U0315194

KINGS FOR BID WINNING —BID PROPOSALS
得标为王—方案篇
2013-2014

中册

龙志伟 编著
Edited by Long Zhiwei

| 城市综合体 |
| Urban Complex |

广西师范大学出版社
·桂林·

序 Preface

“物竞天择，适者生存”这句流传了几千年的警句在当今这个竞争激烈的时代已被奉为至理名言，“优胜劣汰”这一法则在建筑设计领域也同样发挥了指挥棒的作用。如何提高公司的业务水平及竞争力，使己方提出的设计方案赢得招标方及大众的青睐和认可进而得标，是每一个设计师和设计团队都必须深思的问题。一个能在众多方案中脱颖而出、独占鳌头的方案，不仅关系到提案的经济性，更与提案的原创性、创新性、合理性、实用性和完整性息息相关。一个富于想象又不脱离实际、富有创意又经济实用、彰显个性又贴近生活的独创性方案才是招标方心中的首选。

《得标为王——方案篇 2013-2014》是一本大型的设计方案集锦。该书收录了阿特金斯、斯蒂文·霍尔建筑师事务所、Jaspers-Eyers & Partners、3LHD、蓝天组、澳大利亚 SDG、C.F.Møller、UNStudio、深圳天方、北京殊舍等近百家国内外优秀建筑设计公司的知名设计方案。本书精选了度假休闲、商业建筑、酒店建筑、购物中心、文化艺术建筑、城市综合体、办公建筑、医疗建筑、学校建筑、交通建筑、住宅建筑等类别的方案近 200 个。

无论是理念的创新、思维方式及构思角度的转换，还是最新技术的运用、生态节能材料的使用，抑或是独特的造型与外观，它们既使该方案变得独特而唯一，也使之成为备受客户认可和推崇的新设计、新理念、新方案。本书指明了当今建筑设计领域智能、仿生、生态的设计新趋势，阐释了当今备受世人关注的绿色、低碳、以人为本的设计理念，反映了使用者生理上和心理上的需求。这些方案的实施，不仅将促使许多新形式、新类别的建筑的诞生，使人与社会、人与自然和谐发展，同时也将改善人们的生活方式。

“Survival of the Fittest in Natural Selection” has been claimed to be words of wisdom in this fiercely competitive age. “Survival of the Fittest” also plays a leading role in architectural design. How to boost competitiveness and improve professional skills, making the design scheme be favored and accepted by tenderers and the public so as to win the bid has become a thought-provoking issue to every designer and each design team as well. The design scheme, which stands out from numerous proposals, must be not only economical, but also originative, innovative, rational, practical and integrated. Only the most unique proposal that is imaginative, practical, creative, economical, characteristic and close to life can be the priority for tenderers.

Kings for Bid Winning – Bid Proposals 2013-2014 is a large collection of design schemes. Well-known design proposals from nearly 100 national and international renowned architectural design companies, such as Atkins, Steven Holl Architects, Jaspers-Eyers Architects, 3LHD, COOP HIMMELB(L)AU, Shine Design Group, C.F. Møller, UNStudio, Shenzhen TAF Architect and Beijing Shushe Architecture, are included. New projects of about 200 cases involve Resort, Commercial Building, Hotel, Shopping Center, Culture & Art Building, Urban Complex, Office Building, Medical Building, School Building, Transportation Building and Residence.

The innovative concept, thinking mode, design perspective, or the application of latest technology and ecological energy-saving materials, or distinct shape and appearance, make the proposal a special and unique new design, new concept, new scheme recognized and highly appraised by clients. The book implies the architectural design trend for intelligent, bionic and ecological design solutions, interprets a design concept of “Green, Low-carbon and People-oriented” and reflects physical and psychological demands of users. Implementing these proposals not only means the emerging of new forms and new types of buildings, a harmonious development between man and society, man and nature, but also improves people’s lifestyle.

Contents
目录

posten
Radisson SAS

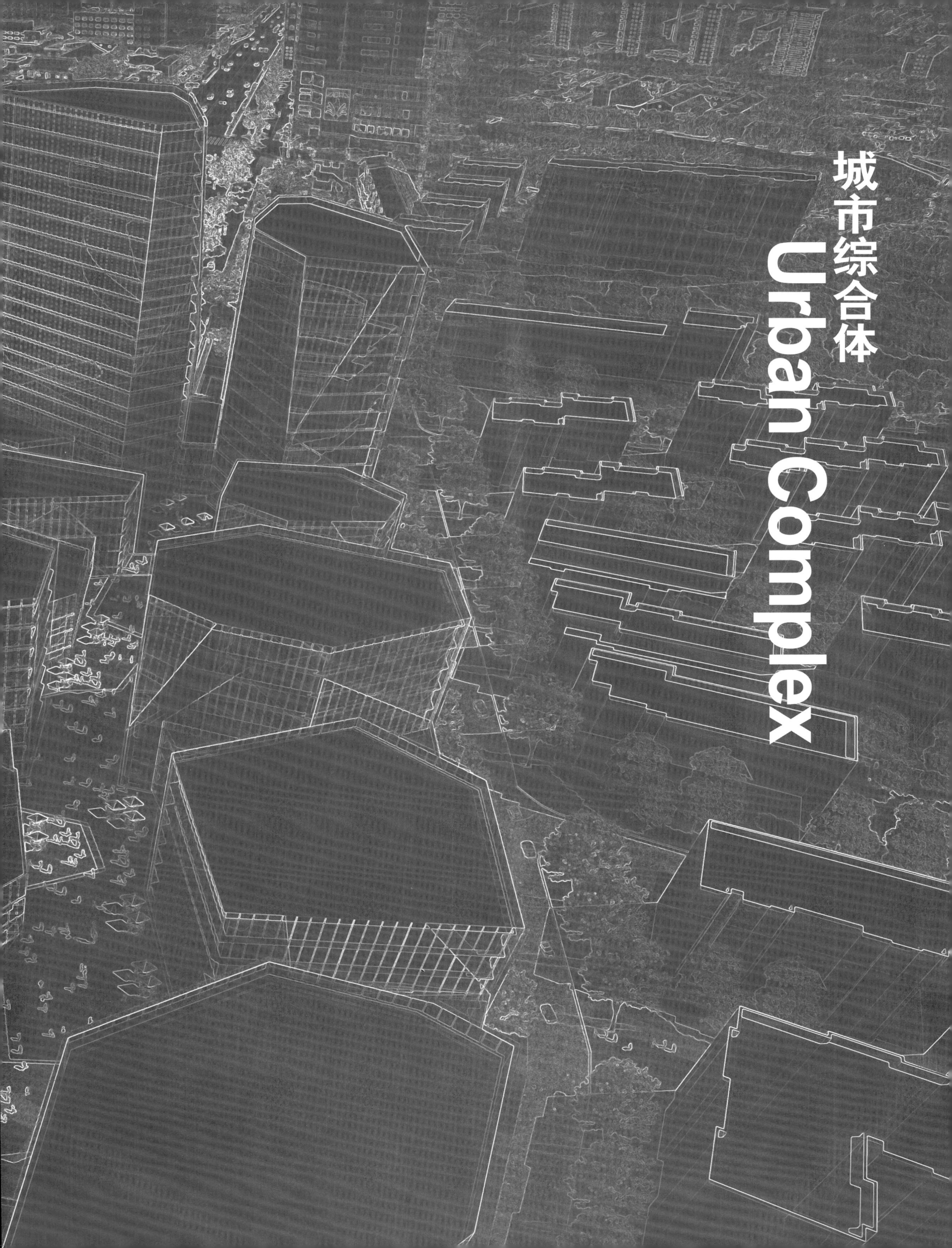
城市综合体
Urban Complex

阿联酋迪拜 The Promenade

The Promenade

设计单位：阿特金斯 ATKINS

开发商：棕榈岛集团

项目地址：阿联酋迪拜

建筑面积：1 060 000 ㎡

Designed by: Atkins

Client: Nakheel

Location: Dubai, UAE

Built Up Area: 1,060,000 m²

项目概况

项目是由棕榈岛集团负责开发，位于迪拜码头区附近的一个专属区域，这一区域也被称为“The Promenade”，它是迪拜码头区最后一个面向大海、可以欣赏到阿拉伯湾风光的地块。

设计理念

设计师将 Promenade 设想为一个集创新住宅、豪华酒店、一级办公空间、充满活力的零售空间及其他公共设施为一体的高端综合社区。方案试图通过标志性的前沿设计，提高全球开发区的标准。

建筑设计

整个项目包含了 7 个主体建筑。Icon 酒店是世界上第一个车轮形态的酒店，高 160 米，位于 Promenade 的尽头，拥有特有的月牙形海滩。Icon 酒店连同左右两侧的两栋翼楼，都建造在 4 层的裙楼上。其他的建筑主体则是三栋高度超过 100 米的服务式公寓住宅和一座办公大楼。

建筑主体的造型并非远眺入海，而是沿海岸线高低起伏；设计遵循半岛风格，场地内无车流通过；建筑之间的宽敞空间及沿海岸线布置的景观，为游客提供了愉悦的徒步之旅。

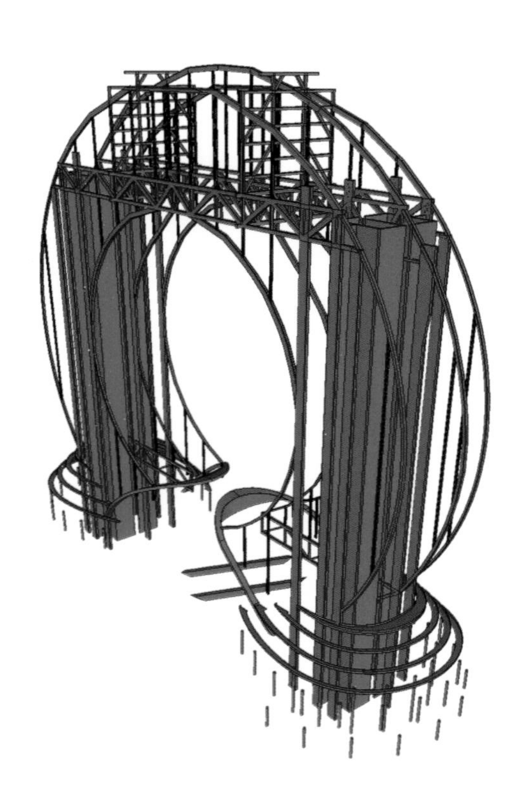

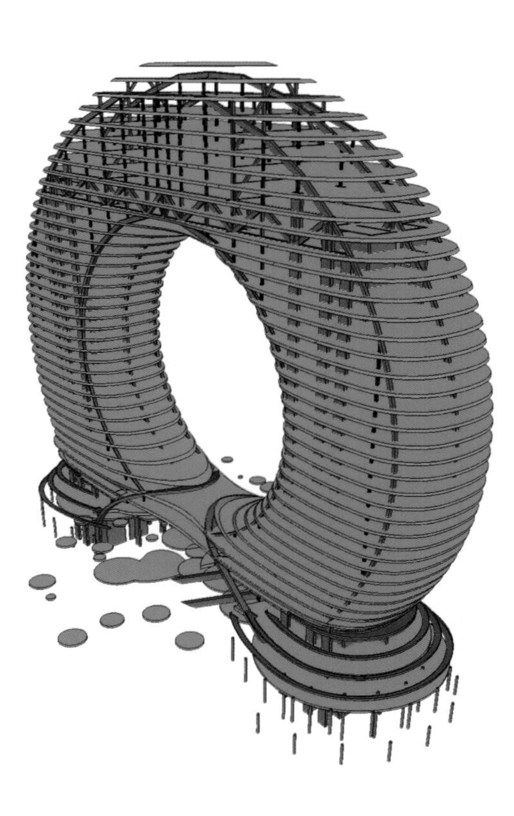

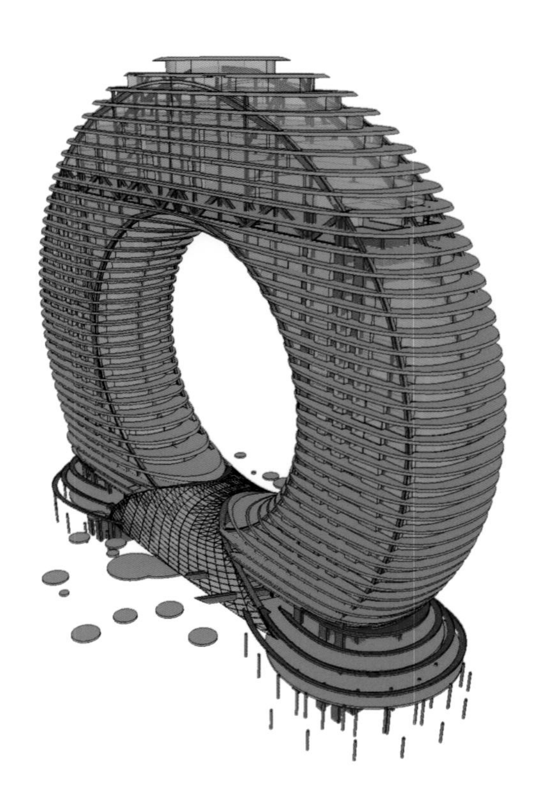

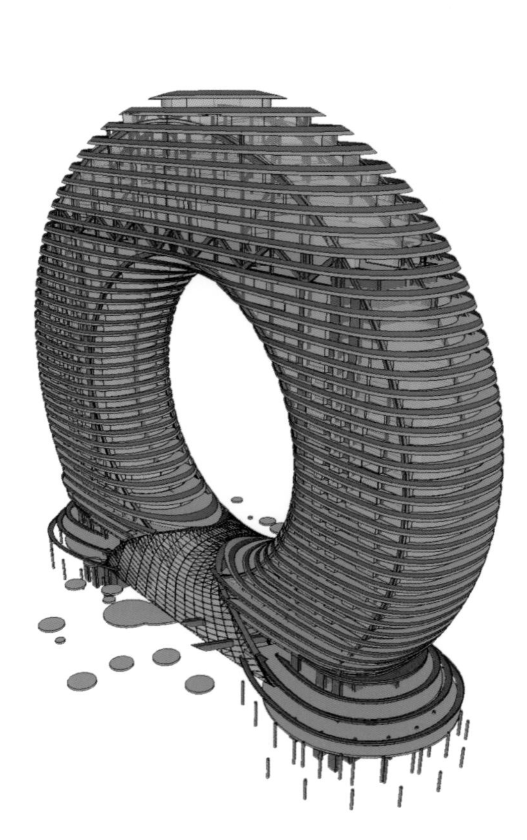

Profile

Nakheel is developing an exclusive area next to Dubai Marina, to be called "the Promenade". The Marina is the world's largest man-made water developments, and the subsequent site of the Promenade is the last plot of land facing the sea with views of the Arabian Gulf.

Design Concept

The Promenade is conceived as a high end, mixed use, integrated community combining innovative residences, luxurious hotels, first class office space and vibrant retail and other public amenities. The architectural aspirations of the Promenade include iconic and cutting edge design that will raise the standard of developments worldwide.

Architectural Design

Atkins is developing seven main structures within the Promenade. The world's first wheel form hotel – the Icon Hotel, with a height of 160 meters, is located at the end of the Promenade with its own private crescent shaped beach. And two further supporting wings, all rest on a four-storey podium. Other buildings include three residential towers of serviced apartments each over 100 meters in height and an office tower.

The shape of the Promenade is not for overlooking the sea but for meandering with undulant coastlines. The design follows a style of peninsula, without vehicles passing through the site. Expansive spaces between buildings and landscapes along coastlines provide tourists with delightful walking experience.

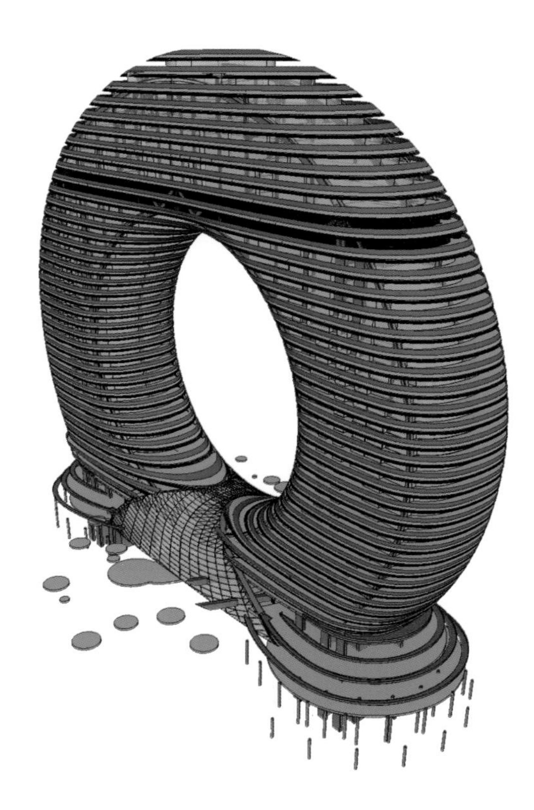

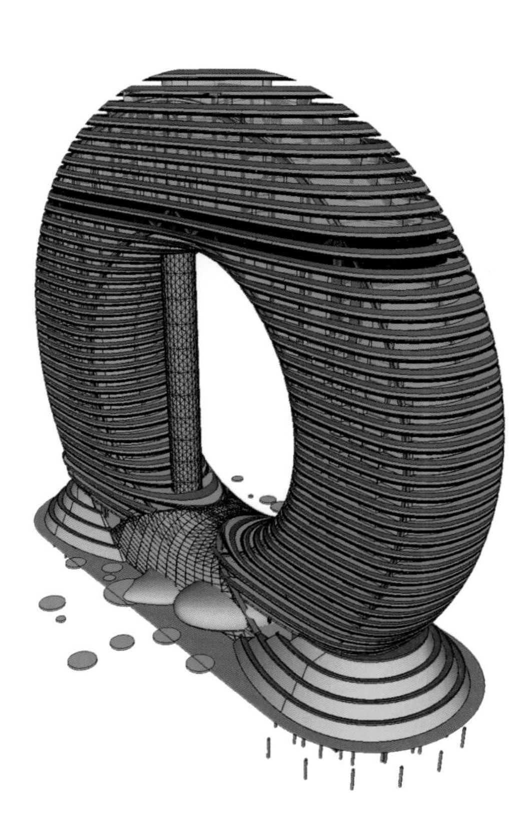

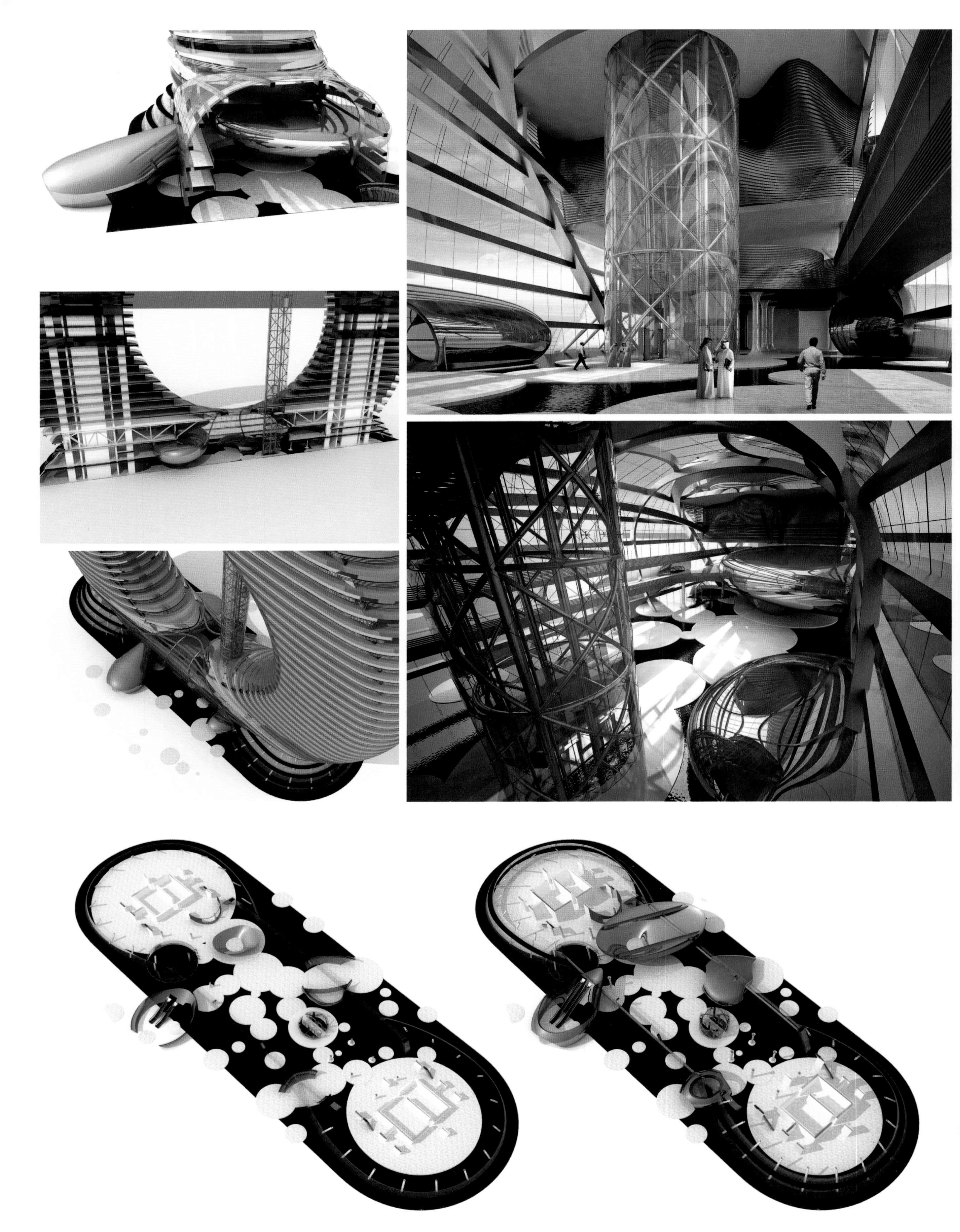

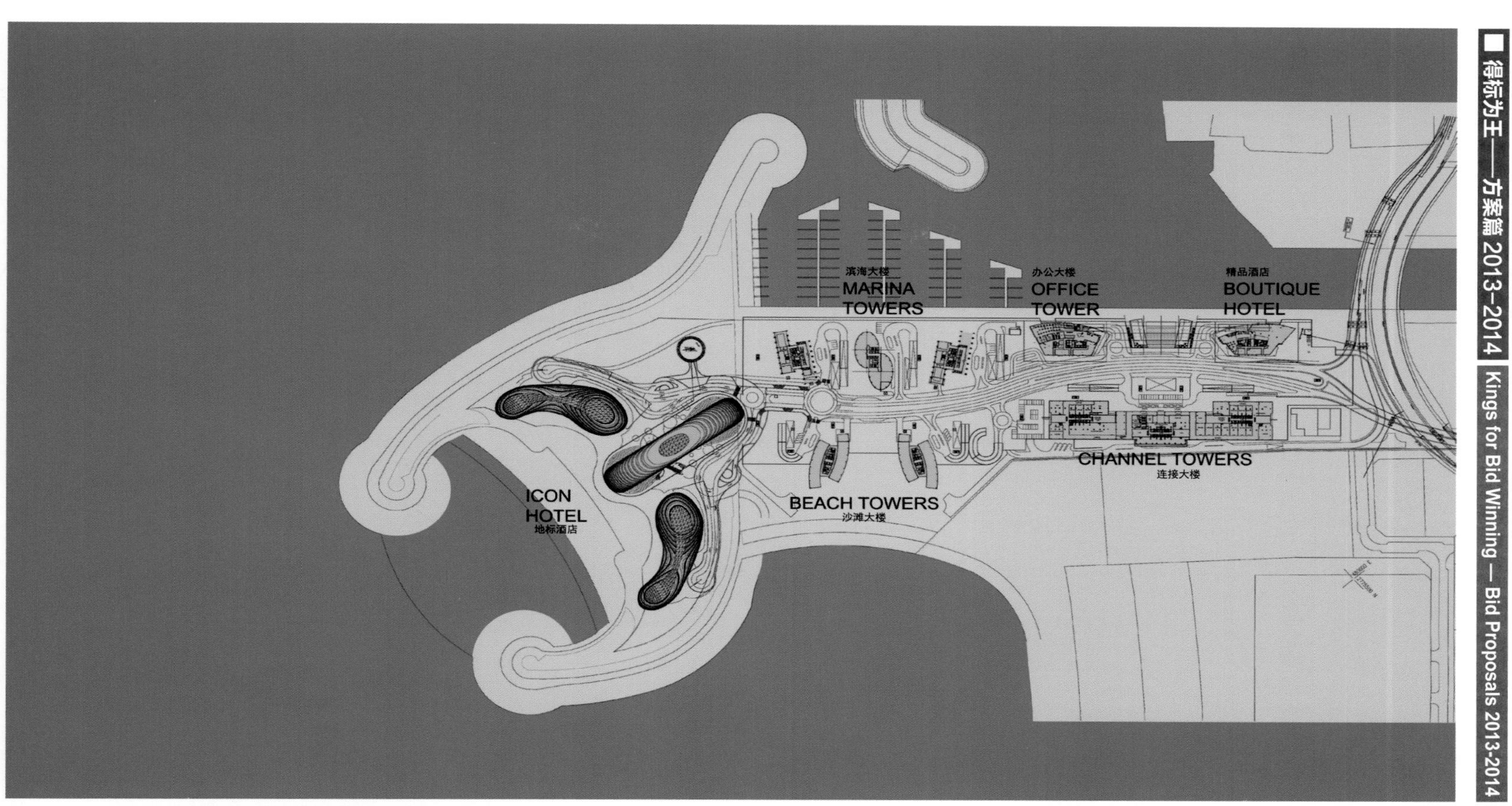
滨海大楼
MARINA
TOWERS
办公大楼
OFFICE
TOWER
精品酒店
BOUTIQUE
HOTEL
CHANNEL TOWERS
连接大楼
BEACH TOWERS
沙滩大楼
ICON
HOTEL
地标酒店

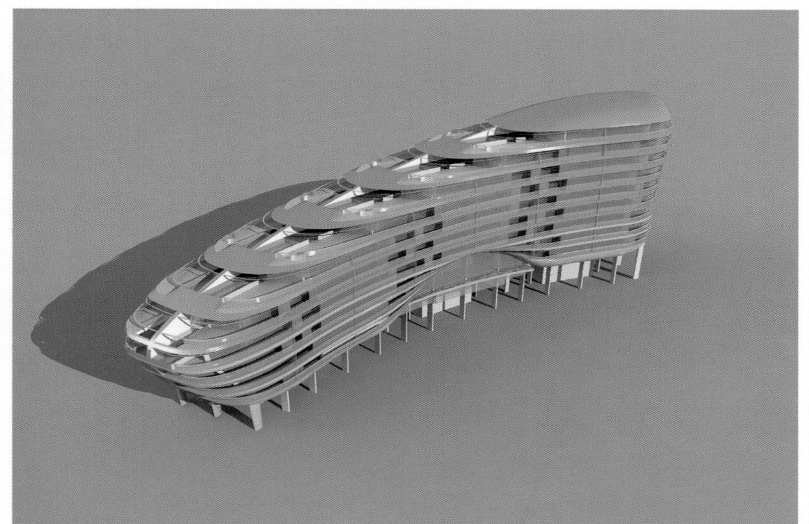

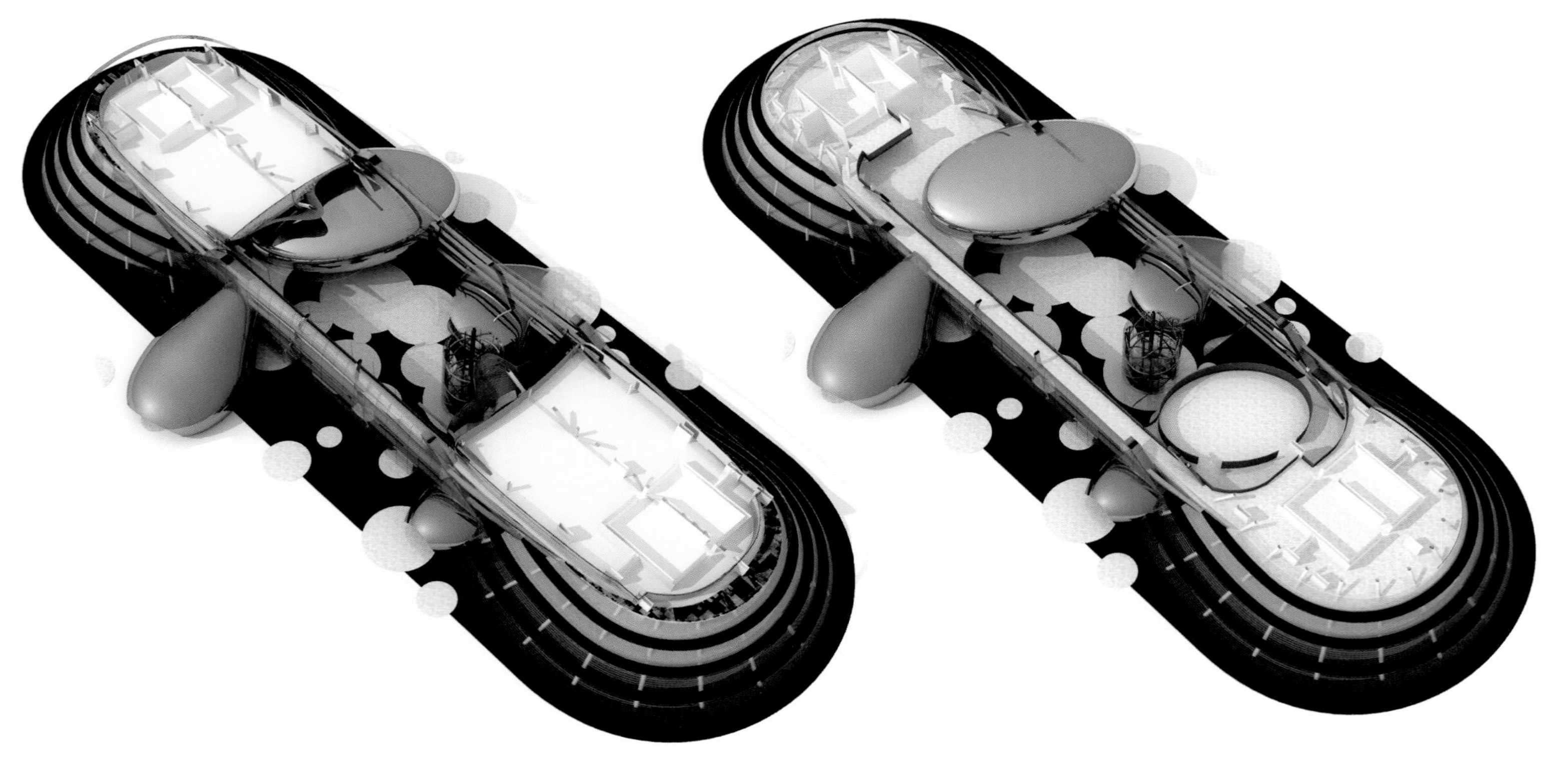

比利时布鲁塞尔 Belair

Belair

设计单位：Jaspers-Eyers Architects

开发商：RAC Development Corp.NV

项目地址：比利时布鲁塞尔

Designed by: Jaspers-Eyers Architects

Client: RAC Development Corp. NV

Location: Brussels, Belgium

项目概况

Belair 是继“城市计划”和“旅游 & 出租车”项目之后，在布鲁塞尔地区进行的第三个大型整修项目。

建筑设计

项目场地位于城市内环的东北端，从植物园路一直延伸至“国会柱”和阵亡的无名战士墓区。项目临近布鲁塞尔历史中心，优越的地理位置使之成为该区域重要的改造项目。

项目主要对 Pacheco 林荫大道、皇家街道及其周围特定的建筑进行整修，在为城市和居民提供公共空间的同时，也将住宅区和市中心连接起来。

在项目构思过程中，设计师设想了多个“点”，通过这些“点”可进入建筑内部，突出了项目的渗透性。同时，设计师设想了两栋三角形的标志性大楼，这两栋大楼的楼层以平板的形式堆叠起来，各楼层有些微的位移，赋予了建筑鲜明的特色。

Profile

Alongside the “Projet Loi Urbain” and “Tour & Taxis” projects, a third major district-wide refurbishment project is taking shape in Brussels.

Architectural Design

The project's location is just inside the city's inner ring road at its north eastern tip, across the road from the Botanical Gardens and stretching to the “Congress Column” and tomb of the unknown soldier. The project with a superior location close to Brussels historic center makes it an important renovation project of this area.

This urban renovation project aims to connect the uptown and downtown districts of the city, provides the public spaces for the city and its inhabitants, and involves the refurbishment of certain buildings on and around the Boulevard Pacheco/Rue Royale.

The architects have designed an overall layout which enables access to the project from a number of points, emphasizing its permeability. The master plan schedules the construction of two new emblematic triangular buildings featuring “offset” floors, as though they were slabs roughly stacked one on top of the other.

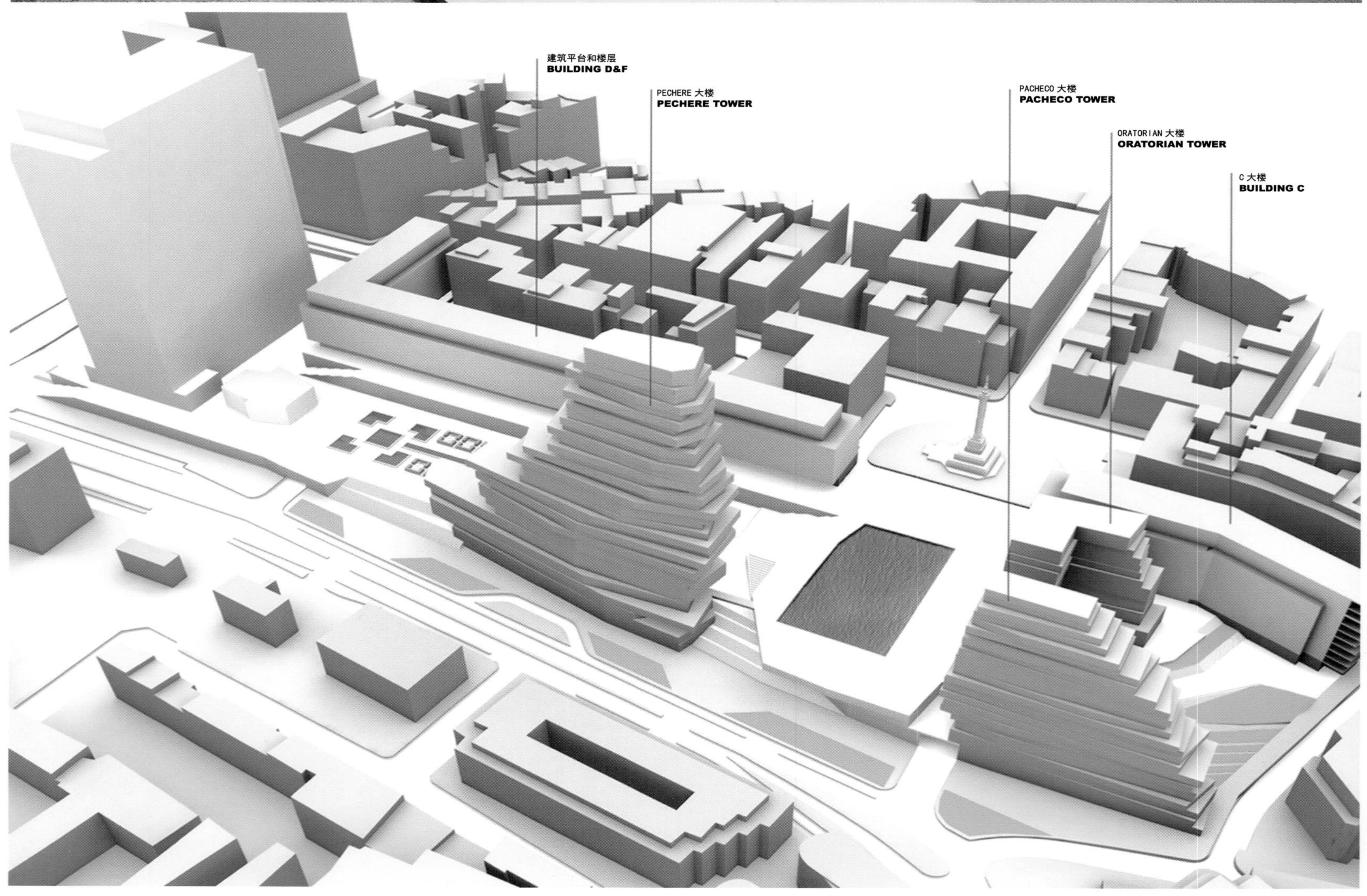
建筑平台和楼层
BUILDING D&F
PECHERE 大楼
PECHERE TOWER
PACHECO 大楼
PACHECO TOWER
ORATORIAN 大楼
ORATORIAN TOWER
C 大楼
BUILDING C

BEYOND TREND
MAGNUM

BEYOND TREND

DESIGN

阿联酋阿布扎比仿生大厦

Bionic Tower

设计单位：LAVA
项目地址：阿联酋阿布扎比

Designed by: LAVA
Location: Abu Dhabi, UAE

项目概况

这栋仿生建筑将未来建筑具体化，它超出了现代主义设计概念，构造了一栋拥有完全智能化外观、可最优化能源效率、符合人性化尺度的大楼。

设计特色

大楼的设计灵感源于自然，试图构建一个超轻、高能效、造型优雅，融合了先进设计技术的建筑。此外，新型材料和尖端技术的使用，提高了建筑的适应能力、反应力，提升了自身的强度和对环境的感知能力。

未来建筑的重点并非是它的外形，而是着重于它的智能系统。整个智能系统以单位的智能化为基础，从而组成庞大的智能系统，通过对行为逻辑进行参数化分析和建模，使该系统不断得到优化。

传统建筑的表面都是被动的，对波动的外部环境缺乏适应能力。该设计的外墙系统背离了这一传统，实现了全面智能化，建筑表皮可以根据内部的需要切实解决诸如通风、采光、排水和储水等问题。正如大自然的有机再生，该建筑采用的智能系统同样能对能源进行调整和重组。

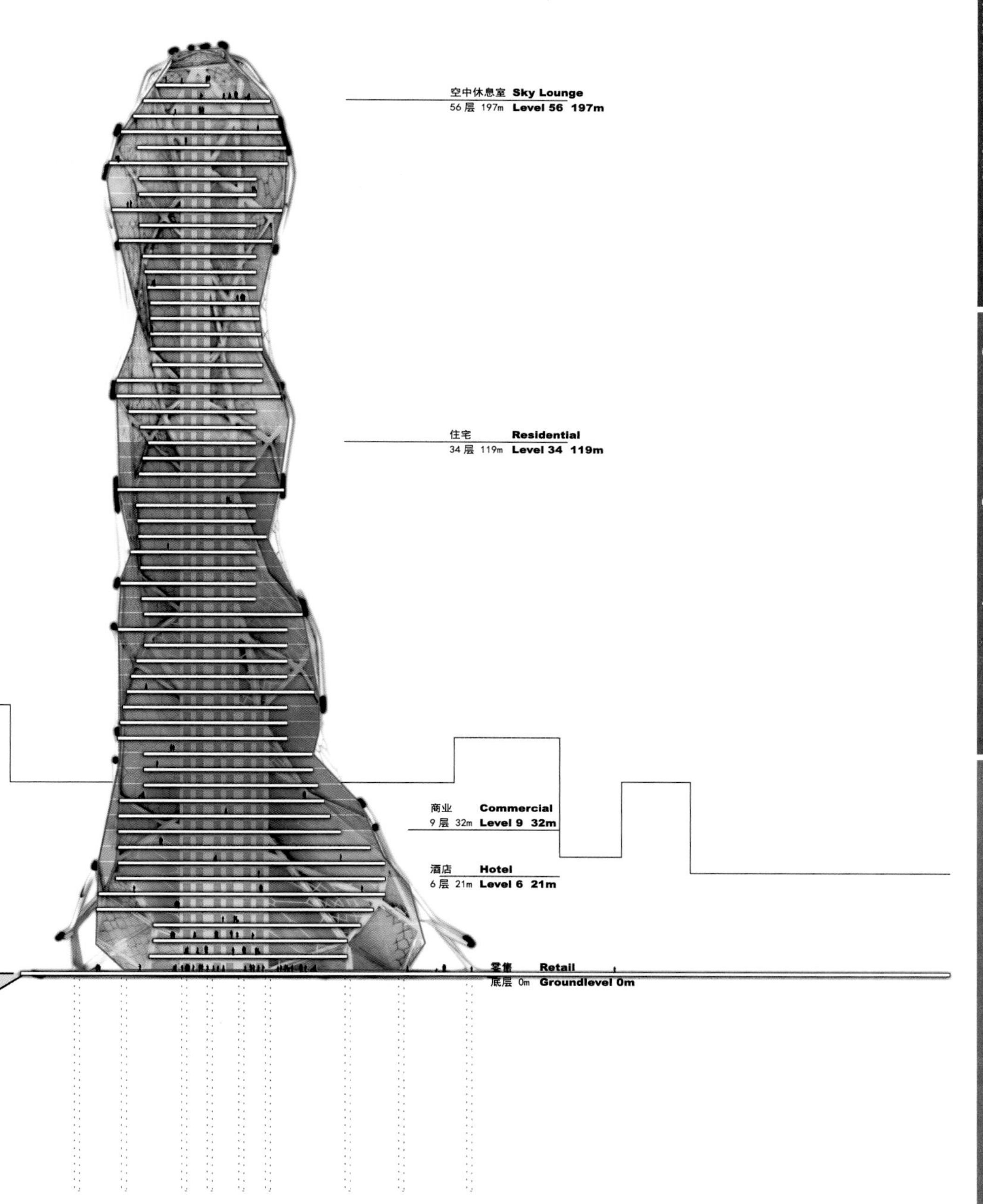

Public Space
公共空间
Information
信息区
大厅
Reception
接待处
Cafe
咖啡区
Relaxation
休息区
Waiting
等候区
Public Space
公共空间
Public Space
公共空间

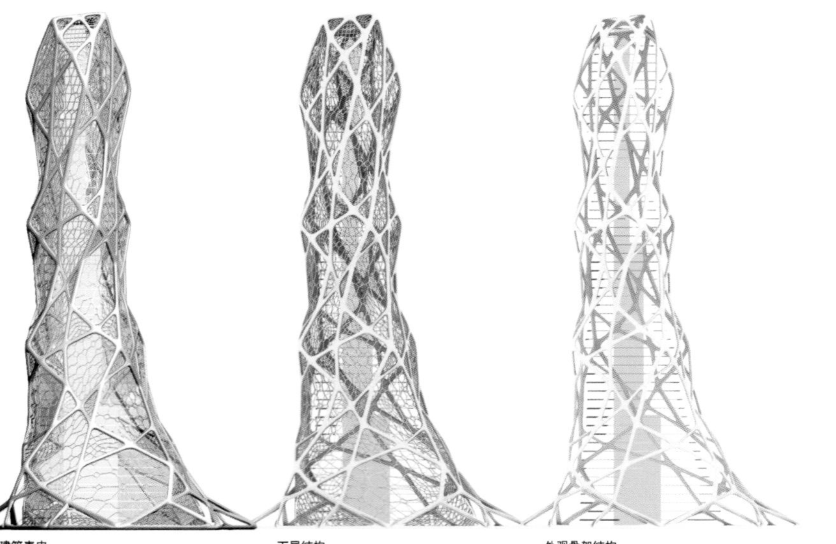

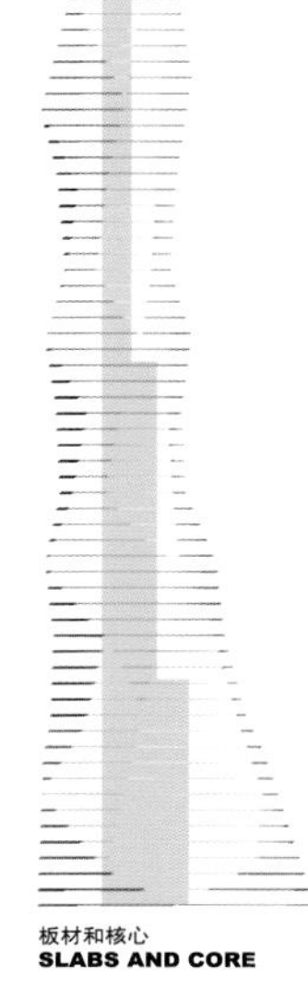

Profile

The Bionic Tower embodies tomorrow's architecture. Its design moves beyond the superseded modernist concept and strives to create a fully integrated intelligent façade, maximum energy efficiency, and humanized scale.

Design Feature

The design inspiration originates from the nature, trying to erect a building that is extremely light, highly efficient, graceful and integrated with advanced technologies. New materials and cutting-edge technologies used in this project have enhanced adaptability, responsiveness, environmental awareness and strength of the project.

Architecture of the future is not about the shape but about the intelligence of the system. The intelligence of the smallest unit results in the intelligence of the overall system. By parametric modeling of the "behavioral logic" the system has been constantly optimized throughout the design process.

The traditional curtain wall is passive, lacking the power to adjust to the fluctuating external environment. The proposed facade is an intelligent automation of surface to address pragmatic issues such as ventilation, solar access and water collection. Just as nature envisions organic regeneration, so the design proposes a natural system of organic structure and organization.

建筑应在城市有机组织中像生态系统一样运作
ACHITECTURE HAS TO PERFORM AS AN ECOSYSTEM
WITHIN THE ORGANIC TISSUE OF THE CITY

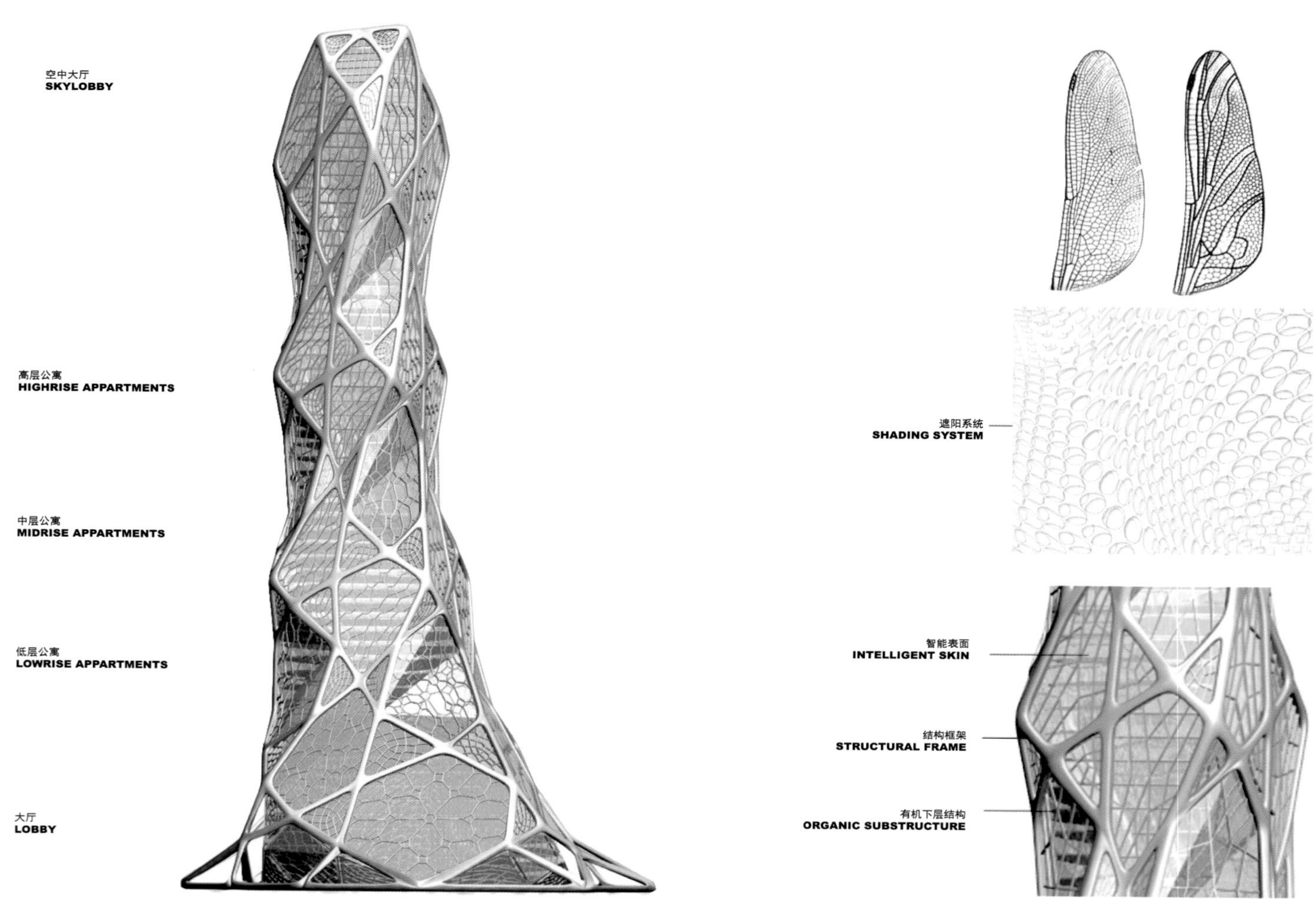
空中大厅
SKYLOBBY
高层公寓
HIGHRISE APPARTMENTS
中层公寓
MIDRISE APPARTMENTS
低层公寓
LOWRISE APPARTMENTS
大厅
LOBBY
遮阳系统
SHADING SYSTEM
智能表面
INTELLIGENT SKIN
结构框架
STRUCTURAL FRAME
有机下层结构
ORGANIC SUBSTRUCTURE

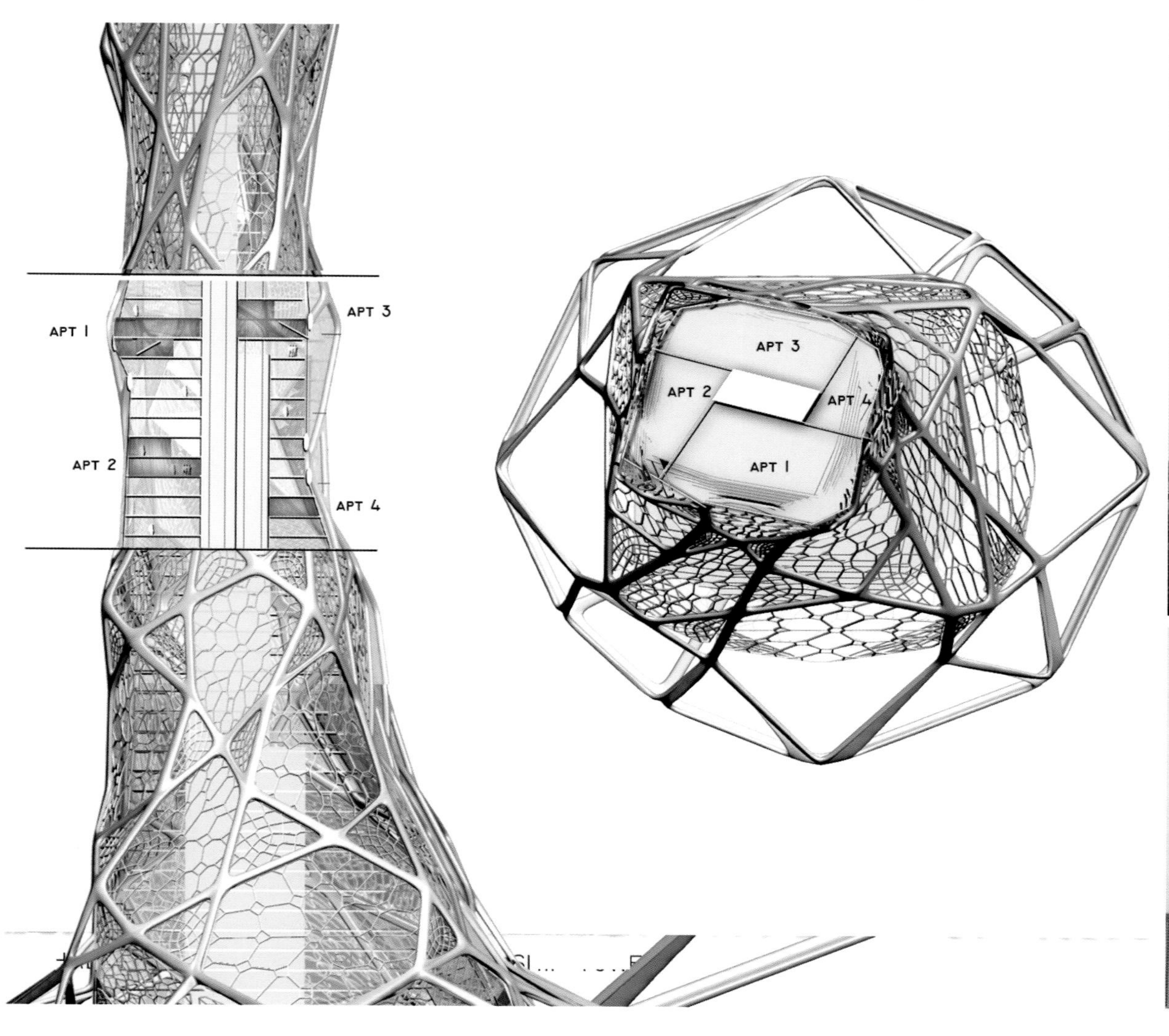
APT 1
APT 2
APT 3
APT 4
APT 3
APT 2
APT 4
APT 1

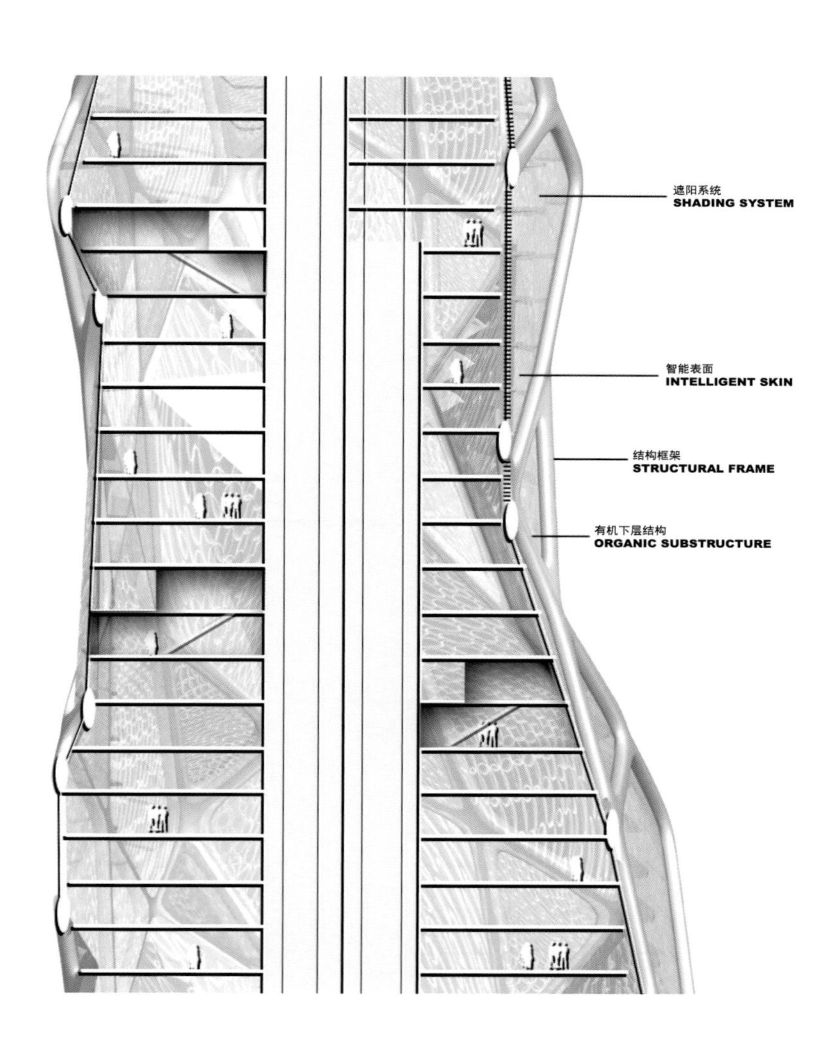
遮阳系统
SHADING SYSTEM
智能表面
INTELLIGENT SKIN
结构框架
STRUCTURAL FRAME
有机下层结构
ORGANIC SUBSTRUCTURE

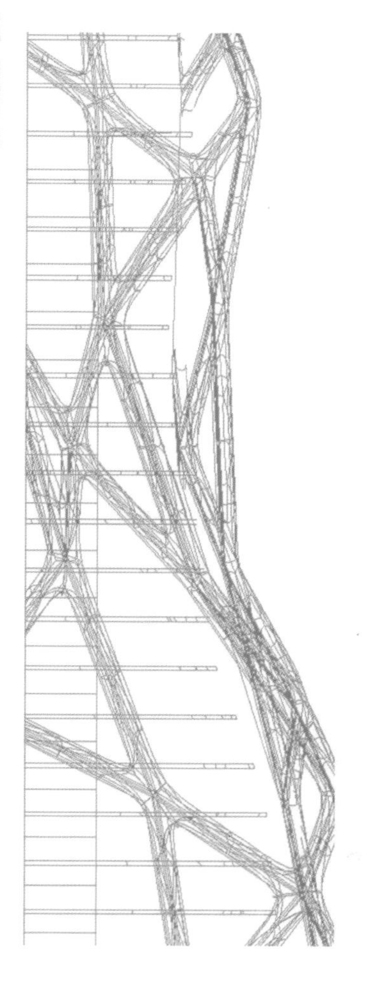

新加坡共和国新加坡市双景坊

DUO

设计单位：奥雷·舍人事务所
开发商：马新私人有限公司
项目地址：新加坡共和国新加坡市
总建筑面积：160 350 ㎡

Designed by: Buro Ole Scheeren
Client: M+S Pte. Ltd.
Location: Singapore City, Singapore
Gross Floor Area: 160,350 m²

项目概况

双景坊是马来西亚首相和新加坡总理联合支持，并由国库控股(Khazanah)和淡马锡控股(Temasek)联合开发的项目之一。设计极具策略性地呈现出这一项目富有象征性意义的合作关系，并注重与周边环境相融合，使其成为新加坡现代大都会的一个新内核。

建筑设计

从热带城市地景中升起的两座雕塑般的塔楼，整合了办公室、住宅、酒店和商业空间等各项功能。双塔的形态设计，巧妙地利用一系列循环的圆形来裁切掉建筑体量，消减建筑体量后形成的空间与外部城市空间相互连接，将双塔融入城市文脉之中。

网状的六角形遮阳系统形成天然的外皮附于凹进的外立面上，动感十足，部分幕墙形状修长，横向延伸至附属的道路。设计师对建筑的悬臂部分和退线部分进行精雕细刻，使之如同一对翩翩起舞的舞伴，又似太极大师在相互切磋。

基地上设计了一处多孔、可渗透的热带景观，消除了大楼直落地面的突兀感。休闲区域和花园作为多个交通枢纽的连接点，形成一个流动的热带花园和商业环境，供公众二十四小时使用。广场镶嵌在大楼中间，设计巧妙地利用周边建筑作为它的边界，形成一个新的公共区域，将 Kampong Glam 历史区域和城市商业延伸地块连接起来。

项目融合了被动节能和积极节能设计来实现其环保功能，通过实现自然通风，来营造自然舒适的热带生活环境，同时最小化能耗。建筑的朝向通过对日照角度和风向的测量得到优化，凹进的建筑体量可以使风在基地内自然流动，在遮阳的室外空间形成凉爽的小气候。

Profile

As part of a historic joint venture collaboration between Khazanah and Temasek supported by the Prime Ministers of Malaysia and Singapore, Ole Scheeren's design for DUO articulates the symbiotic partnership, and actively engages the space of the surrounding city to form a new civic nucleus in Singapore's modern metropolis.

Architectural Design

Featuring two sculpted towers rising from a tropical landscape, the scheme integrates offices, residences, a hotel, and retail space. The two towers are not conceived as autonomous objects, but defined by the spaces they create around them. The design for DUO subtracts circular carvings from the building volumes in a series of concave movements that generate and articulate urban spaces and symbiotically inscribe the two buildings into their context.

Slender facades soar skywards along the adjoining roads, while a net-like hexagonal pattern of sunshades forms a natural texture that reinforces the dynamic concave shapes. The duo of tower volumes is further sculpted to feature a series of cantilevers and setbacks that evoke the kinetic movement of a dancing couple or two sparring tai chi masters.

The buildings dematerialize as they reach the ground

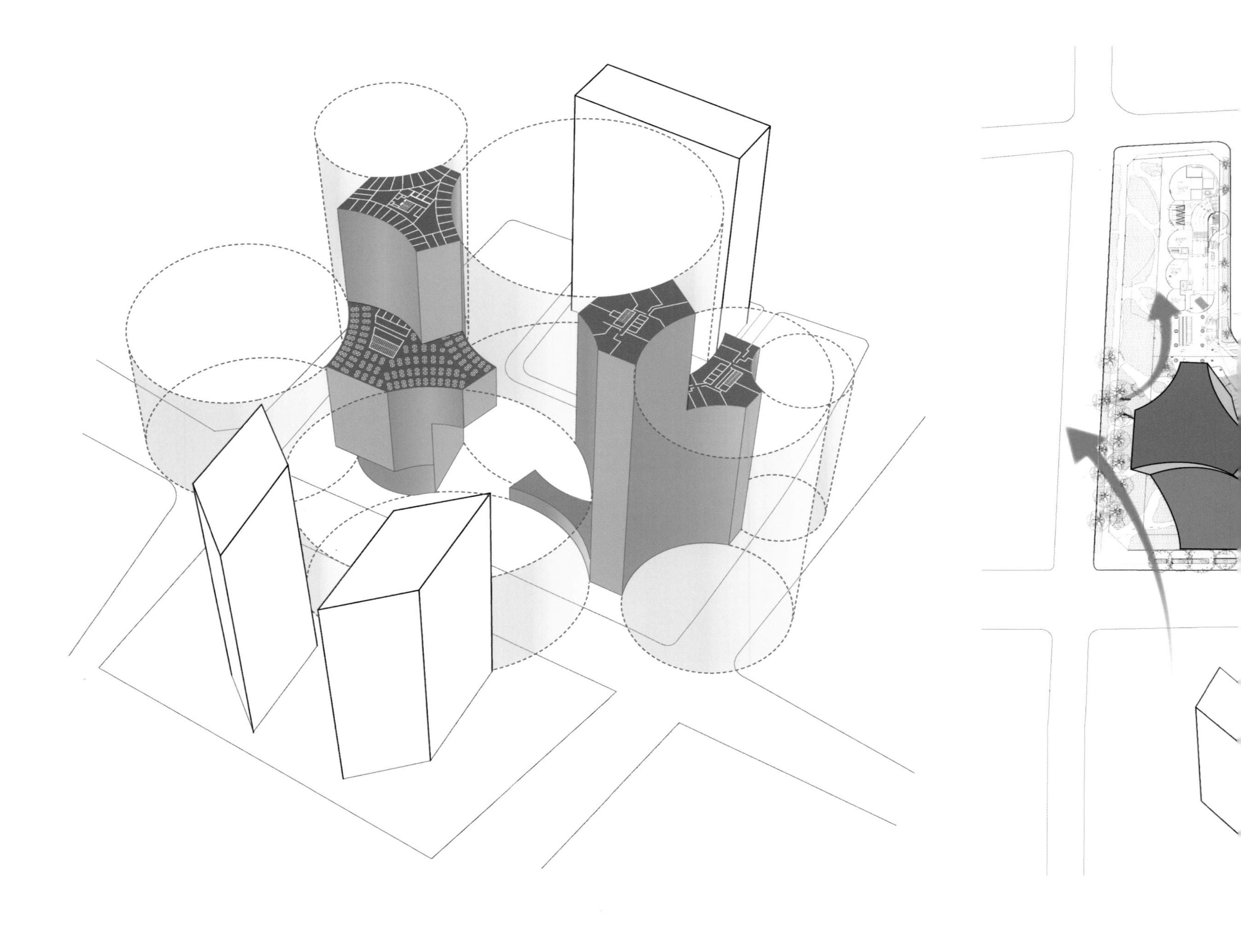

环境水流
ENVIRONMENTAL FLOWS

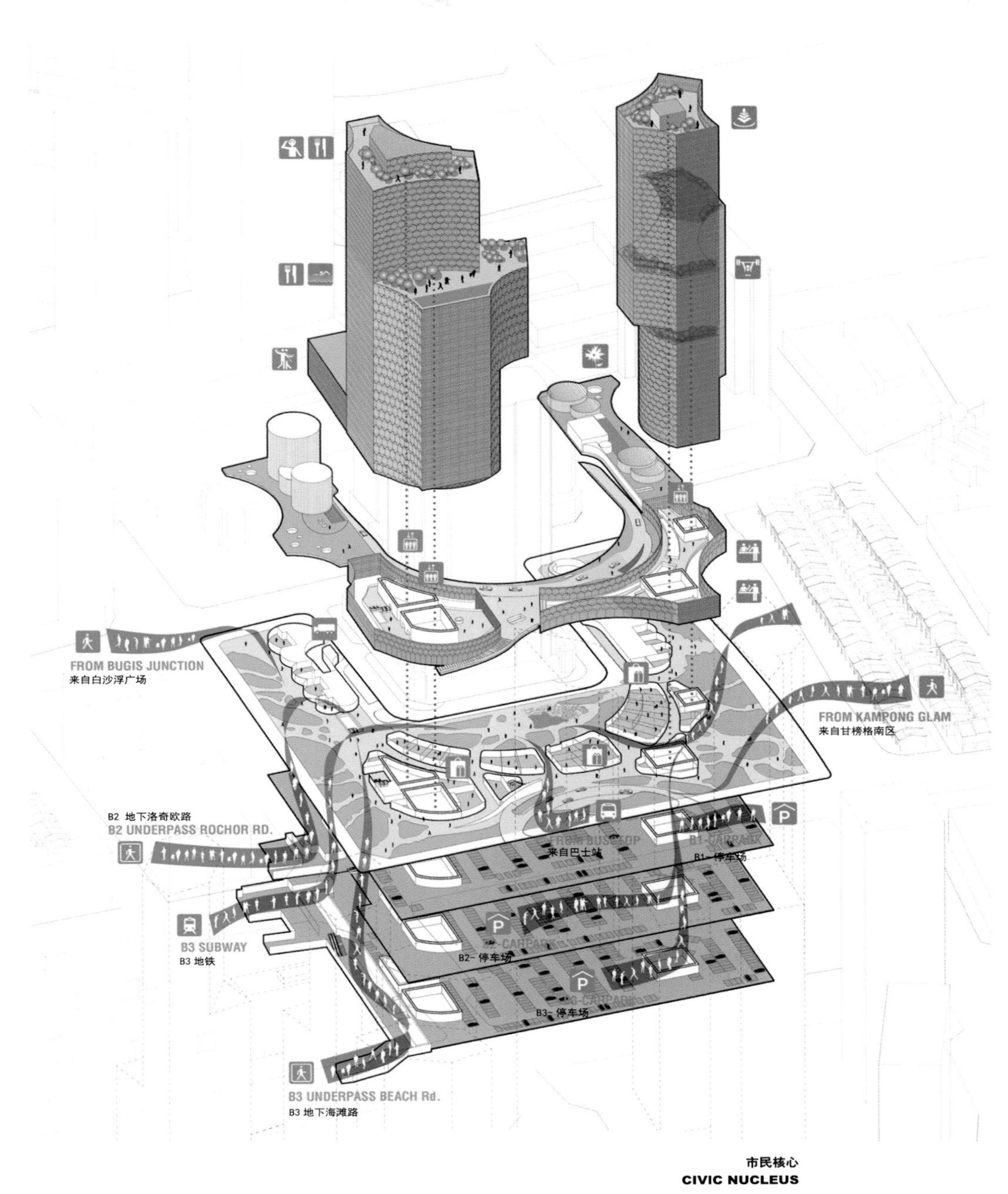

市民核心
CIVIC NUCLEUS

to provide a porous and permeable landscape traversing the site. Leisure zones and gardens act as a connector between multiple transport hubs and establish a flow of tropical green and lively commercial activity, accessible to the public 24 hours a day. A plaza, carved into the center of the towers and strategically incorporating the neighboring building as part of its perimeter, forms a new public nexus between the historic district of Kampong Glam and the extension of the city's commercial corridor.

The development incorporates environmental strategies through passive and active energy efficient design and naturally ventilated spaces. The building's orientation is optimized to prevailing sun and wind angles, while the concave building massing captures and channels wind flows through and across the site, fostering cool microclimates within the shaded outdoor spaces.

挪威奥斯陆“透明水晶”

Crystal Clear Towers

设计单位：C.F.Møller Architects
合作单位：Kristin Jarmund Arkitekter
开发商：KLP Eiendom AS
项目地址：挪威奥斯陆
项目面积：92 000 ㎡

Designed by: C.F. Møller
Collaboration: Kristin Jarmund Arkitekter
Client: KLP Eiendom AS
Location: Oslo, Norway
Area: 92,000 m²

项目概况

“透明水晶”这个独特的设计方案规划了一组同时具备现代元素和北欧元素的高楼，它位于奥斯陆最具开发价值的地段，它的开发将有利于使奥斯陆跻身欧洲最具现代气息的首都行列。

建筑设计

项目被亲切地称为“透明水晶”，包括三个宛若从地面生长起来的、雕塑感十足的塔楼。塔楼婀娜多姿，由众多玻璃体块构成。这三座高楼分别高 110 米、65 米和 55 米，建造在该地段的边缘位置。最高的塔楼与附近的奥斯陆广场和 Postgirobygget 相匹配，较低的两个塔楼则与城市形成链接。

三座塔楼与地段外部建筑有着明显的垂直高差，大型开口式的内嵌窗户，使建筑拥有不同的视角。朝向地段内部的建筑立面由堆叠的棱形玻璃体组成，形成一个透明的棱柱形外观。

在塔楼之间有一个两层高的基地，其内设有商店和餐厅，通过斜坡、台地和楼梯与街面连接，形成波浪式的景观。设计以塔楼为基地的外部边界，将这一基地围合成一个安静的城市花园，为当地的居民提供了一个休闲空间。

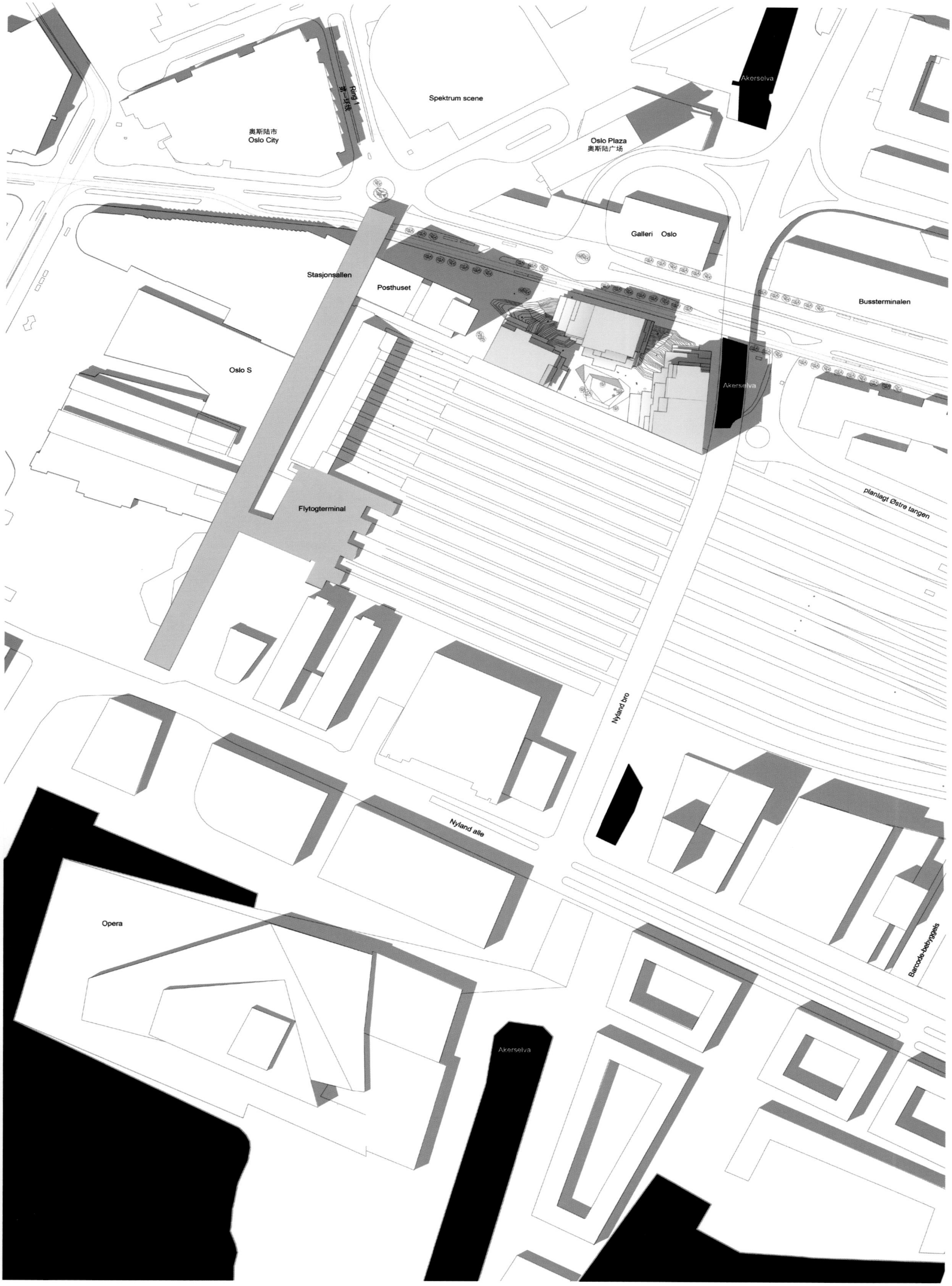

Akerselva
Spektrum scene
Ring 1
第一环线
奥斯陆市
Oslo City
Oslo Plaza
奥斯陆广场
Galleri Oslo
Stasjonsallen
Posthuset
Bussterminalen
Oslo S
Akerselva
planlagt Østre tangen
Flytogterminal
Nyland bro
Nyland alle
Opera
Barcode-bebyggels
Akerselva

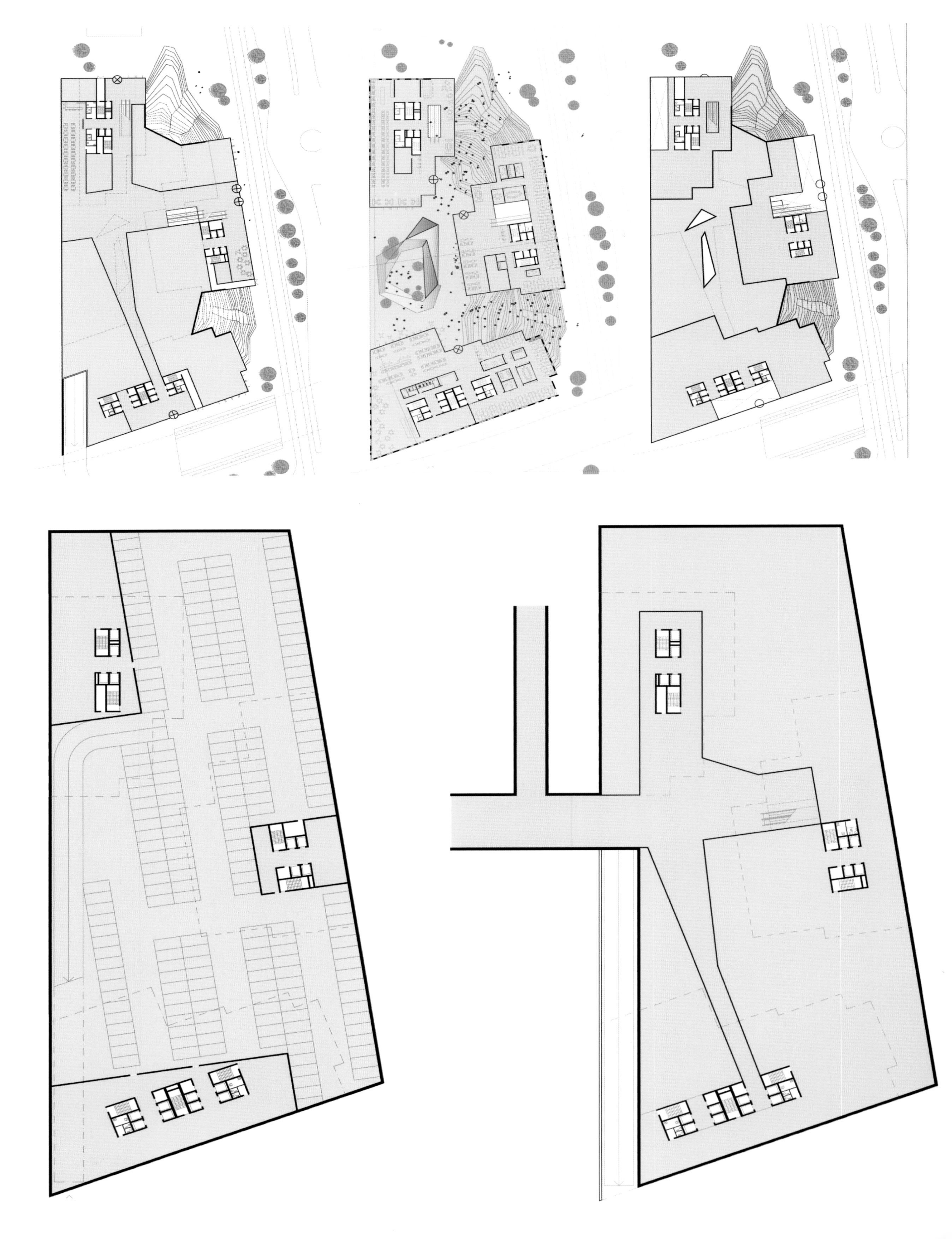

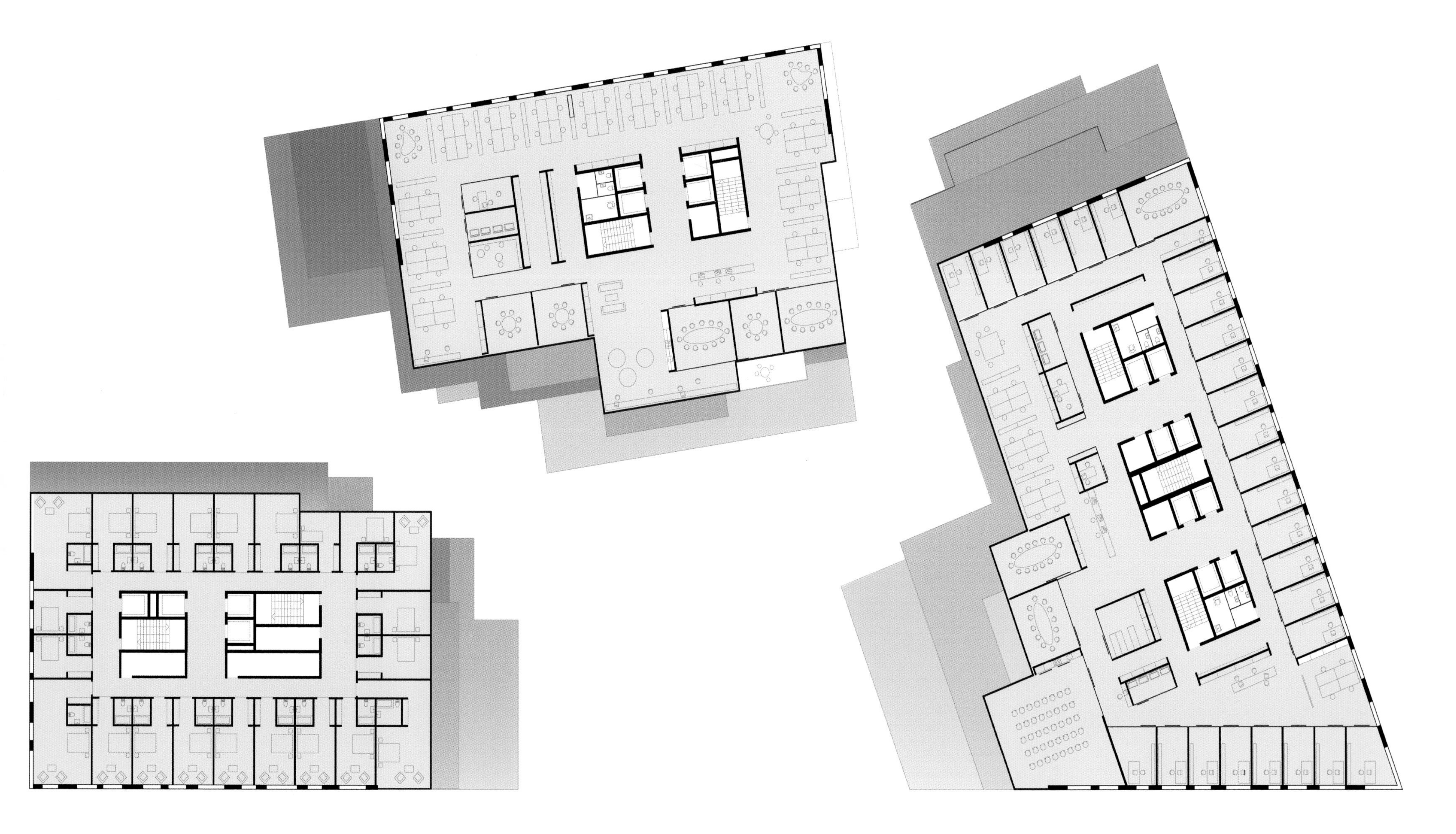

entra
posten
steria
OSLO
CopyCat
Radisson

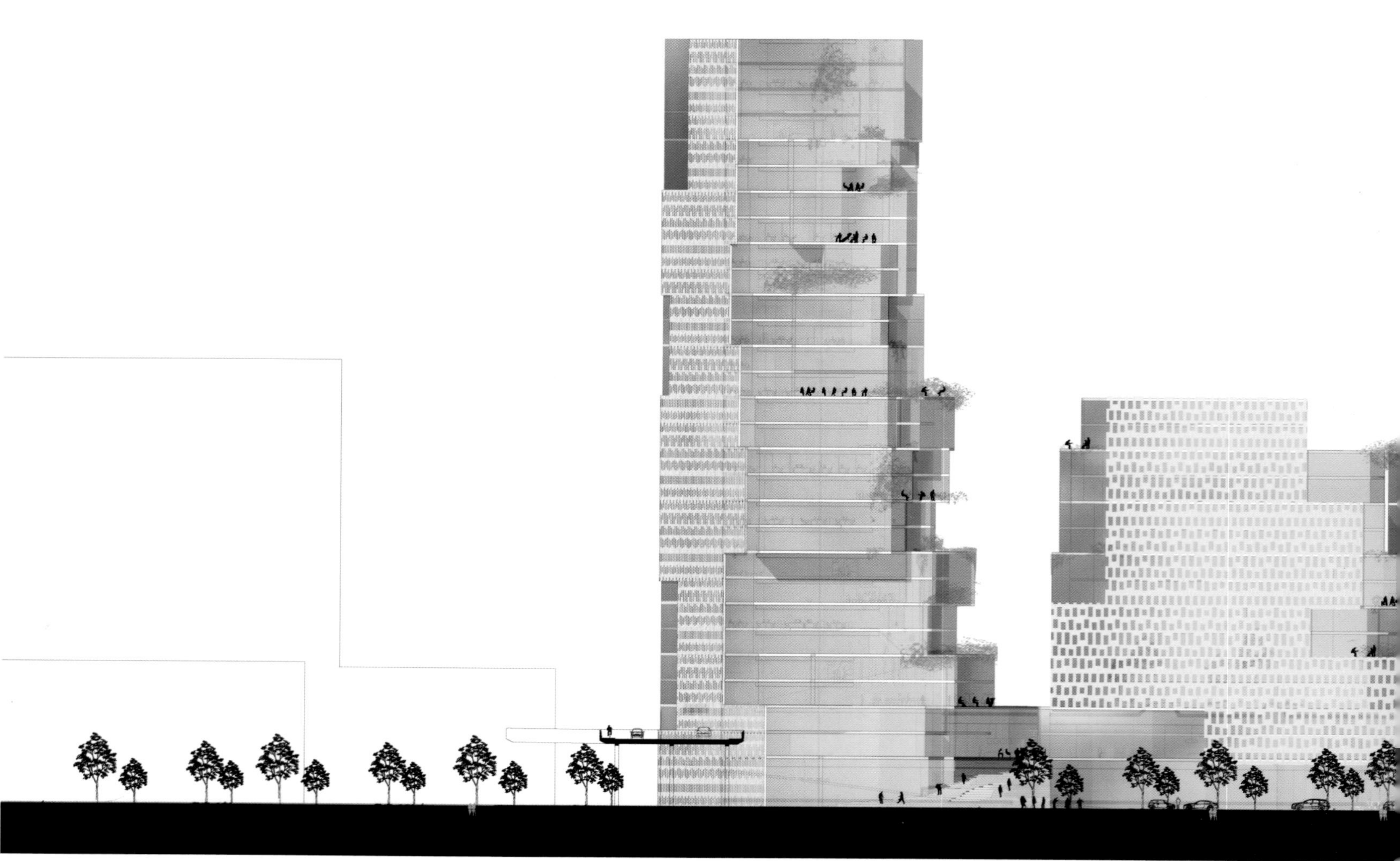

Profile

"Crystal Clear" is a unique proposal for a modern, Nordic cluster of towers .It is located at one of Oslo's most valuable sites. Its development will help turn Oslo into one of Europe's most modern capitals.

Architectural Design

The project, which has been dubbed "Crystal Clear", consists of three towers, which grow organically from the ground to form a sculptural cluster, and are composed of stacked, prismatic volumes. The three towers of approx. 110, 65 and 55 meters height, are arranged along the edges of the site, and the tallest tower is aligned with the existing nearby Oslo Plaza and Postgirobygget towers, while the lower buildings form the link to the city.

The three towers have clear-cut and vertical elevations to the exterior of the site, with large openings and setbacks forming windows to selected viewpoints. In contrast, the elevations towards the interior of the site are composed of stacked, glazed volumes, freely arranged to form a prismatic and crystalline appearance.

In between the towers, a two-storey base containing shops and restaurants forms an undulating landscape that connects to street level via ramps, plateaus and stairs. This base creates a calm urban garden, framed by the tall buildings, with recreational space for the city and the buildings occupants.

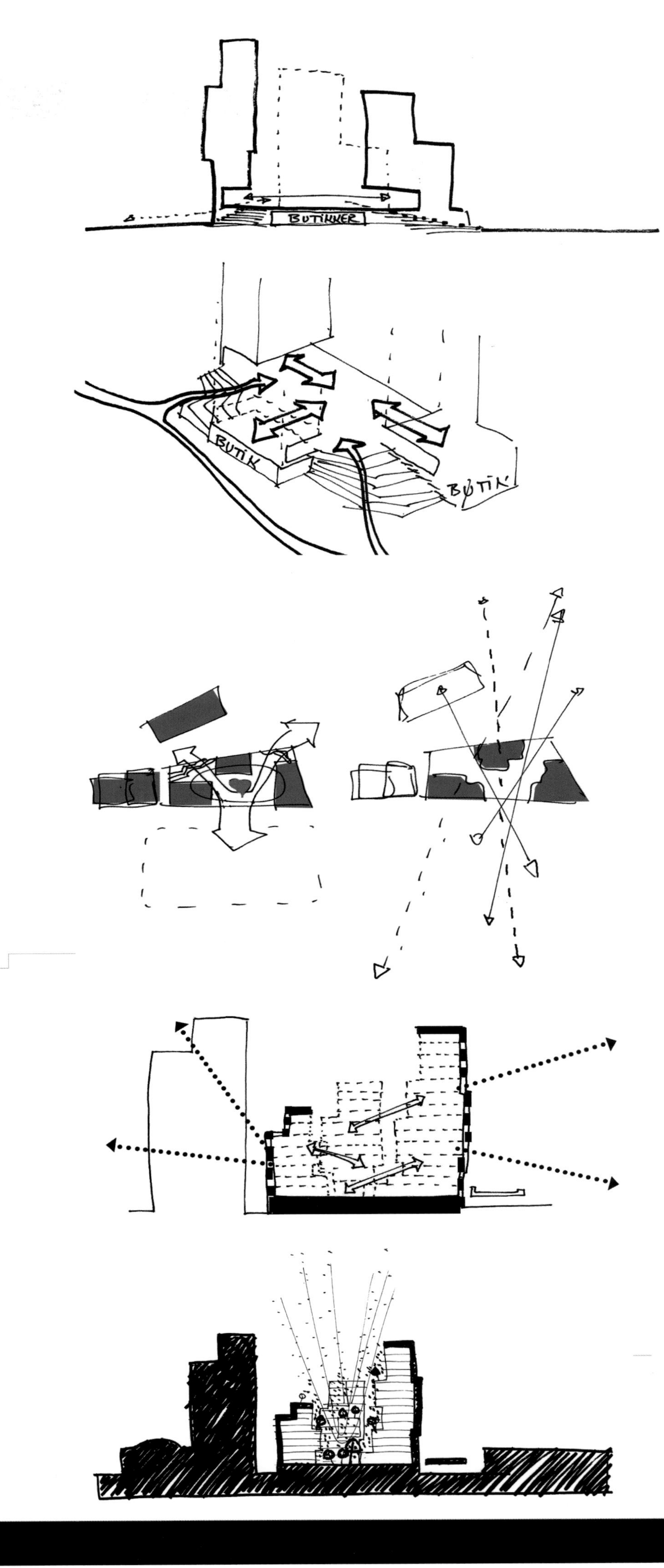

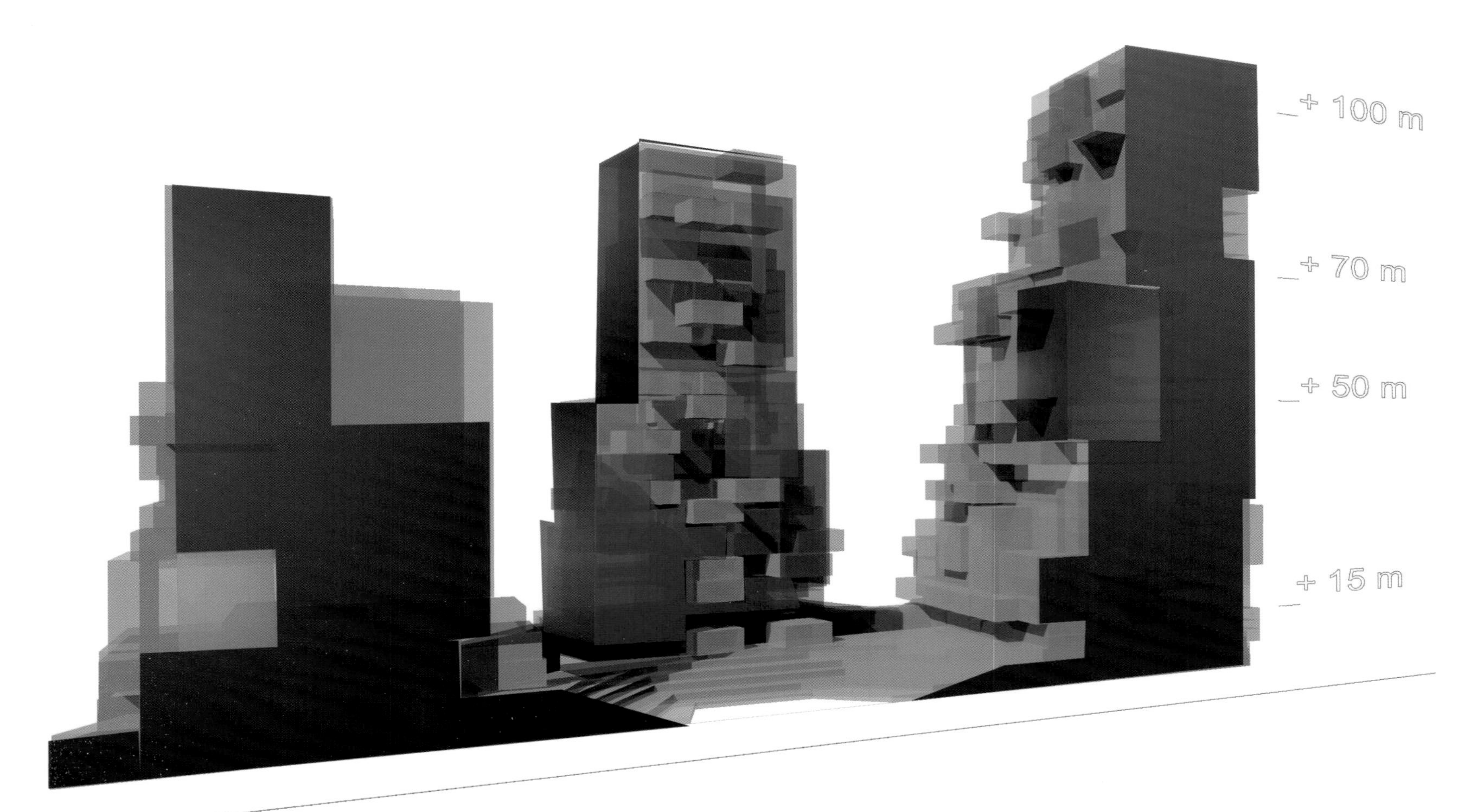
+ 100 m
+ 70 m
+ 50 m
+ 15 m

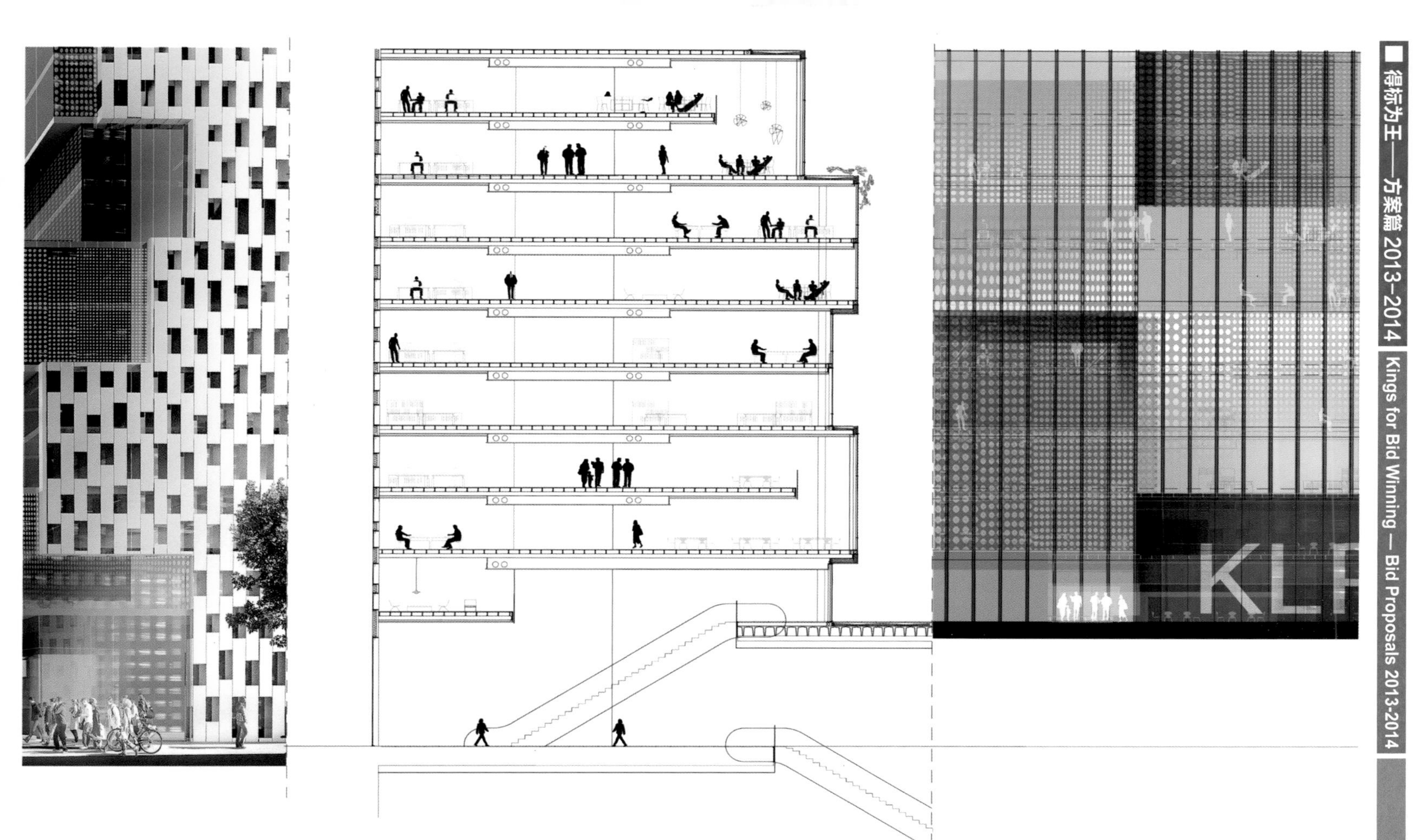
KLF

CAFE BABYLON

哥伦比亚波哥大 Torre Bacata

Torre Bacata, Colombia

设计单位：Alonso Balaguer y Arquitectos Asociados

项目地址：哥伦比亚波哥大

Designed by: Alonso Balaguer y Arquitectos Asociados

Location: Bogota, Colombia

项目概况

Bacata 大楼是一个大胆的、决断的、有政治支撑的项目，设计旨在创造一个新的城市标志，使之成为推动城市改造和重建的引擎。

设计构思

设计师认为，与世界其他城市的逻辑性可持续增长不同，这一城市的增长就像“油渍”一样，势不可挡地向周围蔓延，产生难以预料的社会性和功能性影响。去往城市郊区既费能源又费时，私家车车主们也难以维持这一模式，那么，紧凑的城市建设则成为一种选择。

在紧凑城市的发展模式中，土地利用的一个有效因素就是建筑高度。这并不是漫无目的地建造高楼，而是根据城市的具体情况构建在特定的区域里。

建筑设计

这个正在施工的、高 250 米的建筑将涵盖混合用途的功能区，包括酒店、住宅、办公、商业、运动中心和停车场。建筑内的台阶式景观区域可供社区人员自由使用，同时也成为了一个交流、聚会、休闲的户外空间。

建筑大楼将成为城市垂直空间的一部分，建筑呈阶梯状向空中延伸，逐层后退的建筑结构赋予了其独特的外观，同时也保证了各个小空间的视野和采光条件。

Profile

Bacata tower is a bold, determined project supported from the political sphere. The project design aims to create a new urban reference, a real engine of transformation and regeneration of the area.

Design Concept

Designers believe that the growth of this city is unlike the logical sustainable growth of large cities worldwide, it grows like “Oil Spot” that expanding inexorably toward peripheries and generating unpredictable functional results. The high energy consumption and travel time when banished to the outskirts of the city make private vehicle owners unable to maintain such a mode, therefore compact urban construction becomes a priority.

To the development mode of a compact city, one of the resources is the building height, not indiscriminately, but in specific areas to be fitted and referencing to the city.

Architectural Design

With its 250 meters high under construction, the project will house an intense mix of uses: hotel, residential, office, commercial, sports and parking. The terraced and landscaped areas are free for the community, providing an outdoor space for communication, gathering and leisure activities.

The tower is proposed as a part of the vertical space of the city. Stepped extension of the building together with retreated building structure makes a unique architectural appearance while ensures enjoyable views and lighting conditions of each small space.

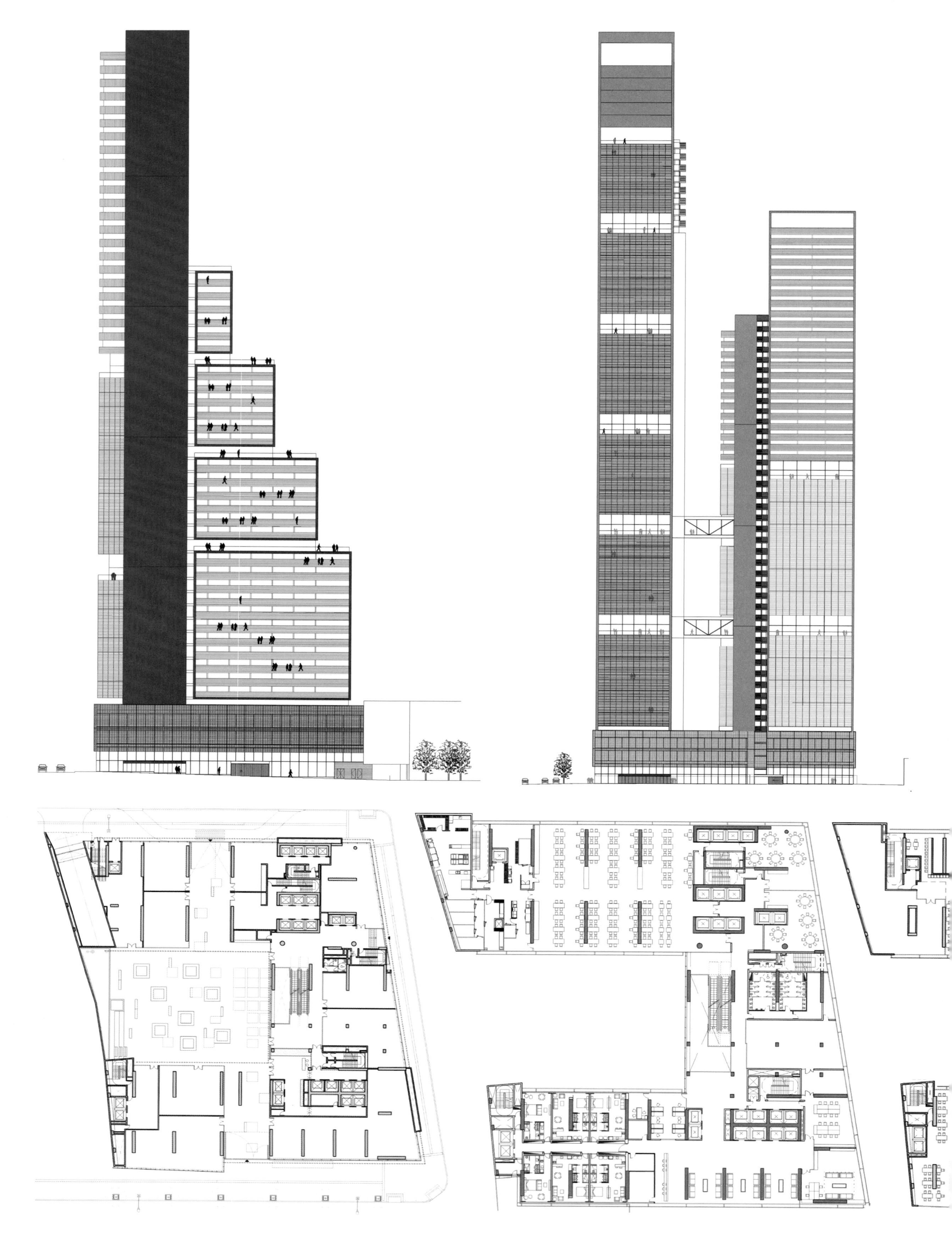

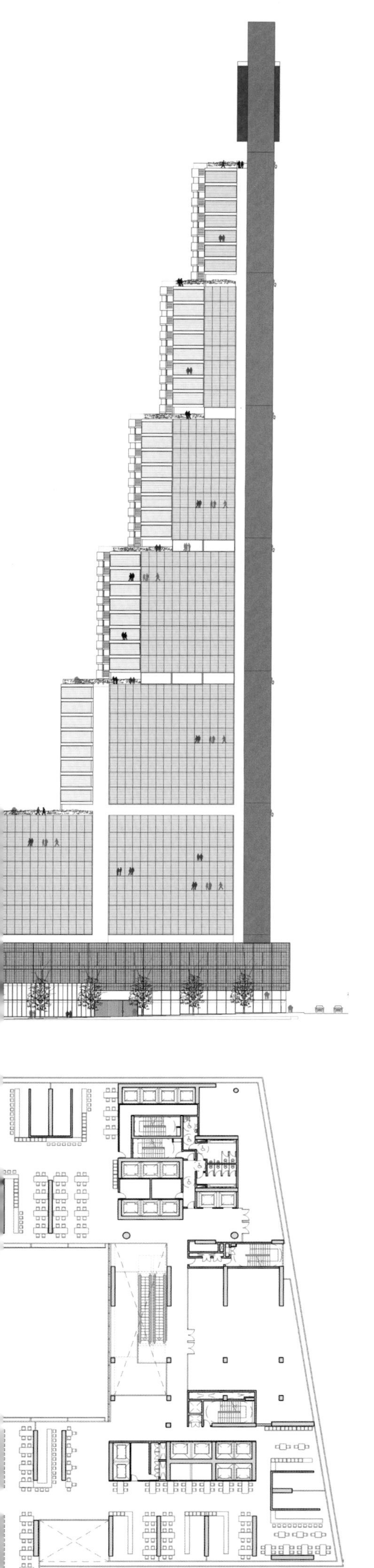

重庆弹子石商业零售娱乐城

Danzishi Retail and Entertainment District

设计单位：10 DESIGN（拾稼设计）
开发单位：重庆规划局
重庆雷士照明
重庆能投置业
华发集团
重庆中誉房地产开发有限公司
项目地址：中国重庆市
占地面积：360 000 m²
建筑面积：800 000 m²
设计团队：Gordon Affleck Ross Milne
Brian Fok Jamie Webb
Ryan Leong Nigel Heigh
Frisly Colop Morales Jason Easter
Lukasz Wawrzenczyk Colin Aston
Francisco Fajardo Mike Kwok
Dan Narita Caterina Choi
Laura Rusconi Clerici Kevis Wong

Designed by: 10 DESIGN
Client: Chongqing Planning Bureau; NVC-Lighting; Chongqing Energy Investment Real Estate; Huafa Group; Zhongyu Real Estate Development
Location: Chongqing, China
Site Area: 360,000 m^2
Gross Floor Area: 800,000 m^2
Design Team: Gordon Affeck, Ross Milne, Brian Fok, Jamie Webb, Ryan Leong, Nigel Height, Frisly Colop Morales, Jason Easter, Lukasz Wawrzenczyk, Colin Aston, Francisco Fajardo, Mike Kwok, Dan Narita, Caterina Choi, Laura Rusconi Clerici, Kevis Wong

项目概况

这个多功能综合发展项目位于长江沿岸，处于中国最大的直辖市——重庆境内，是为该市力争区域经济强化发展的三大热点项目之一。这个以高端商业与零售为主导功能的项目将创建基地面积达 13 万平方米的城市新区。

建筑设计

项目总建筑面积为 80 万平方米，包括 15 万平方米的专属高端零售及娱乐区，配备有文化、体育、酒店及娱乐功能，这些功能占据了裙楼的主要空间。其余区域则用于商务办公、酒店及酒店式公寓等，容纳在高度为 100-250 米的系列塔楼中。

项目设计旨在为开发区提供既实用又具备创新性的整体解决方案。台阶造型的零售庭院围绕荫庇式花园广场而建，其间由高架行人道相互连接。各地块的共同建筑特征建立了彼此间的整体联系，而独特建筑语言的融入，又突出了单体建筑的个性特征。

设计特色

基地地势陡峭，向下延伸的坡道直通长江江岸，这不仅体现出重庆典型的地形特征，也呼应了传统的楼梯街和面朝长江的旧时建筑。70 米的地势高差贯穿整个开发区，为打造多重零售层、由沿街面激发活力的商业业态创造了得天独厚的条件。

设计利用地形优势，使每一零售层与上、下层面形成互动，提高具有多重标高的商业裙楼的价值。延伸至江边的坡道以及与轻轨站高架连接桥将商业人流源源不断地引入建筑内，进一步激活了多层次的零售业态。这些通路均延伸自阶梯造型的商业零售裙楼，它们形成的无缝商业动线，不仅激活了该区域的活力，为购物者带来多维度体验，还营造出方便日常居住的高效商业环境。

Profile

Located on the banks of the Yangtze River in China's largest municipality, Chongqing, this mixed use development is one of the three major focal projects in the city seeking to enhance the economic development of the region. The project focusing on high-end business and retail will create a new urban district with a total site area of 130,000 square meters.

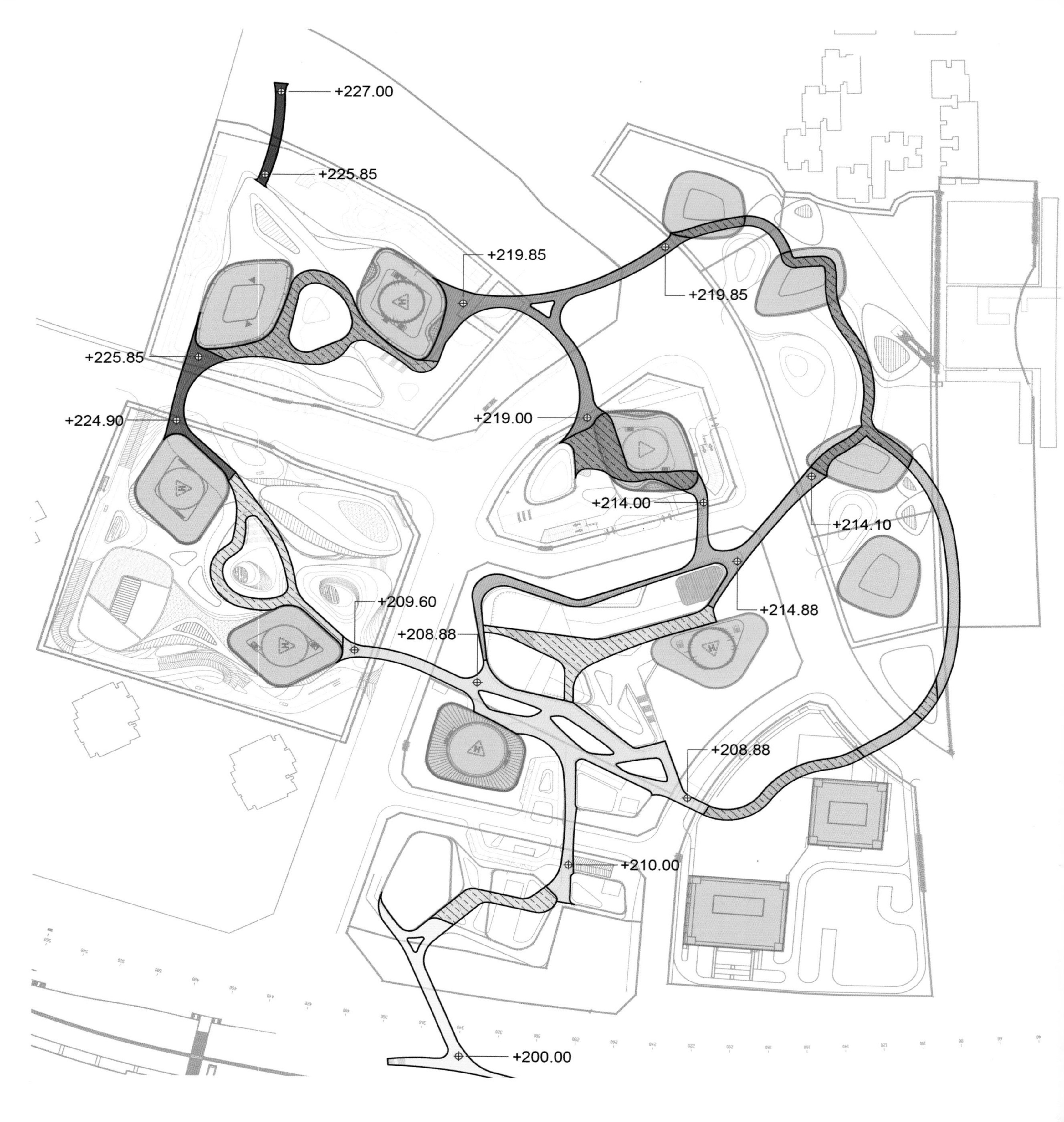

Architectural Design

Comprising a total of 800,000 square meters of accommodation, the development includes 150,000 square meters of dedicated high-end retail and entertainment area including cultural, sport, hospitality, and entertainment functions. This accounts for the majority of podium level accommodation, whilst the remaining area is given to commercial office space, hotel and serviced apartments, all accommodated in a series of towers ranging between 100-250 meters in height.

The design aims to provide a cohesive solution that is both pragmatic and architecturally innovative within the context. By creating a series of terraced retail courtyards surrounding sheltered garden plazas and linked together by elevated walkways, the plots are connected by common features whilst allowing for distinct architectural languages to differentiate their respective identities.

Design Feature

The steep site topography sloping down to the Yangtze River is characteristic of Chongqing and reminiscent of the city's traditional stepped streets and buildings fronting onto the river. A 70 meters level difference across the development area provides a unique opportunity to create multiple retail levels each activated by street frontage.

By taking advantage of this key topographical feature, the design multiplies the commercial value of the various podium levels significantly. This multi-level retail diagram is further activated by footfall along the riverside promenade at the lower end, and a light rail station at the top of the site from which pedestrians are drawn into the core of the development by means of the elevated link bridges. These walkways extend from the terraced retail podiums to provide seamless retail circulation such that the district becomes both a dynamic multi-dimensional experience for the occasional retail visitor, and an efficient business environment for the day-to-day inhabitant.

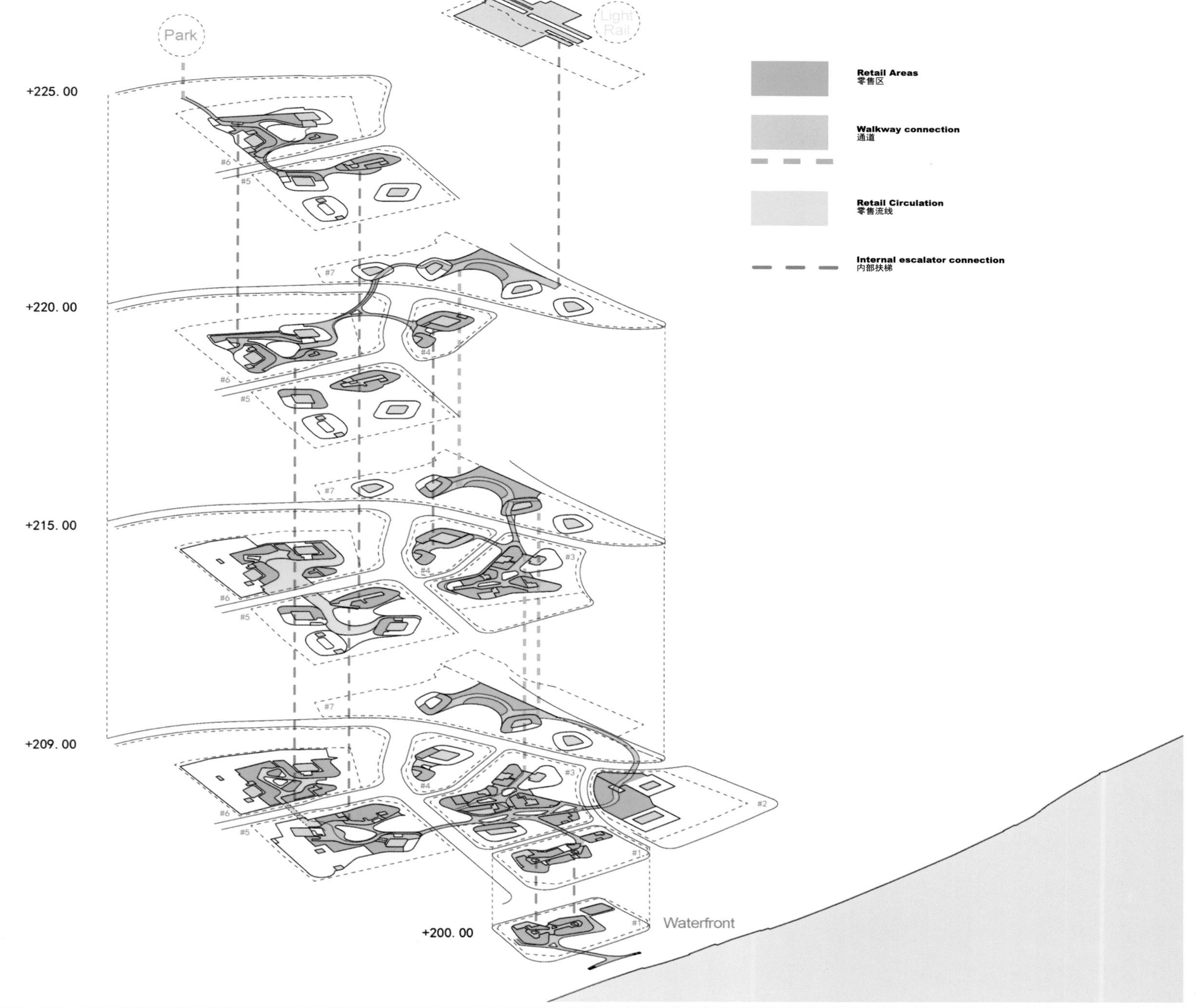
Park
Light Rail
+225.00
+220.00
+215.00
+209.00
+200.00
Waterfront
#1
#2
#3
#4
#5
#6
#7
Retail Areas
零售区
Walkway connection
通道
Retail Circulation
零售流线
Internal escalator connection
内部扶梯

NVC LIGHTING (C5-1)
NVC LIGHTING (C5-4)
ZHONGXUN PLAZA (C5-4)
ENERGY (C5-4)
ENERGY (C4-2)

江苏苏州高铁新城

Suzhou Hi-speed Railway New City

设计单位：RTA-Office 建筑事务所

开发商：苏州市高速铁路新城建设投资有限责任公司

项目地址：中国江苏省苏州市

总建筑面积：150 000 m²

摄影：RTA-Office 建筑事务所

Designed by: RTA-Office

Client: Suzhou Hi-speed Railway New City Construction & Investment Co., Ltd.

Location: Suzhou, Jiangsu, China

Total Building Area: 150,000 m²

Photography: RTA-Office

项目概况

这是由 RTA-Office 建筑事务所设计的苏州高铁新城地标塔楼项目，这个项目将成为北京—上海高铁沿线苏州站的地标性建筑。

设计特色

设计师不仅是在构建建筑，同时也是在营造空间，通过营造一个出人意表的新空间来丰富人们的生活体验。设计师着重突出苏州这座城市的特征，通过水、花园、窗户等细节，流露出这座城市独有的气质和韵味。同时，设计赋予了建筑现代的特征，使之既具备了当地的历史感，又增添了一抹现代都市的时尚感，恰如其分地将古典与现代两种气息糅合在一起。

整个方案基于两个独立建筑的设计：一个为酒店，一个为办公商业建筑。这两个大型的如雕塑般的建筑体表面覆盖有特殊的纹理，构成了建筑易于识别的肌理。这是一种技术性的网面，由激光切割玻璃和金属质地的元素构成，它裹覆着塔楼，为其增添了现代感和活力。覆盖酒店、商业区、办公区的材质都不一样，但是却表达着相同的特性。塔楼底座的纹理紧凑封闭，越向上越轻盈，直至最后消失不见。

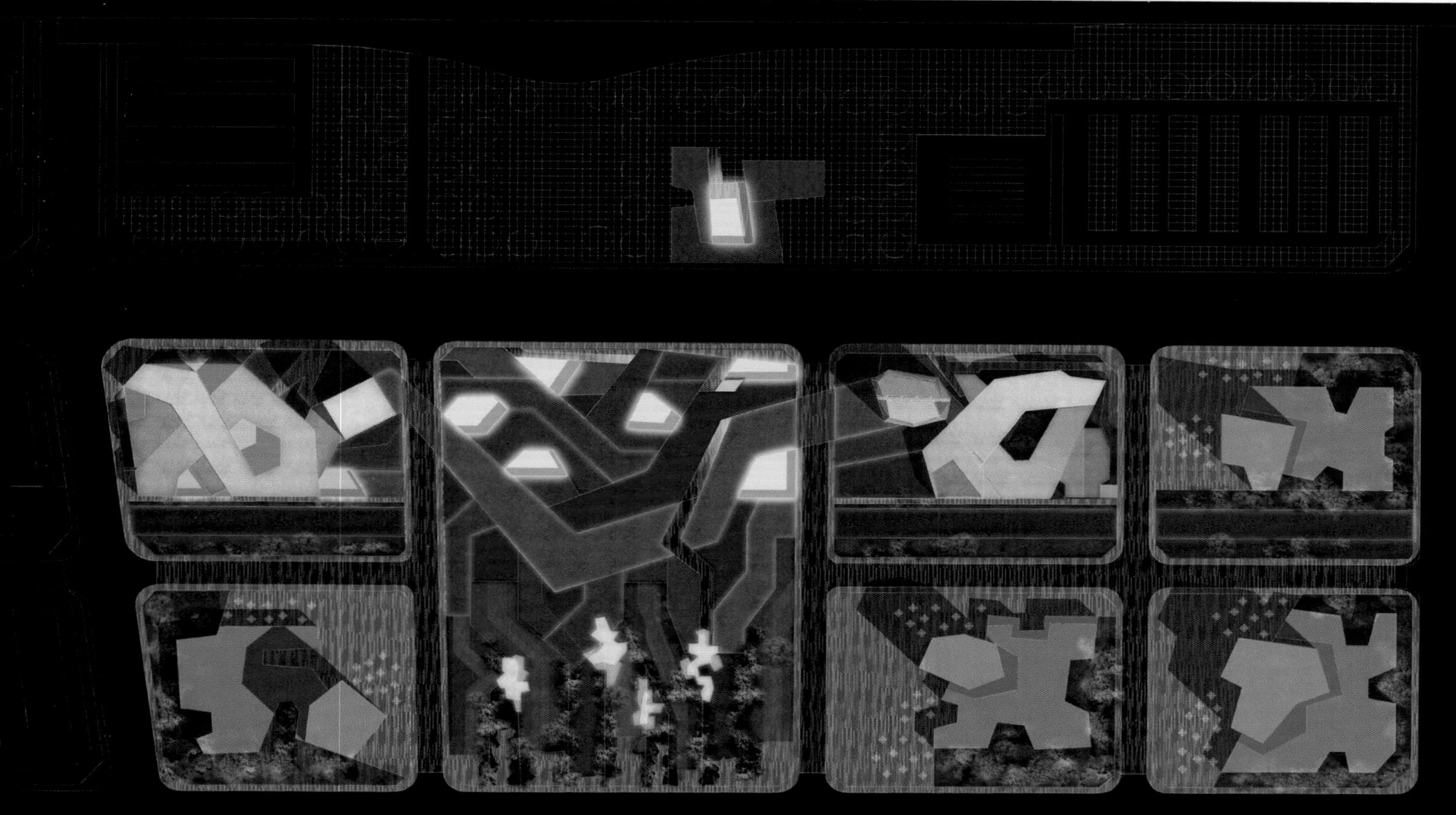
0 5m 10m 20m
N

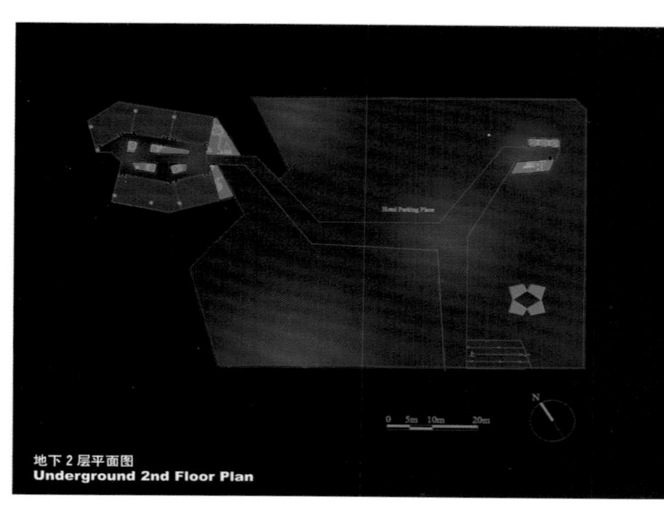

地下 2 层平面图
Underground 2nd Floor Plan

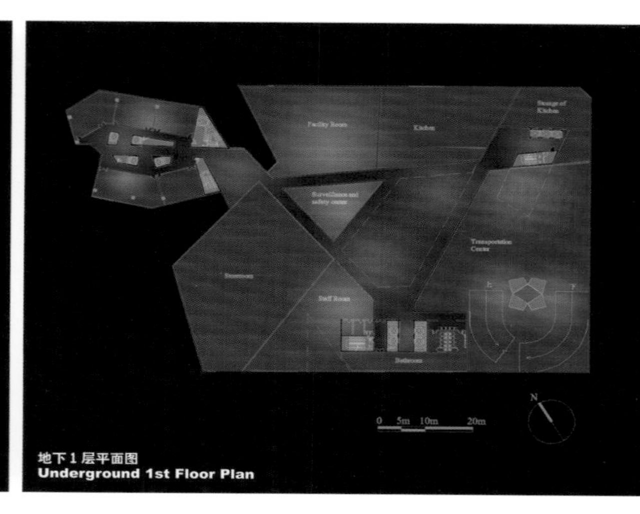

地下 1 层平面图
Underground 1st Floor Plan

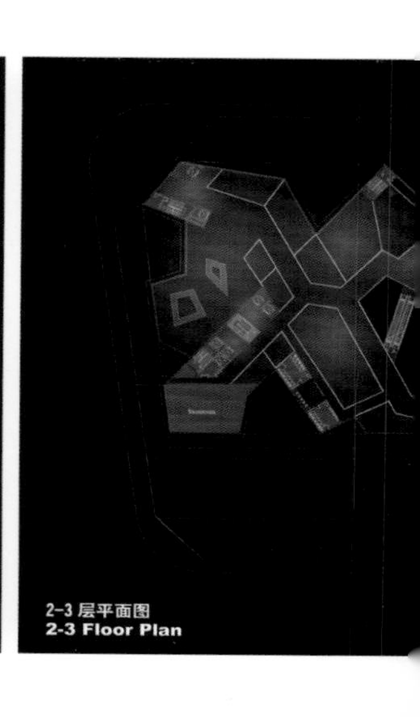

2-3 层平面图
2-3 Floor Plan

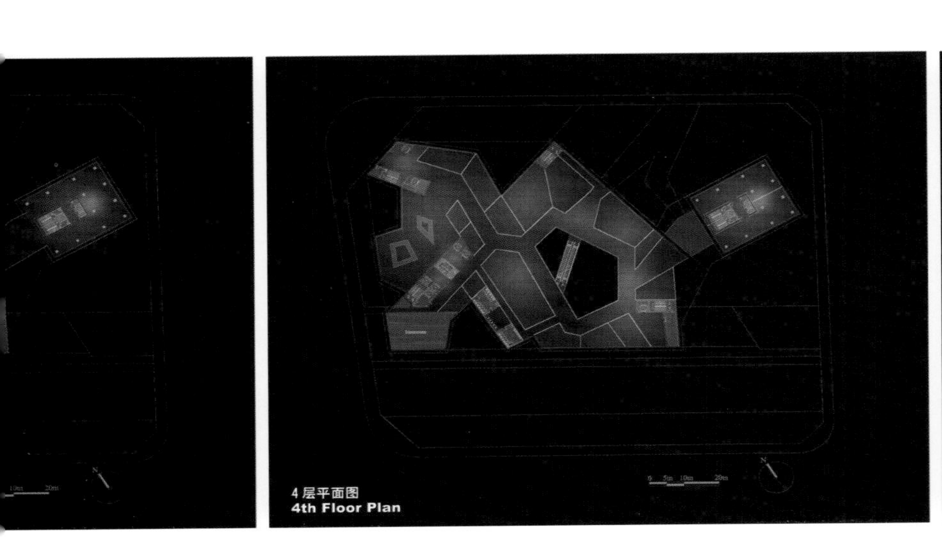

4 层平面图
4th Floor Plan

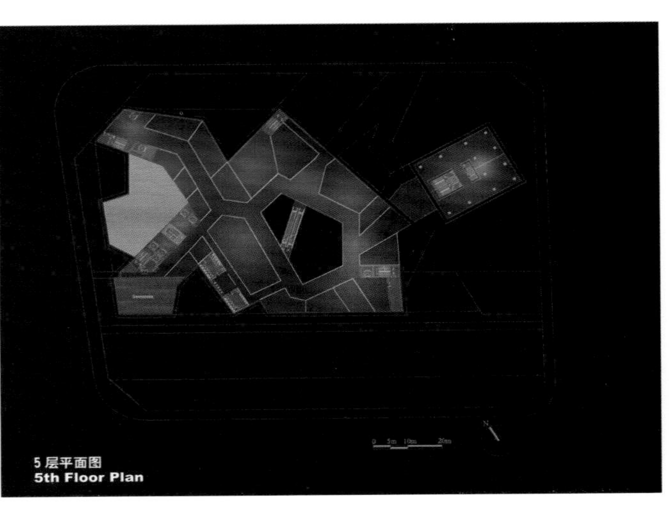

5 层平面图
5th Floor Plan

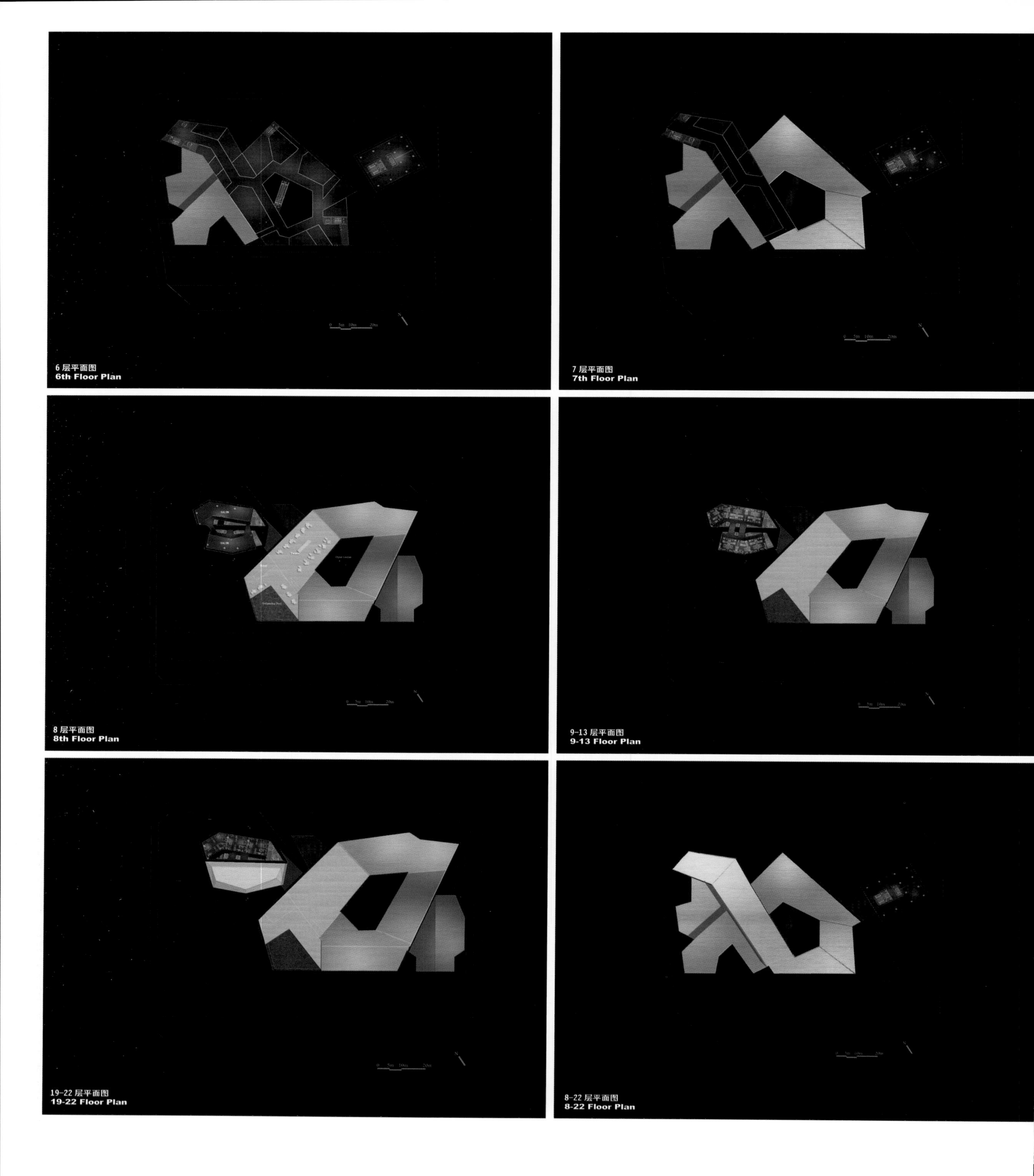

6 层平面图
6th Floor Plan

7 层平面图
7th Floor Plan

8 层平面图
8th Floor Plan

9-13 层平面图
9-13 Floor Plan

19-22 层平面图
19-22 Floor Plan

8-22 层平面图
8-22 Floor Plan

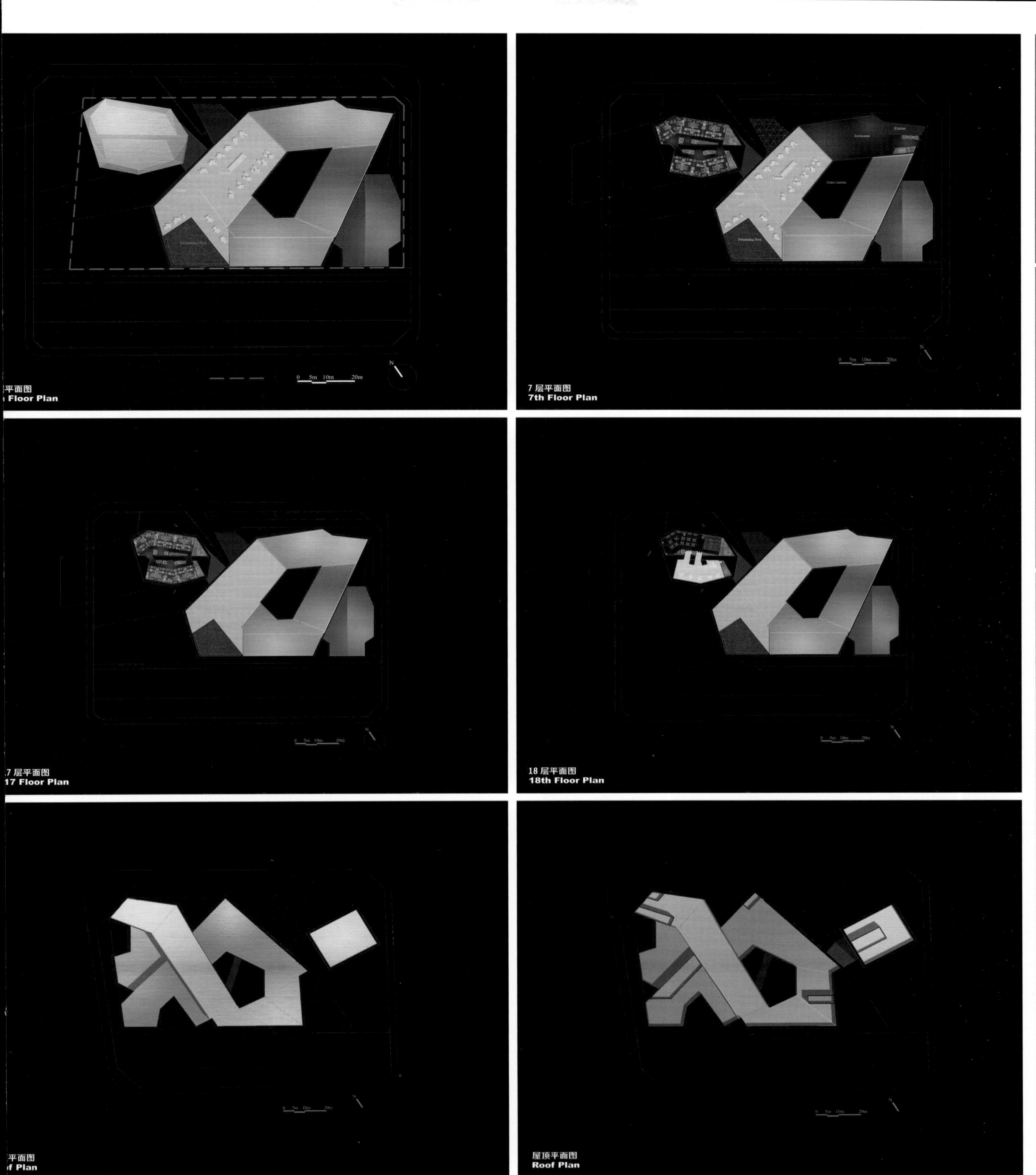

层平面图
Floor Plan

7 层平面图
7th Floor Plan

17 层平面图
17 Floor Plan

18 层平面图
18th Floor Plan

平面图
f Plan

屋顶平面图
Roof Plan

Profile

It is a landmark tower project of Suzhou Hi-speed Railway New City designed by RTA-Office. The project will be the landmark of Beijing-Shanghai Express Railway Suzhou Station area.

Design Feature

Designers are more than constructing buildings, but creating spaces. Creating unexpected new spaces will enhance the life experience of man. Designers also try to emphasize the main characters of Suzhou. Water, gardens, windows and other details reveal tremendous charm and appeal of this city. In the meantime, the design imposes modern architectural features on these towers, which enable them to possess sense of history and sense of modern fashion as well. Classical and modern styles are perfectly integrated.

The proposal is based on the design of two singular objects, a hotel and an office & commercial building. These two large sculptures wear a special dress with recognizable texture, a technical mesh composed of glass and metal texture elements cut by laser. It covers the tower and gives them both modernity and vitality. The texture covering the buildings is different in the hotel, in the shopping and in the office building. But they have the same property. In the bottom of the towers, the texture is more compact and closed while at the top of the building becomes lighter until even disappearing.

广西贵港盛世华府

Guigang Shengshi Mansion, Guangxi, China

设计单位：广州市纬纶建筑设计公司
开发商：贵港市远辰中海房地产开发有限公司
项目地址：中国广西壮族自治区贵港市
占地面积：51 485 m²
总建筑面积：486 357 m²
容积率：7.52
设计团队：许佐锐 朱溢钊 杜淑莹
郑燕畅 江启冬 张健玮

Designed by: Win-land Architecture Design Co., Ltd.
Developer: Guigang Yuanchen Zhonghai Real Estate Development Co., Ltd.
Location: Guigang, Guangxi, China
Site Area: 51,485 m²
Floor Area: 486,357 m²
Plot Ratio: 7.52
Design Team: Xu Zuorui, Zhu Yizhao, Du Shuying,
Zheng Yanchang, Jiang Qidong, Zhang Jianwei

项目概况

项目位于贵港市新城区的中心地带，处于城市发展的中轴线上，拥有良好的区位发展优势。设计旨在深度挖掘地块价值，将项目打造成区域范围内的城市地标。

建筑设计

该项目是由办公酒店主楼、商业区、住宅区三大部分组成的城市综合体，办公酒店主楼为 270 米的超高层，商业区由集中商业和商业街两部分组成，住宅区主要为大户型超高层。

在结构布局上，设计通过整合办公、酒店、商业购物、影视娱乐、健身休闲、餐饮等功能空间的布局，提高客群的广域性和循环性，创造高效的商业价值。在外观上，设计通过统一外立面风格，来协调整个社区的形象。

分区设计

办公区注重办公空间的实用性，通过合理组织垂直交通以及新材料的运用，打造布局灵活、高效节能的办公空间。集中商业区明确集中商业的动线，使主轴贯穿整个商业空间，同时，合理布置立体交通，引导人流快速便捷地出入商场。沿街商业沿城市主干道分布，布局连续，并强调与集中商业的衔接。住宅区注重平面与景观的关系，室外大型的中央园林等开放空间的设计，营造了高品质的生活环境。

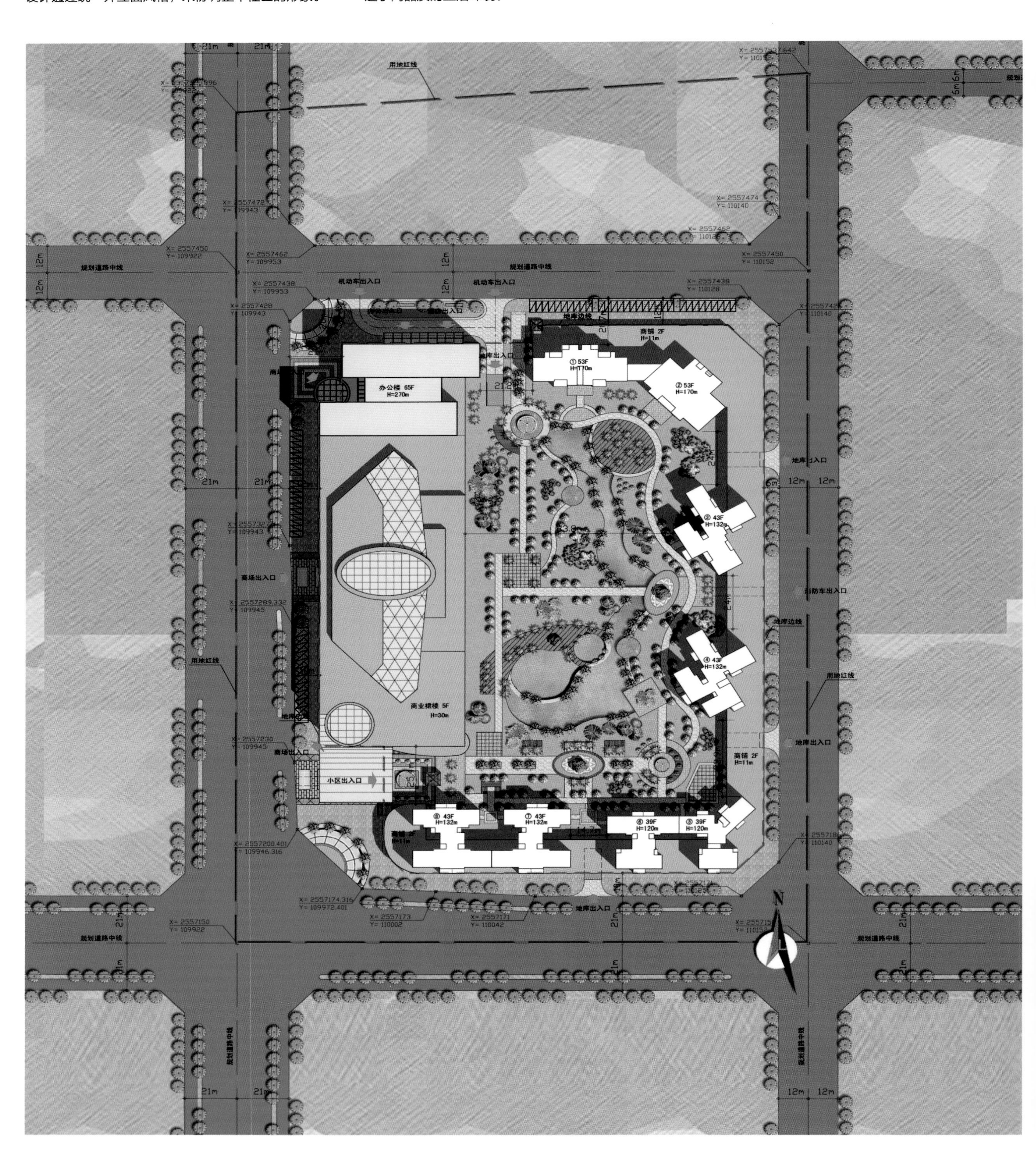

规划道路
Planning Road

可达步行人流路线
Accessiable Pedestrian Route

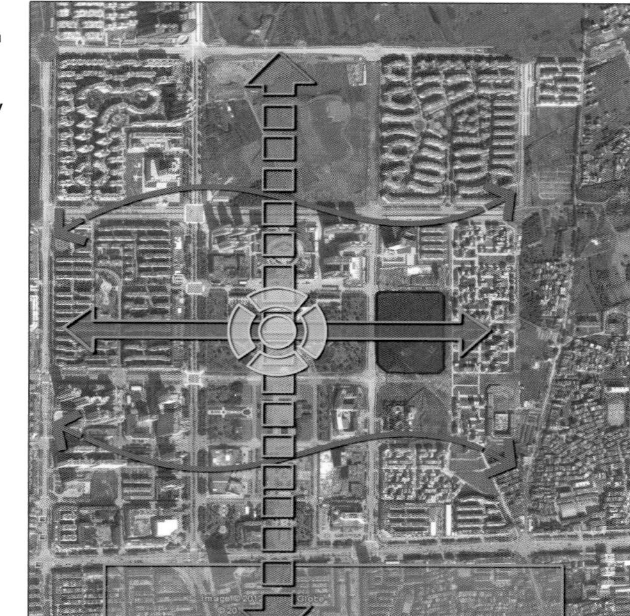

功能与景观主轴
Function & Landscape Main Axis

功能与景观次轴
Function & Landscape Secondary Axis

城市发展带
Urban Development Area

城市中心
City Center

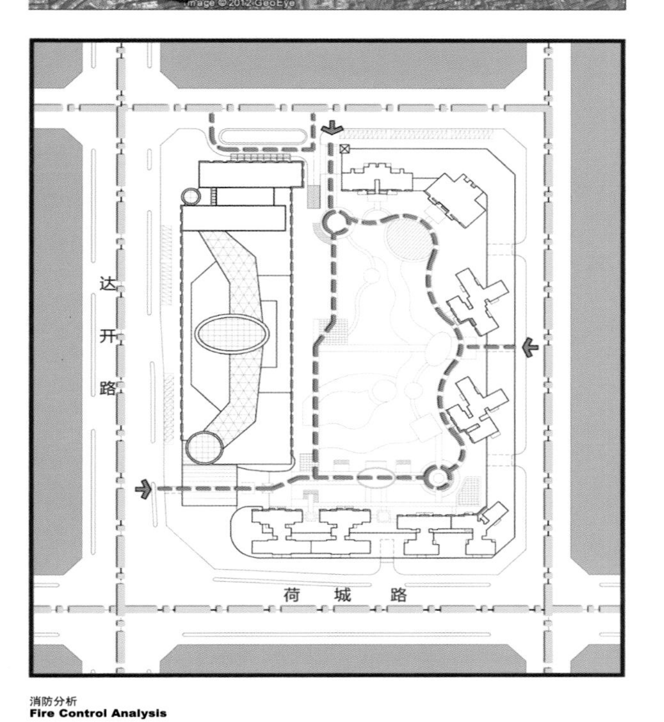

消防分析
Fire Control Analysis

城市路网
Urban Road Network

消防扑救面
Fire-fighting Plane

消防车道
Fire-fighting Lane

消防出入口
Fire-fighting Access

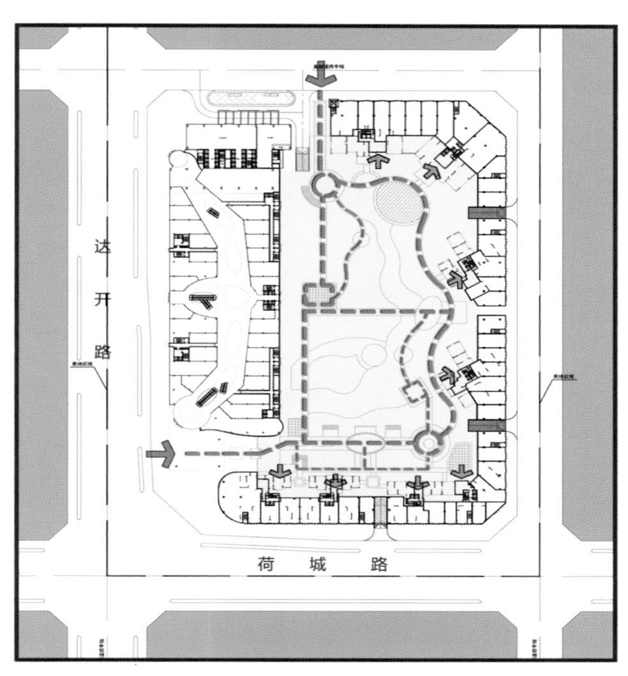

住宅流线
Residence Circulation

住宅区步行道
Residential Area Footpath

住宅大堂出入口
Residential Lobby Entrance & Exit

地库出入口
Garage Entrance & Exit

住宅区出入口
Residential Area Entrance & Exit

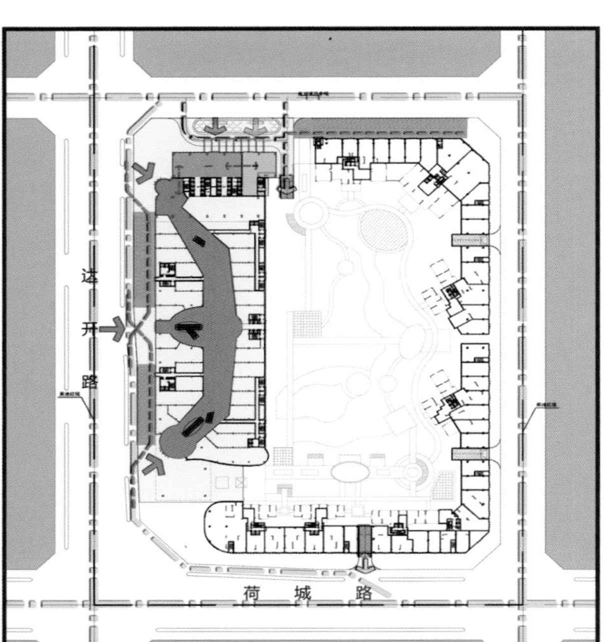

商业、酒店办公流线
Commerce, Hotel, Office Circulation

商场流线
Mall Circulation

商业出入口
Commercial Access

办公大堂
Office Lobby

人流渗透
Pedestrian Flow

酒店大堂
Hotel Lobby

地面停车
Ground Parking

地库入口
Garage Entrance

停车路径
Parking Path

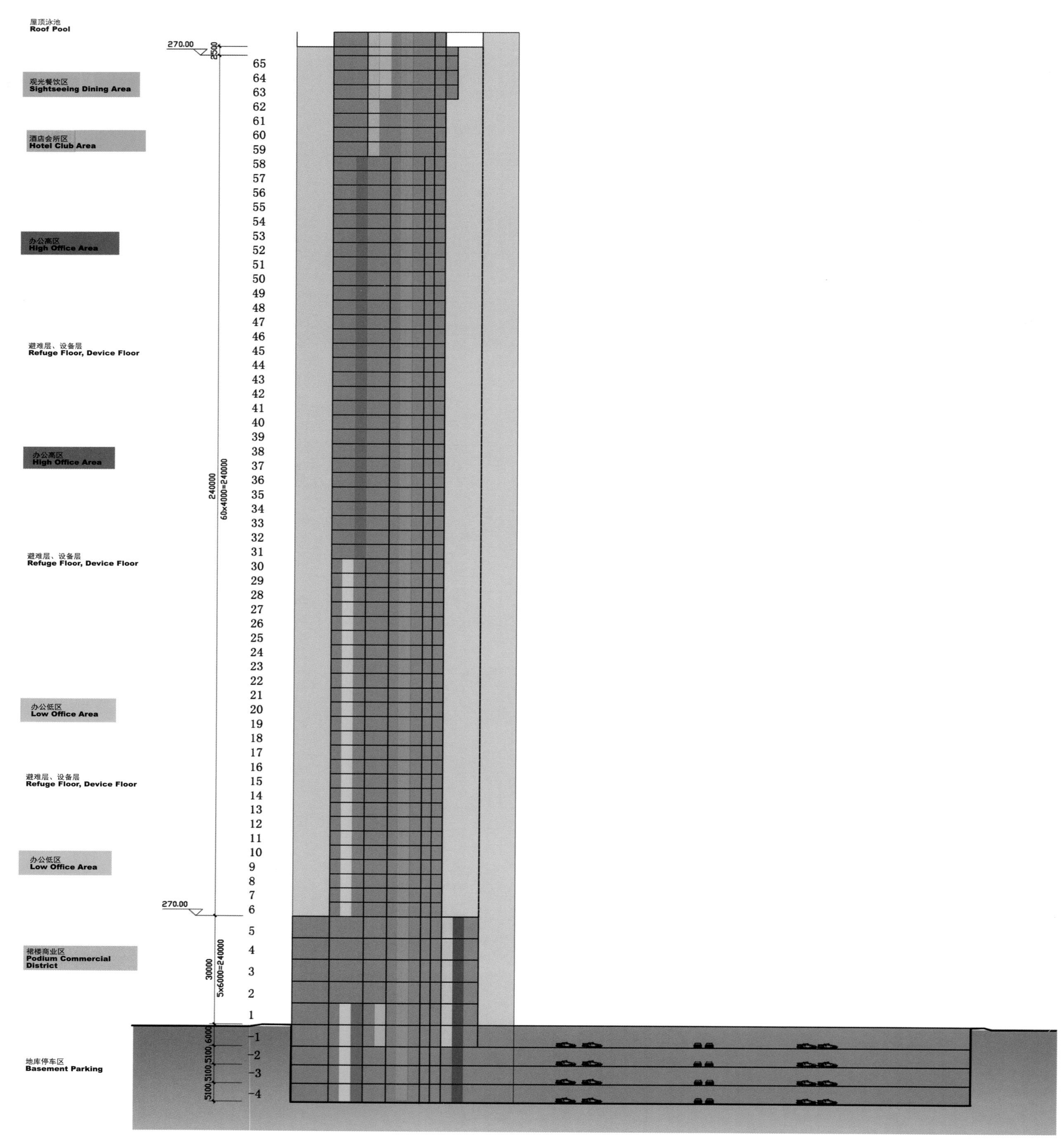

Profile

The project is located in the center of Guigang New City Region along urban development axis with favorable development advantages. The design aims to further explore the site's value and build an urban landmark of this region.

Architectural Design

The project is an urban complex composed of office & hotel main building, commercial block and residential block. The main building is 270 meters high comprising intensive commercial district and commercial street. Residential block is mainly occupied by large-unit super high-rises.

On structural layout, the design has integrated office, hotel, shopping, theater, fitness, restaurant and other functional spaces to expand scope of customers, increase circulation and elevate commercial value. On architectural appearance, façade style is unified to coordinate overall community image.

Subarea Design

The design emphasizes practicality of office space and rational organization of vertical transportation. Through application of new materials, a flexible high-efficient and energy-saving office space is created. Specify concentrated commercial lines, ensuring principal axis to penetrate the whole commercial space. Reasonably organize vertical traffic and lead customers to get in or out of the mall conveniently. Street commerce is distributed along main urban roads. Continuous layout emphasizes the connection with intensive commerce. Residential block pays attention to the relationship between planes and landscape. Large outdoor central garden and other open spaces contribute to the creation of high quality living environment.

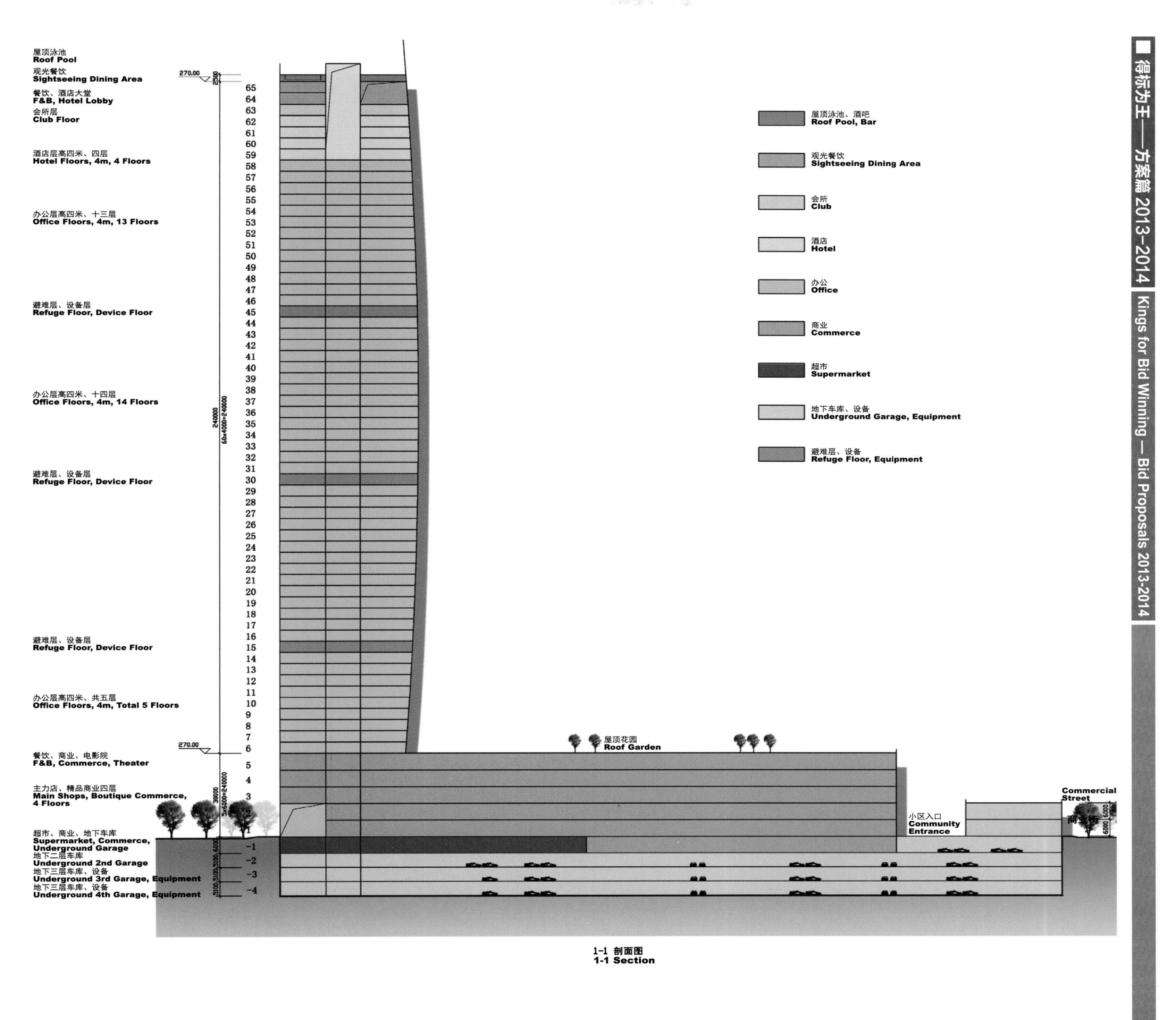

1-1 剖面图
1-1 Section

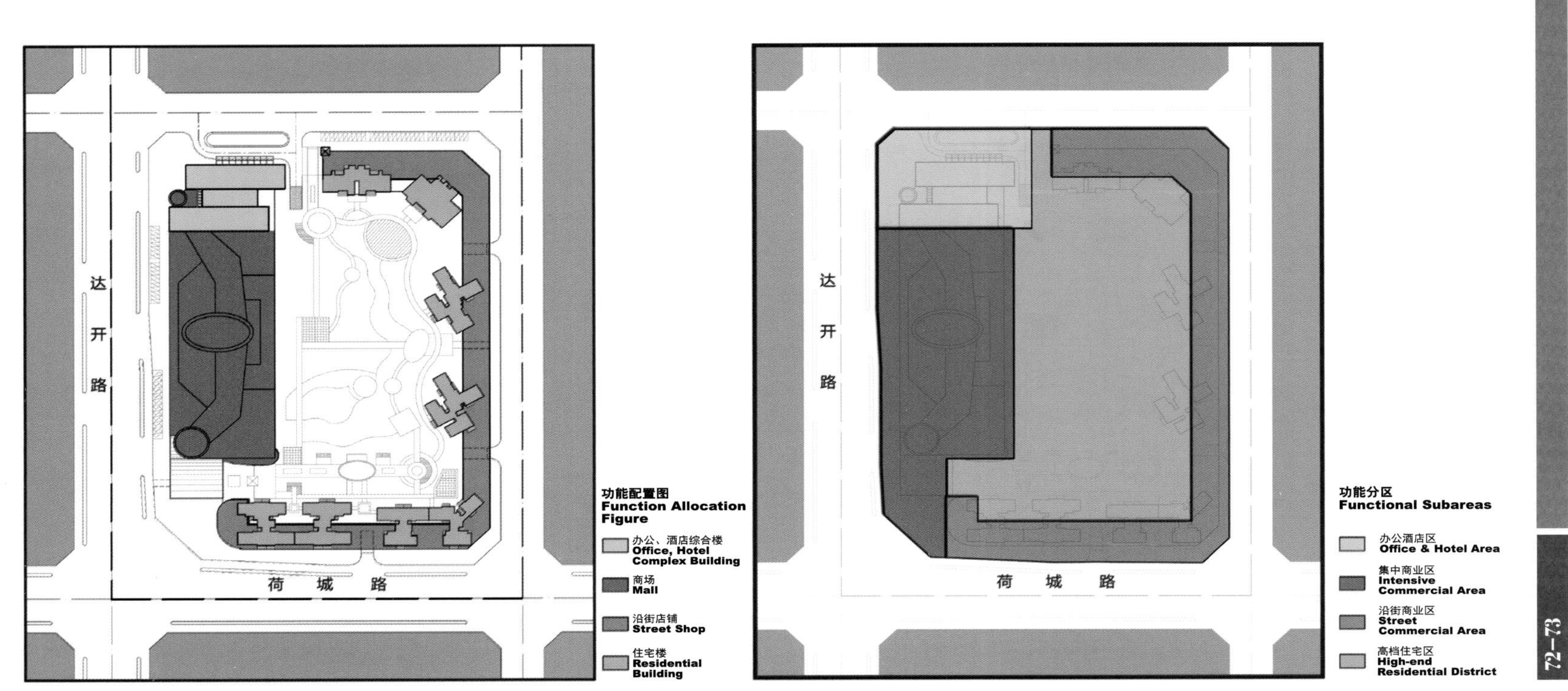

广东深圳创维公明新城

Skyworth Gongming Mixed-use

设计单位：阿特金斯 ATKINS
开发商：新创维电器（深圳）有限公司
项目地址：中国广东省深圳市
建筑面积：422 990 ㎡

Designed by: Atkins
Client: New Skyworth Electrical (Shenzhen) Co., Ltd.
Location: Shenzhen, Guangdong, China
Area: 422,990 m²

项目概况

项目位于深圳西部光明新区，是深圳首个集居住、购物、休闲、娱乐、办公于一体的中高端绿色城市综合体。

建筑设计

项目总建筑面积达 422 990 平方米，涵盖了 17 栋高 100 米的高层住宅、3 栋弧形塔楼及 61 000 平方米购物中心。3 栋塔楼分别为高 85 米、135 米的公寓以及高 250 米的酒店 + 办公混合体，它们沿着轻轨线分布，呼应了城市的天际线。购物中心由 5 层高的大型商场及 3 层高的零售商业区构成，内置餐饮、KTV 和电影院。

绿色节能设计

绿色节能设计是该项目的设计重点之一。塔楼的南北向采用横向遮阳，东西向则采用竖向遮阳。购物中心的裙房顶部采用了弧形大格栅，既可遮阳又不影响自然采光和通风，同时，在屋顶设置的空中花园向社会开放，有利于将项目打造成该区域的城市大客厅。

Profile

This landmark mixed-use project in Western Shenzhen consisting of 3 interconnected blocks of residential, shopping, leisure, entertainment, and office uses aims to become the first eco-friendly development in Guangming District.

Architectural Design

The total GFA of the project is 422,990 square meters covering 17 high-rise residences of 100 meters, three curvilinear towers and 61,000 square meters shopping center. The three curvilinear towers including two apartment buildings respectively 85 meters and 135 meters high, and a 250 meters hotel & office mixed building dominate the skyline along the lightrail. The commercial program weaves shops, restaurants, KTV and cinemas into a 5-storey shopping mall and 3-storey shopping street complex.

Energy-saving Design

Energy-saving design is the key of the project design. Horizontal shading is adopted in north-south direction of the towers while east-west direction applies vertical shading. Arc-shaped grid used at the top of the podium of shopping center plays a role in sun shading while without affecting ventilation and natural lighting. Sky garden set on the roof opens to the public, which facilitates to be built an "urban living room concept" for the new CBD .

BRYKOR
DAIRY
BOUCHERON

总体经济技术指标

		7#	8#	9#	合计
总用地面积		18747	34417	33212	86376
总建筑面积		79460	216709	128340	424509
其中	住宅	69380		113340	182720
	商业	5800		15000	20800
	产业用房		151820		151820
	产业配套用房		61739		61739
	公共配套	4280	3150		7430
建筑覆盖率		28.00%	60%	28%	
绿化覆盖率		68.8%	60.2%	70%	
容积率		4.24	6.30	3.86	
停车位	（地上）	30		20	
	（地下）	800	2700	1400	4900

分项经济技术指标

	7#		8#			9#	
总用地面积	18747		34417			33212	
总建筑面积	79460		216709			128340	
其中	住宅	69380	产业用房	写字楼	30400	住宅	113340
	街铺	5800		酒店	33400	街铺	15000
	幼儿园	3600		公寓	72000		
	物业管理用房	680		街铺	16020		
			产业配套用房	商场	61739		
			公共配套	文化室	2500		
				居委会	200		
				服务站	300		
				邮政所	150		
容积率	4.24		6.30			3.86	

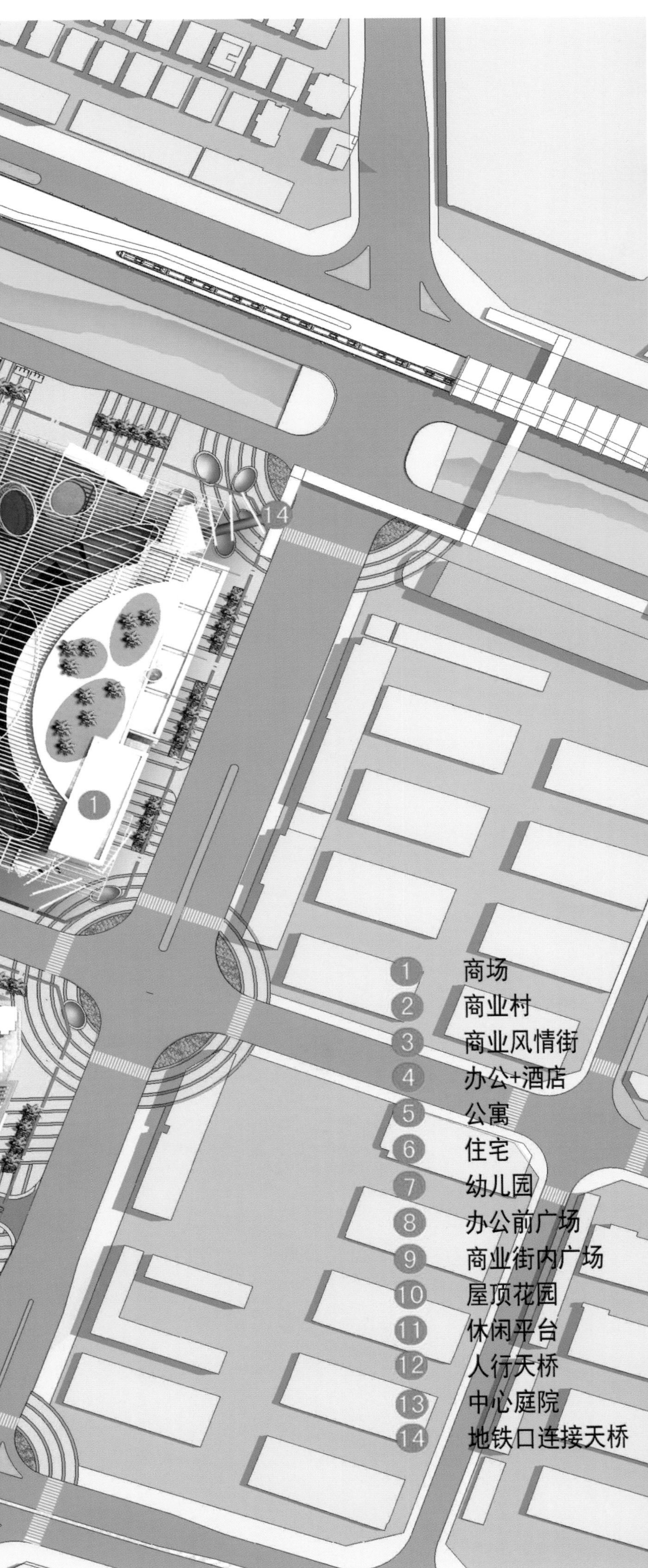
14
10
1
1 商场
2 商业村
3 商业风情街
4 办公+酒店
5 公寓
6 住宅
7 幼儿园
8 办公前广场
9 商业街内广场
10 屋顶花园
11 休闲平台
12 人行天桥
13 中心庭院
14 地铁口连接天桥

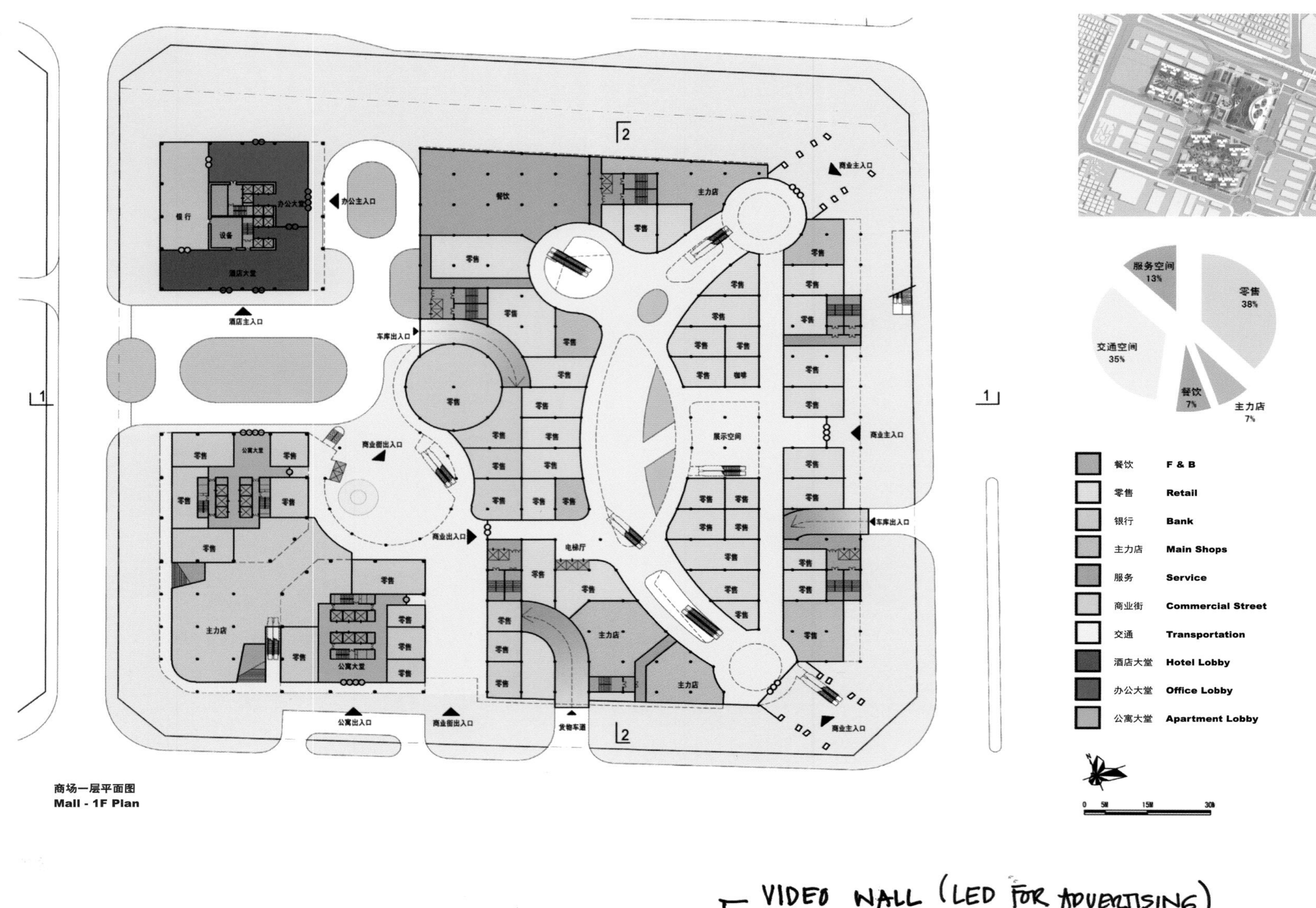

商场一层平面图
Mall - 1F Plan

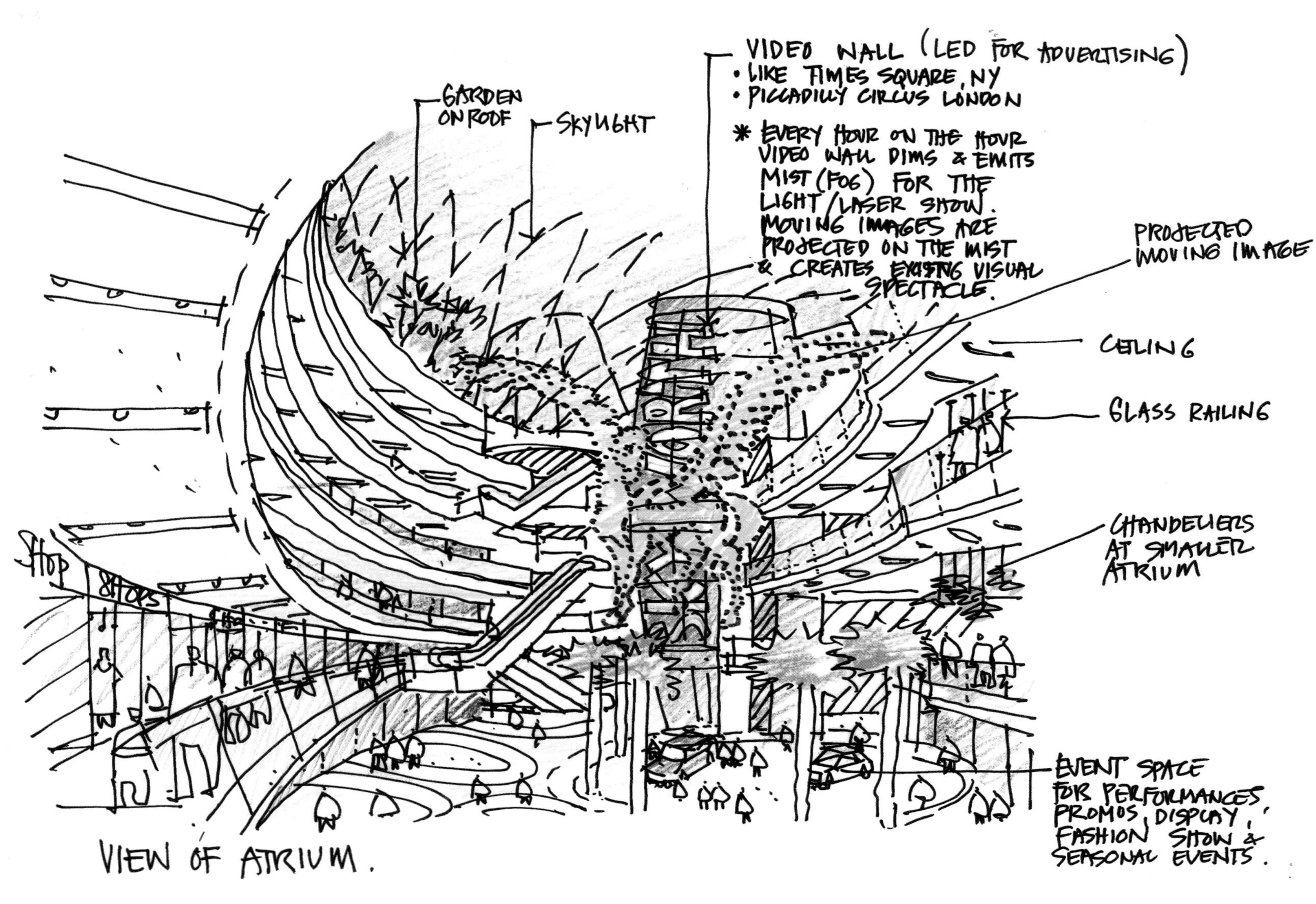

VIEW OF ATRIUM.

西立面
West Elevation

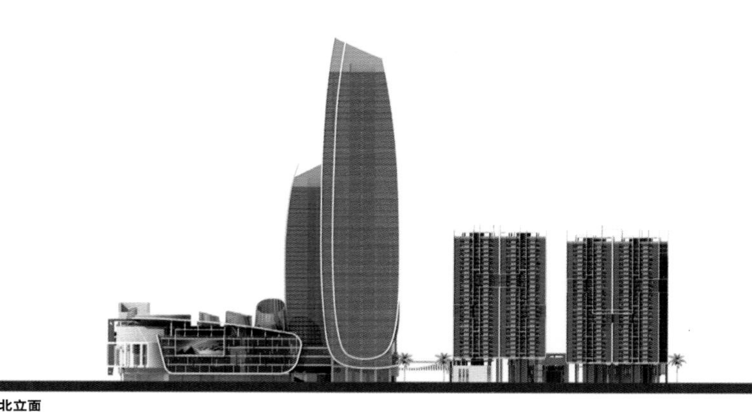

北立面
North Elevation

东立面
East Elevation

南立面
South Elevation

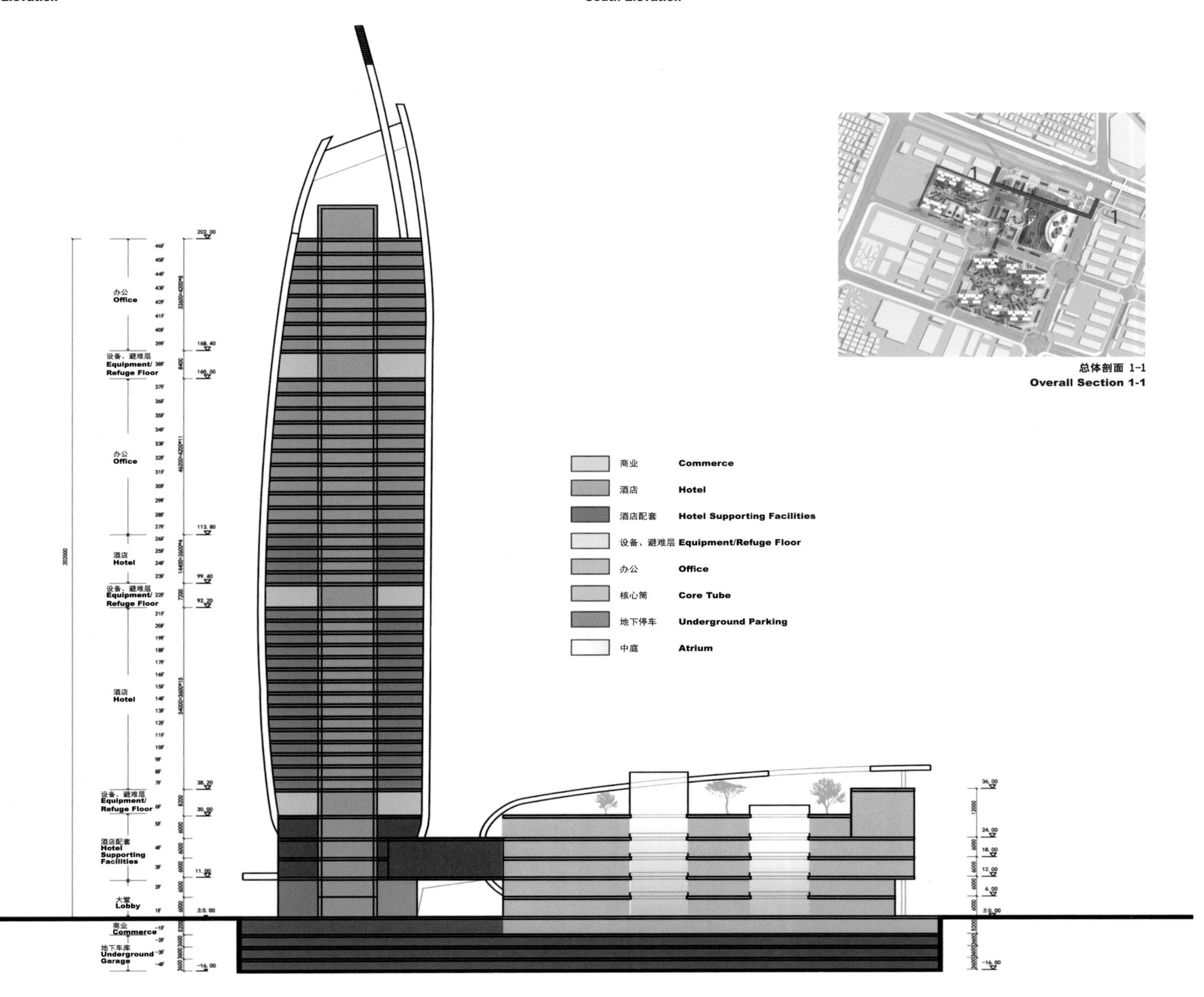

总体剖面 1-1
Overall Section 1-1

阿联酋迪拜 Al Salam Tecom 塔楼

Al Salam Tecom Tower

设计单位：阿特金斯 ATKINS
开发商：Abdulsalam M Rafi Al Rafi
项目地址：阿联酋迪拜
总建筑面积：94 500 m²

Designed by: Atkins
Client: Abdulsalam M Rafi Al Rafi
Location: Dubai, UAE
Total Built Up Area: 94,500 m²

项目概况

Al Salam Tecom 塔楼位于迪拜科技、经济、贸易和媒体自由区，建筑高 195 米，是一栋涵盖了零售商业、酒店式公寓和办公空间的多功能建筑。

建筑设计

这个 47 层的建筑建造在 28 米高的裙房之上，建筑的 1 层由零售区和通向公寓和办公区域的门厅组成，2 层的地下室和 5 层的裙楼层可提供专用的停车区域。美食广场、商务中心和远处的零售区位于 1 层位置。底部 15 层为酒店式公寓，上部 23 层为办公室楼层。裙房的露天层可以提供各种娱乐设施，包括两个游泳池和两个健康俱乐部，其中一个为酒店式公寓的住户专用，而另一个为办公区的租户服务。

倾斜的立面表现形式是建筑最大的特色，突出了由两个三角形图案组成的建筑主立面。反光的银色和蓝色玻璃与灰色玻璃混合使用，有利于区分和界定各功能区间。住宅区的立面采用了预制板这一表现元素，而对窗体和阳台的处理，则将建筑内部的混合用途元素凸显并分离出来。

Profile

The Al Salam Tecom Tower in the Dubai Technology, Electronic, Commerce and Media Free Zone on Sheikh Zayed Road stands at 195 meters. This mixed use building comprises a retail area, serviced apartments and offices.

Architectural Design

The 47-storey building rests on a 28 meters high podium. The ground floor comprises a retail area and the entrance lobby to the apartments and office area. The two basement levels and five podium levels are dedicated parking spaces. The food court, business centre and further retail areas are located at first floor level. This mixed use building comprises serviced apartments on the lower 15 floors and offices on the upper 23 floors. The podium accommodates recreational facilities at deck level, including swimming pools and health clubs - dedicated for the serviced apartment tenants and for office space tenants.The most significant feature of the building is the inclined expression which emphasizes the two triangular masses composing the tower's main elevation. Reflective silver and blue curtain glazing is used along with tinted grey glazing helping to define areas on the structure. The residential component is expressed by the use of precast panels with window treatment and balconies to highlight and separate the mixed use elements of the building.

SHOPPING
HALL
ALSALAM

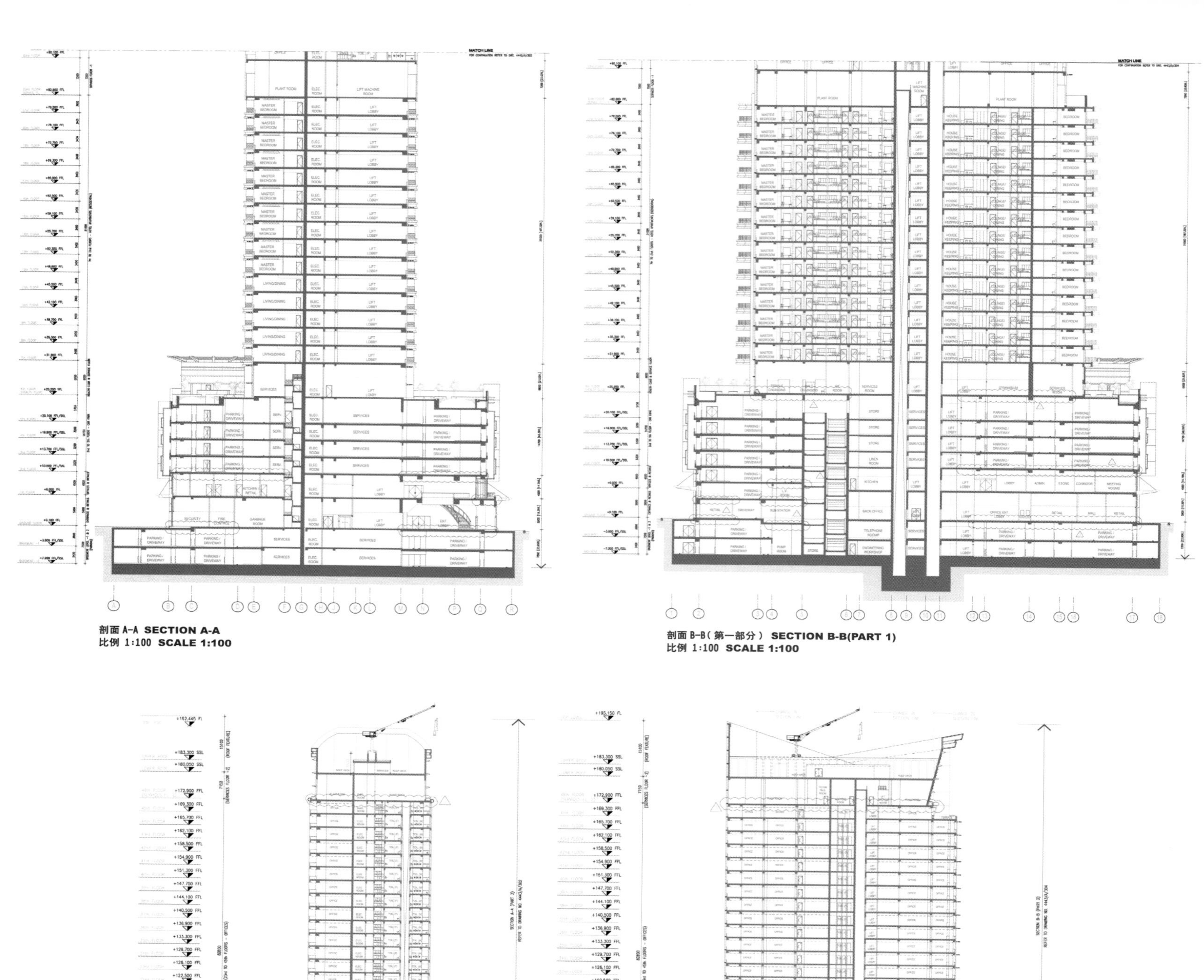

剖面 A–A **SECTION A-A**
比例 1:100 **SCALE 1:100**

剖面 B–B（第一部分） **SECTION B-B(PART 1)**
比例 1:100 **SCALE 1:100**

剖面 A–A **SECTION A-A**
比例 1:100 **SCALE 1:100**

剖面 B–B **SECTION B-B**
比例 1:100 **SCALE 1:100**

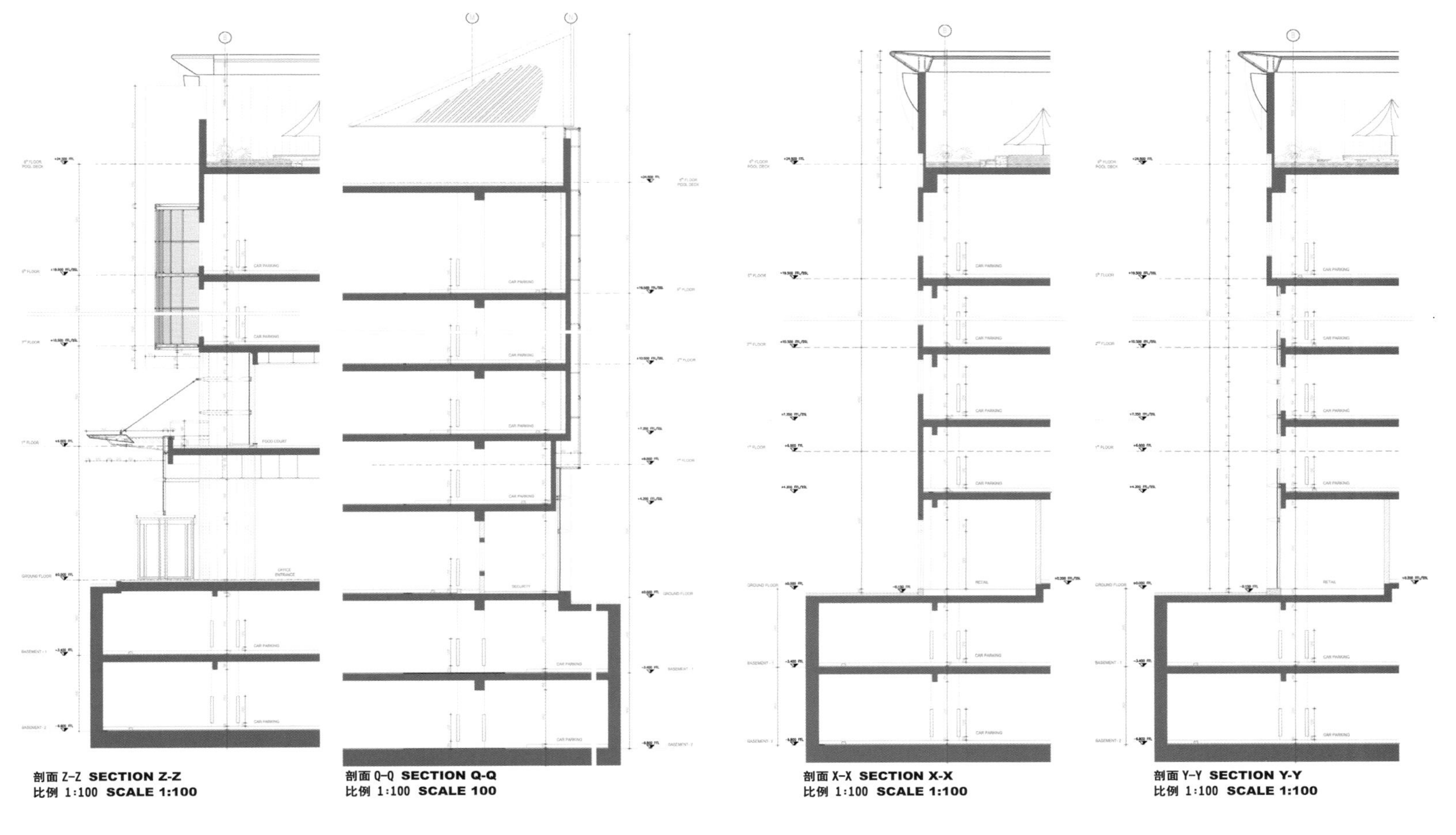

剖面 Z-Z SECTION Z-Z
比例 1:100 SCALE 1:100

剖面 Q-Q SECTION Q-Q
比例 1:100 SCALE 100

剖面 X-X SECTION X-X
比例 1:100 SCALE 1:100

剖面 Y-Y SECTION Y-Y
比例 1:100 SCALE 1:100

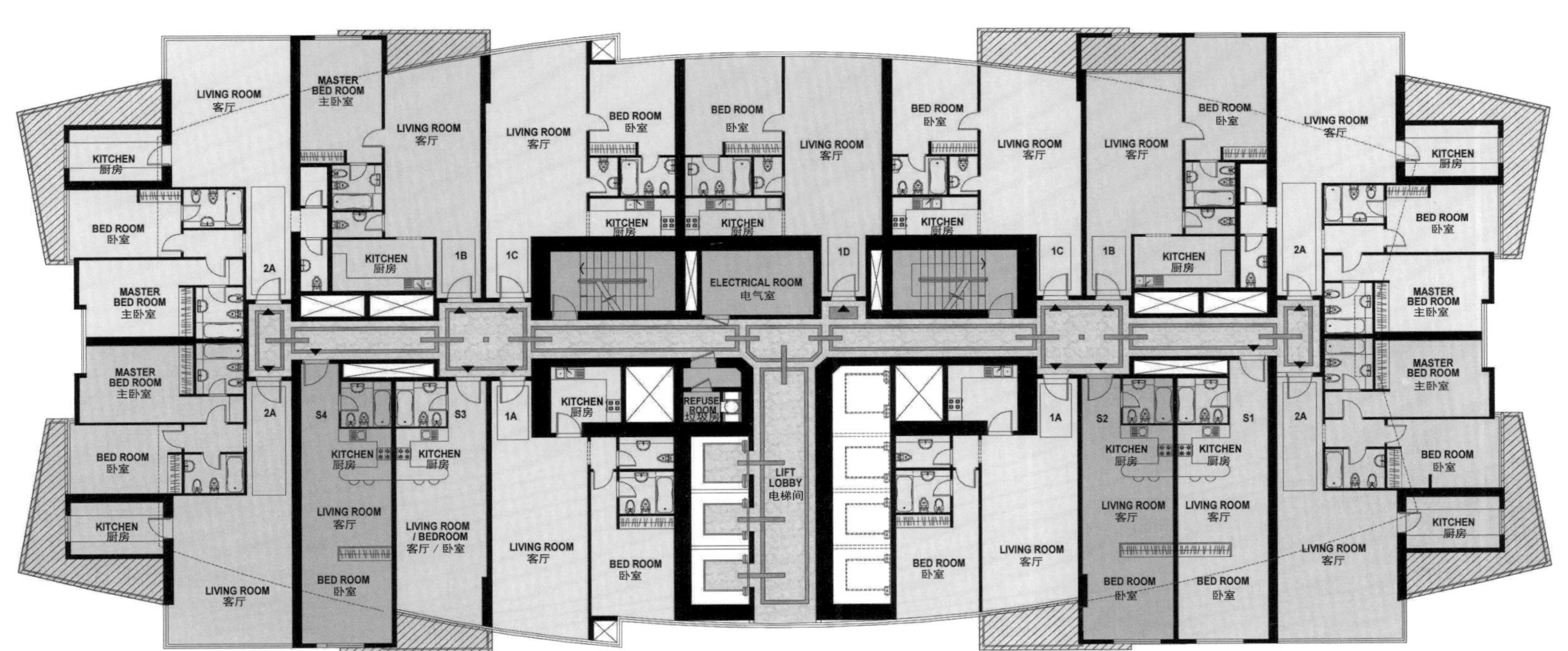

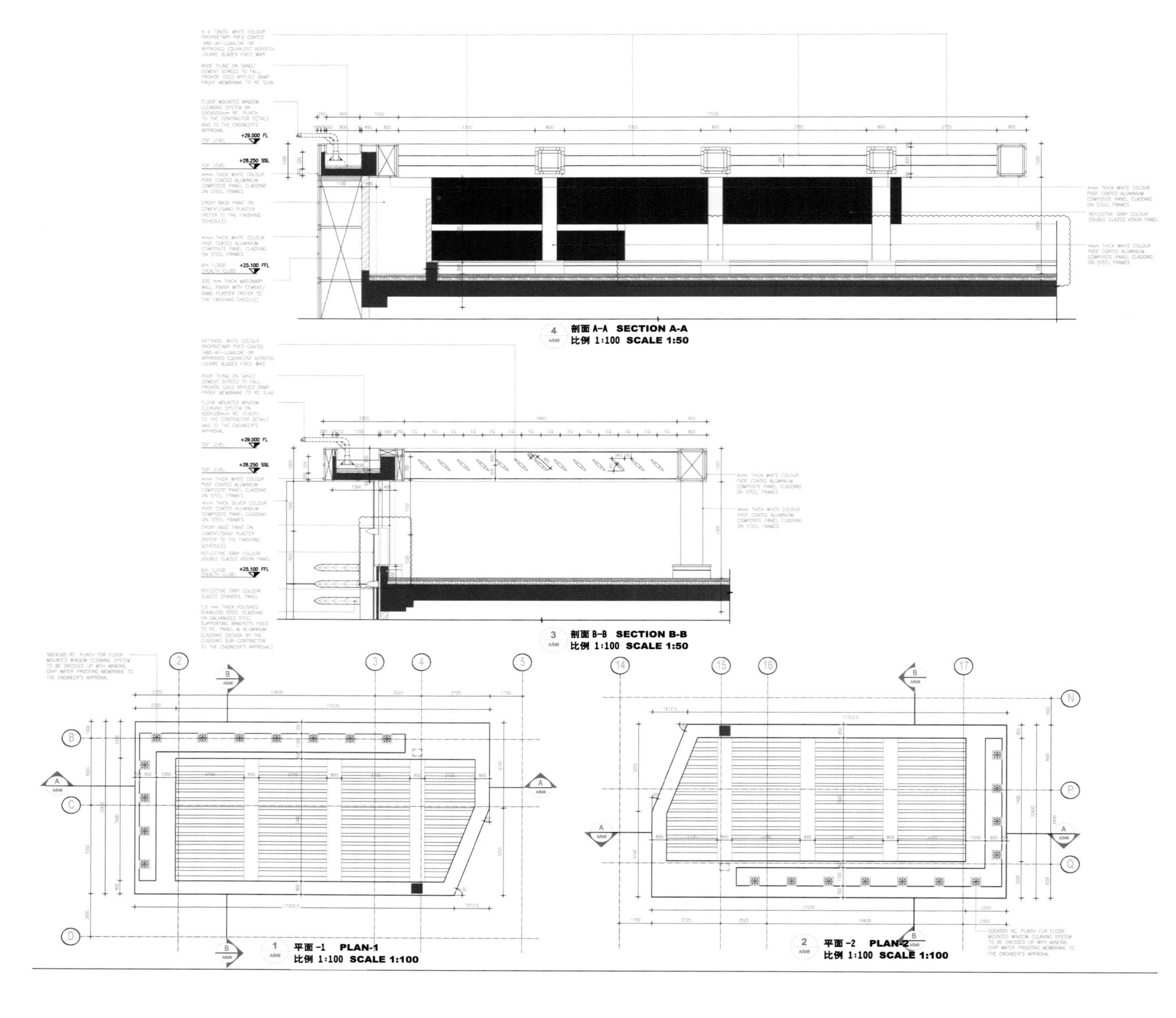

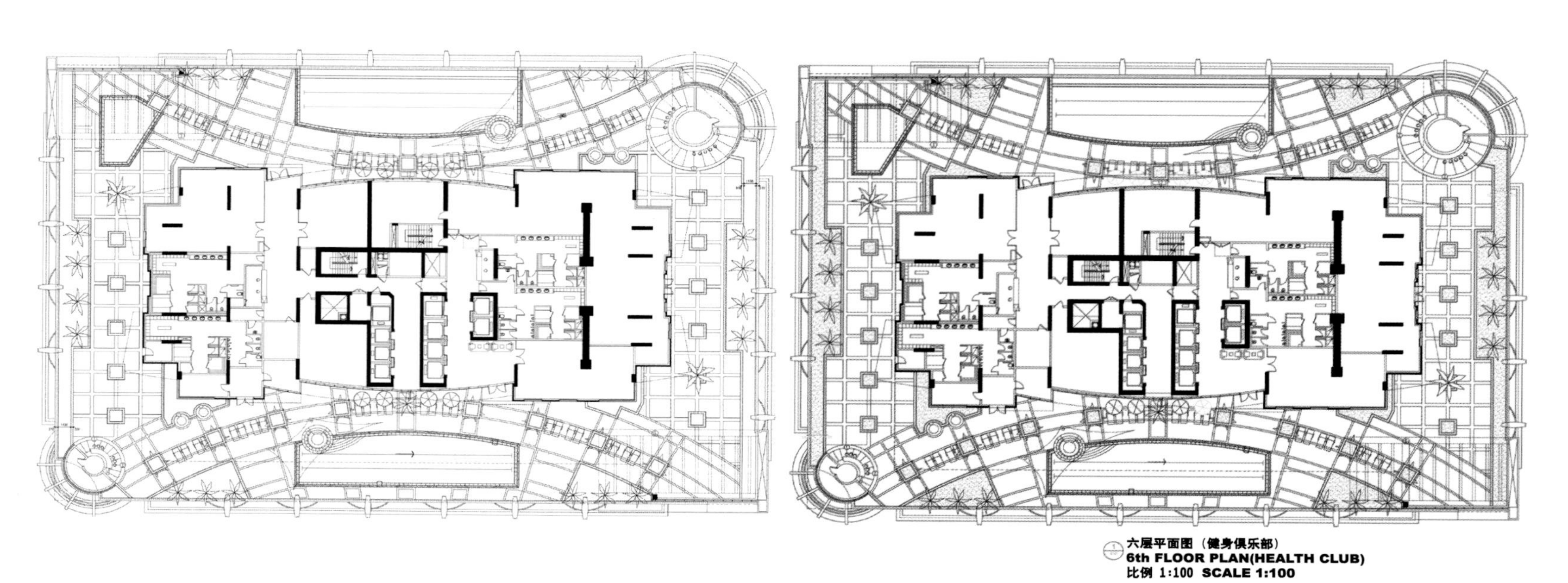

六层平面图（健身俱乐部）
6th FLOOR PLAN(HEALTH CLUB)
比例 1:100 SCALE 1:100

1 剖面 C-C SECTION C-C
比例 1:50 SCALE 1:50

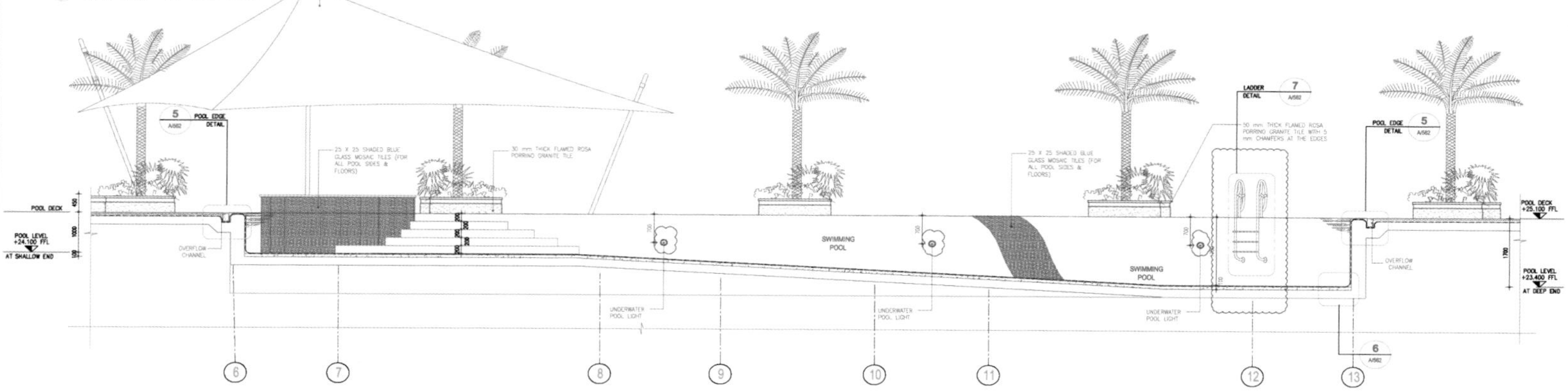

2 剖面 B-B SECTION B-B
比例 1:50 SCALE 1:50

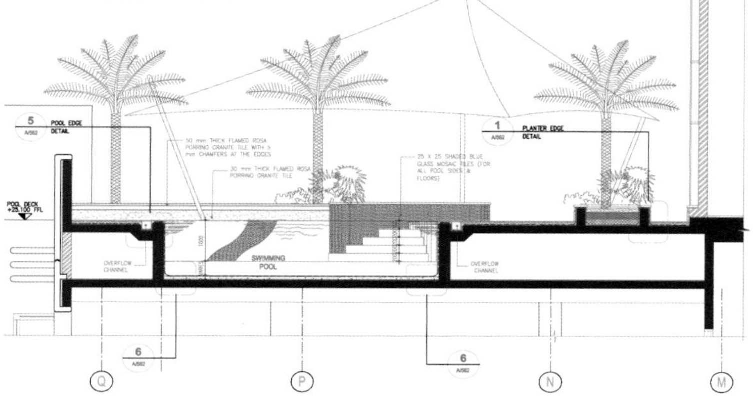

1 剖面 A-A SECTION A-A
比例 1:50 SCALE 1:50

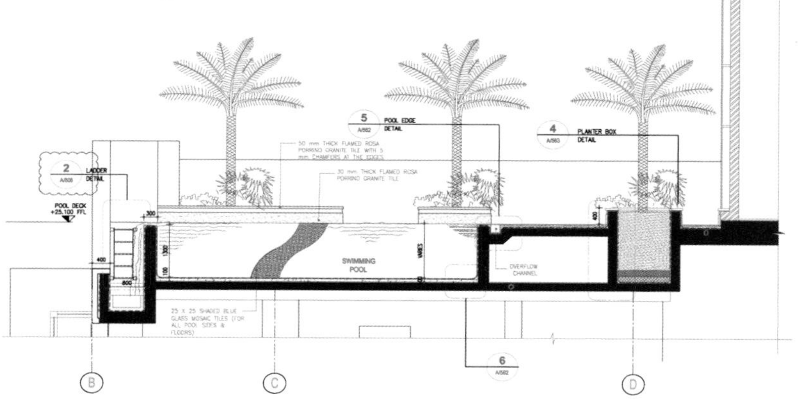

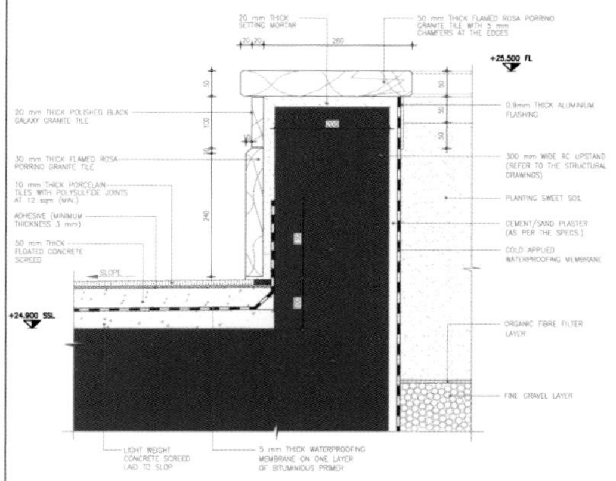

1 花盆边缘细部 PLANTER EDGE DETAIL
比例 1:5 SCALE 1:5

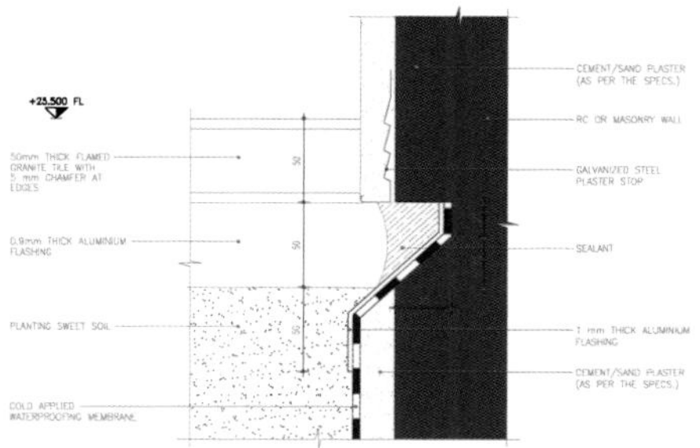

2 花槽墙细部 PLANTER WALL DETAIL
比例 1:2 SCALE 1:2

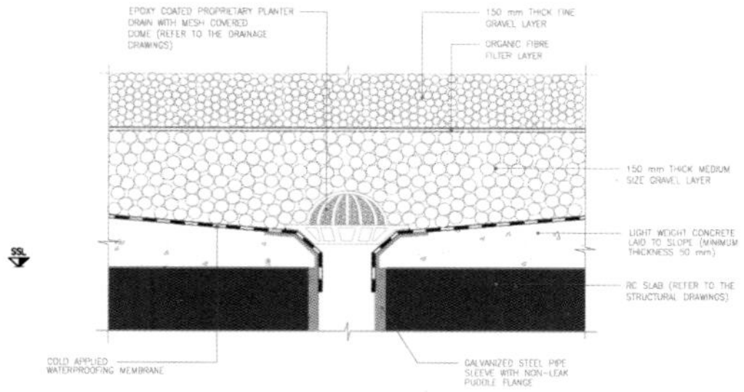

3 花槽排水细部 PLANTER DRAIN DETAIL
比例 1:5 SCALE 1:5

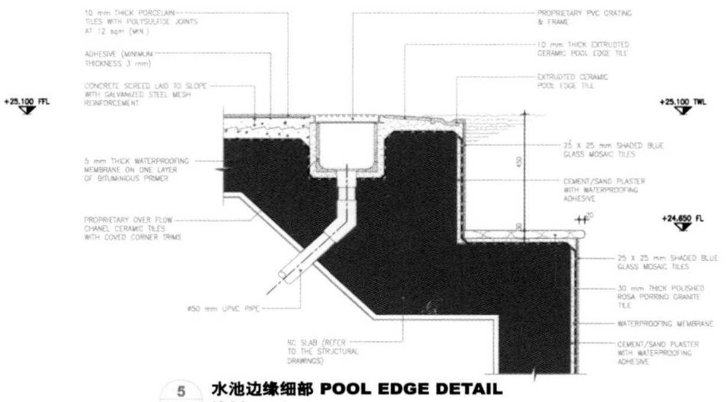

5 水池边缘细部 POOL EDGE DETAIL
比例 1:10 SCALE 1:10

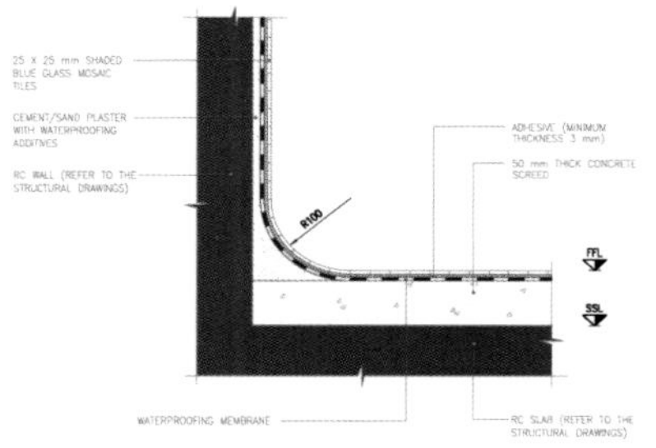

6 池壁细部 POOL WALL DETAIL
比例 1:5 SCALE 1:5

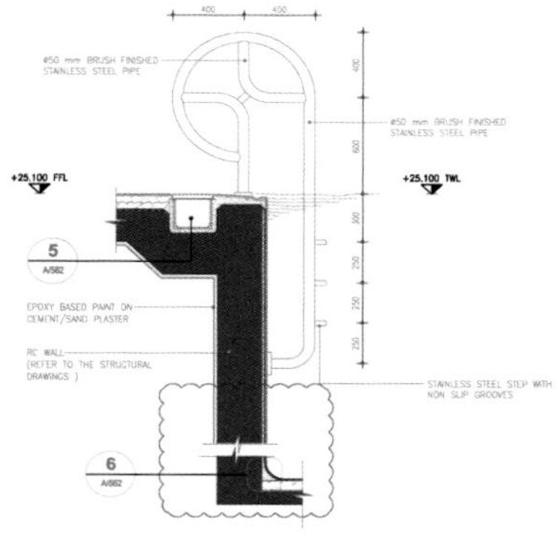

7 水池阶梯细部 POOL LADDER DETAIL
比例 1:20 SCALE 1:20

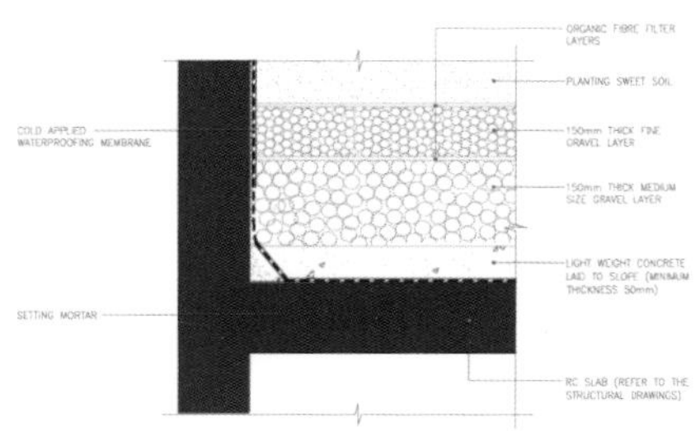

8 花盆细部 PLANTER DETAIL
比例 1:10 SCALE 1:10

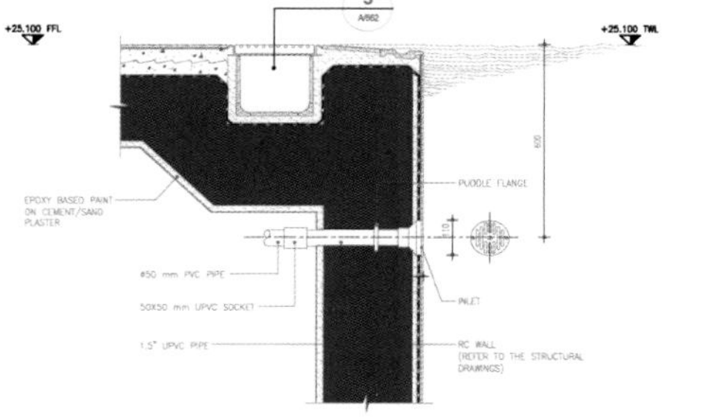

4 入口细部 INLET DETAIL
比例 1:10 SCALE 1:10

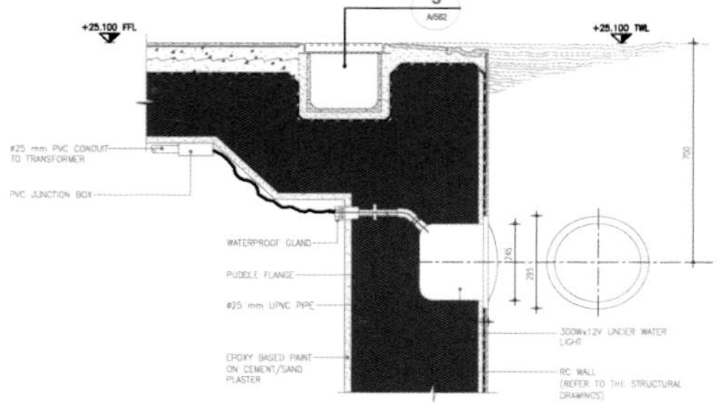

9 水下灯具细部 UNDER WATER LIGHT FITTING DETAIL
比例 1:10 SCALE 1:10

陕西西安信德中心

Shaanxi Xi'an Xinde Center

设计单位：深圳市天方建筑设计有限公司
开发商：西安信德投资有限公司
项目地址：中国陕西省西安市
用地面积：4 622.76 ㎡
总建筑面积：24 668.42 ㎡
绿化率：30%
容积率：4.59

Designed by: Shenzhen TAF Architectural Design Co., Ltd.
Developer: Xi'an Xinde Investment Co., Ltd.
Location: Xi'an, Shaanxi, China
Site Area: 4,622.76 m^2
Gross Floor Area: 24,668.42 m^2
Greening Ratio: 30%
Plot Ratio: 4.59

项目概况

项目位于西安北郊，处于汉城南路、红光路以及丰镐西路的交界处。地块北面有汉长安遗址，西面现为红光小区，南接红光路，东面有西安协和医院和利君教育公司，东南面为西安市自来水西郊总公司。该地块交通便利，文化环境良好，周边各种配套设施完善，为项目的建设提供了优越的条件。

建筑设计

信德中心由 5 层裙房和 16 层的塔楼组成，主要功能为商业、酒楼、酒店和办公。塔楼以长方形的形式布局，可以有效提高塔楼的使用率，方正的平面布局形式也有利于将来的改造。塔楼南侧临红光路布置，这既使建筑在城市界面上具备了良好的标识性，而且交通流线也更为便捷顺畅。

建筑立面采用了 Art Deco 风格。简洁流畅、挺拔向上的线条，金字塔状的台阶式退台和极具对称性的构图，使建筑在转折、进退和高低错落中显得动感而高雅，流露出优雅而内敛的文化气息。

Profile

The project is located in the northern suburb of Xi'an, the intersection of South Hancheng Road, Hongguang Road and West Fenghao Road. The site is close to Chang'an historic site of Han Dynasty to the north, Hongguang Community to the west, Hongguang Road to the south, Xi'an Xiehe Hospital and Lijun Education Company to the east, Xi'an water supply company to the southeast. Convenient transportation, enjoyable cultural environment and complete supporting facilities provide favorable conditions for the project.

Architectural Design

Xinde Center is composed of a five-floor podium and a 16-floor tower. Its main functions include business, restaurant, hotel and office. Rectangle layout of the tower can effectively elevate its usage rate. The square plane layout also facilitates future reconstruction. The south side of the tower facing Hongguang Road offers favorable iconic image for the building and also makes traffic lines smooth and convenient.

The building elevation adopts Art Deco style. Concise, smooth, straight and upward lines, pyramid-shaped stepped terrace and symmetric composition make the building dynamic and elegant, revealing graceful unpretentious cultural atmosphere.

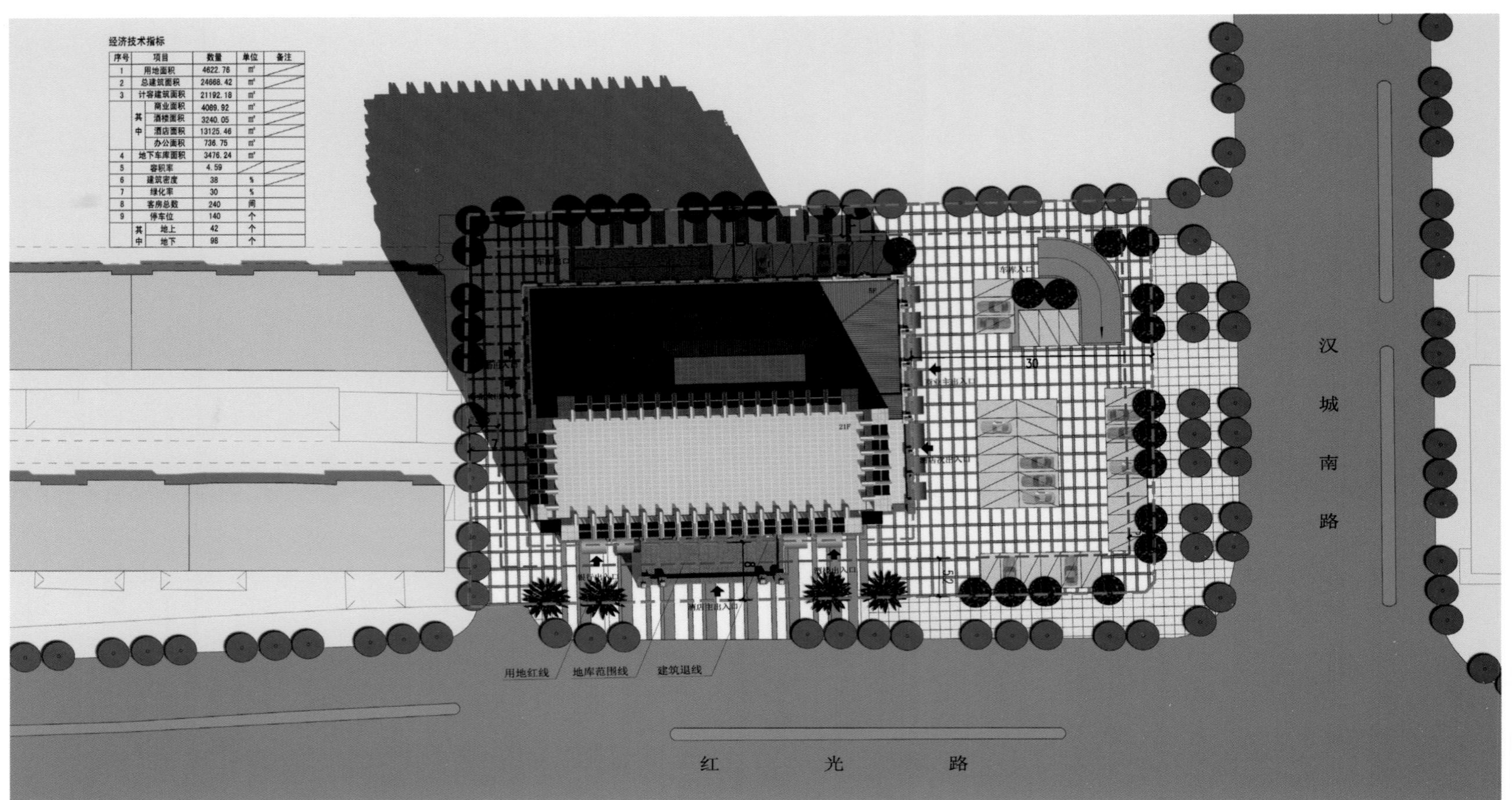

经济技术指标

序号	项目		数量	单位	备注
1	用地面积		4622.76	㎡	
2	总建筑面积		24668.42	㎡	
3	计容建筑面积		21192.18	㎡	
	其中	商业面积	4089.92	㎡	
		酒楼面积	3240.05	㎡	
		酒店面积	13125.46	㎡	
		办公面积	736.75	㎡	
4	地下车库面积		3476.24	㎡	
5	容积率		4.59		
6	建筑密度		38	%	
7	绿化率		30	%	
8	客房总数		240	间	
9	停车位		140	个	
	其中	地上	42	个	
		地下	98	个	

■ 总平面图

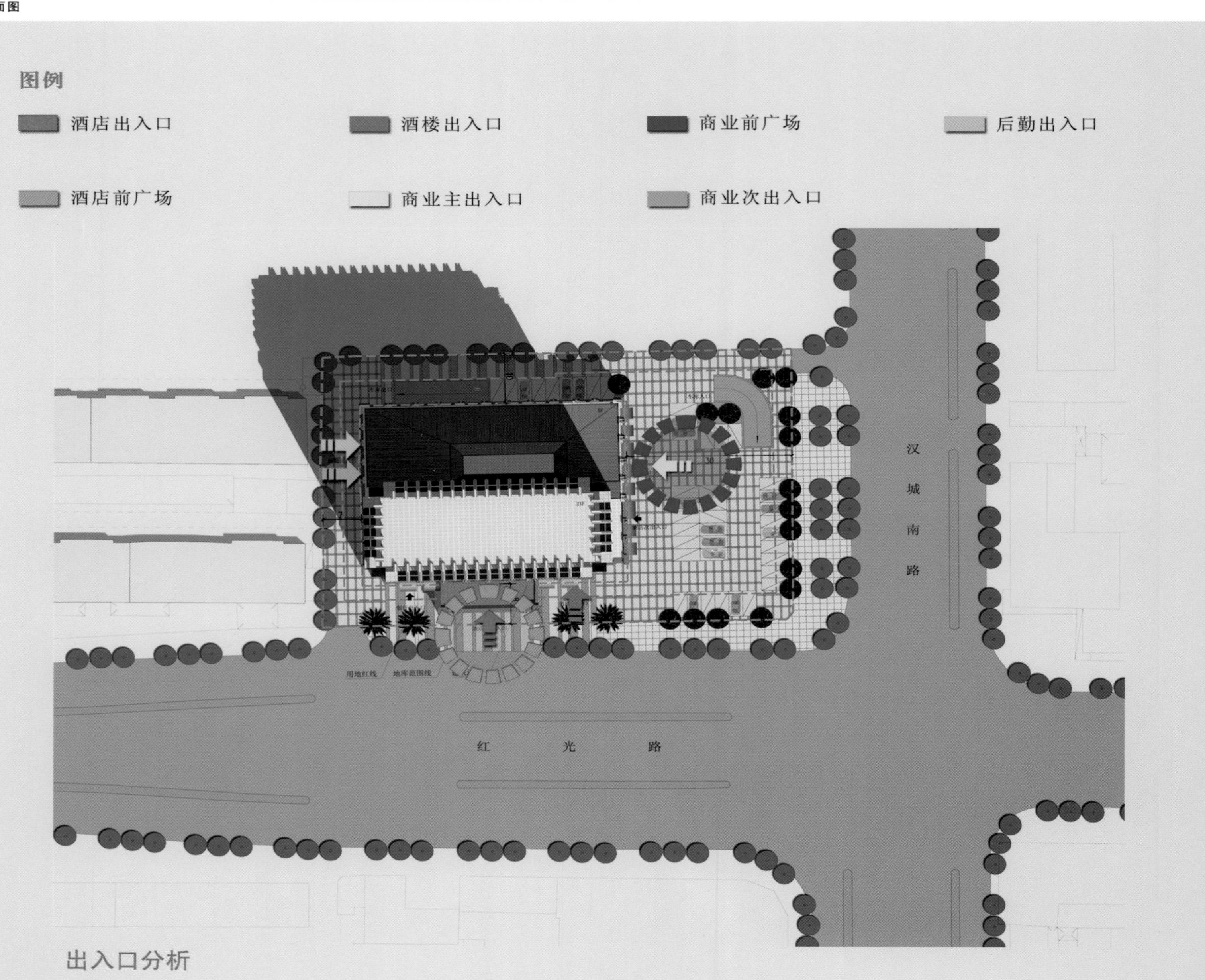

出入口分析

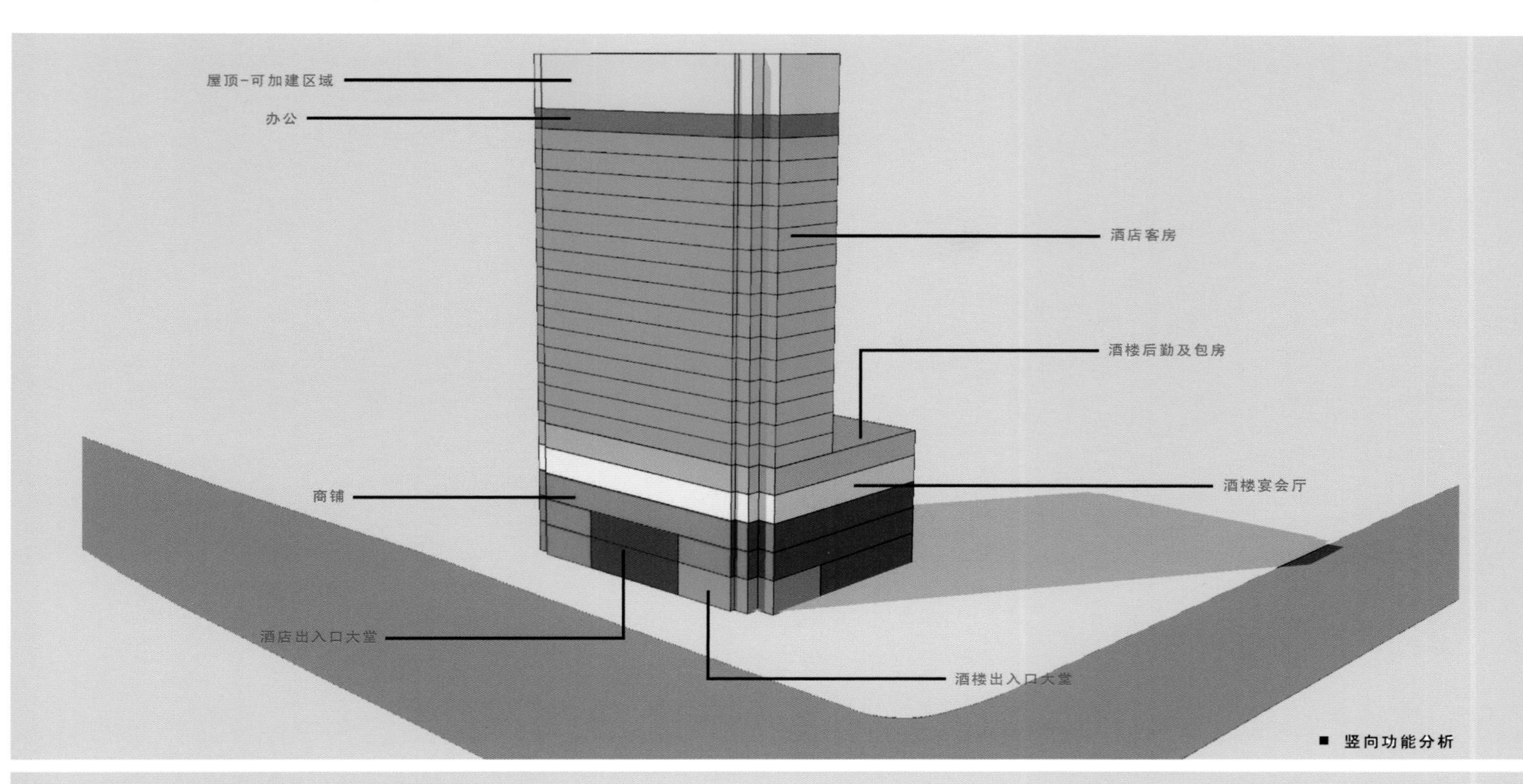

■ 竖向功能分析

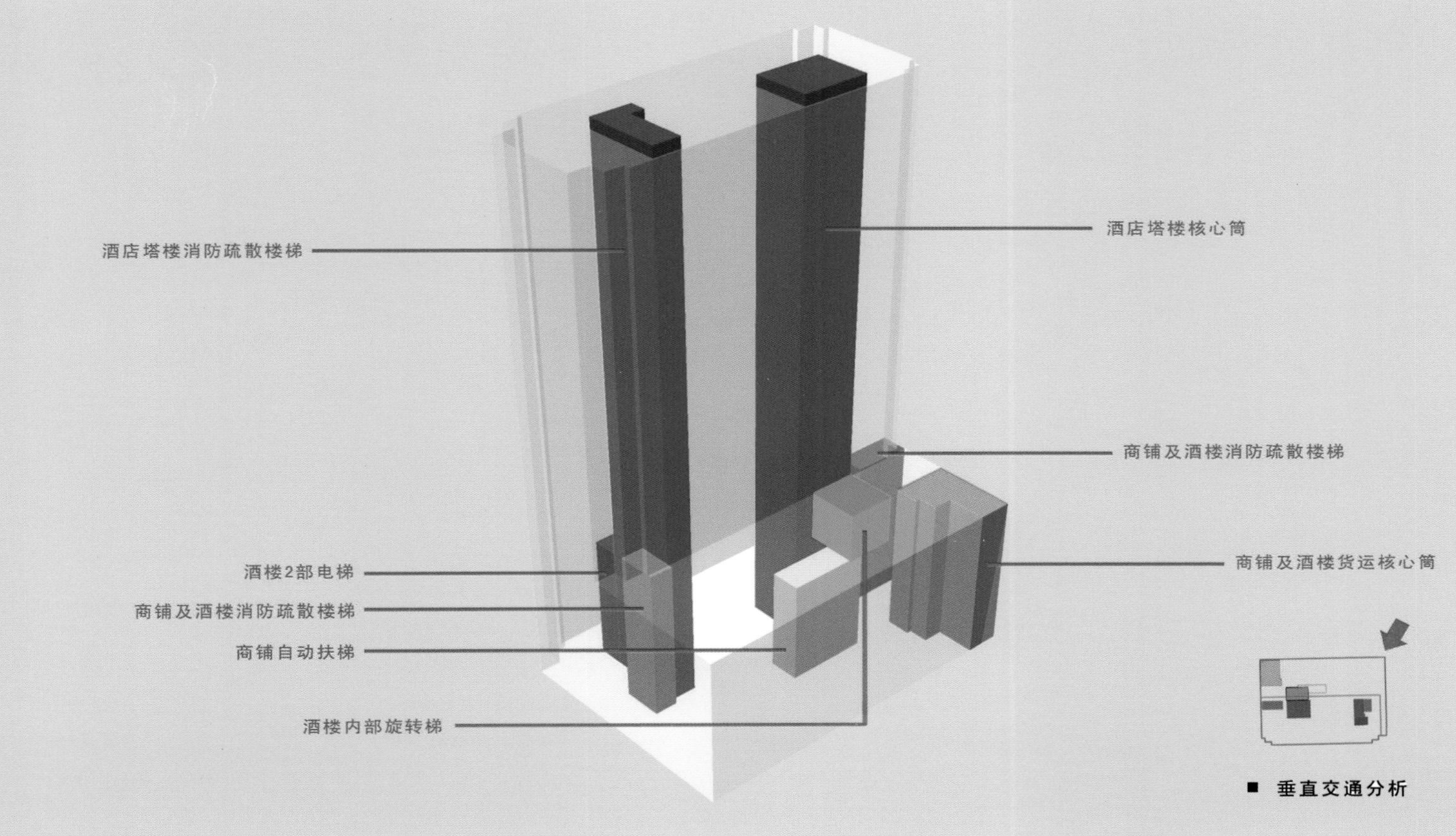

■ 垂直交通分析

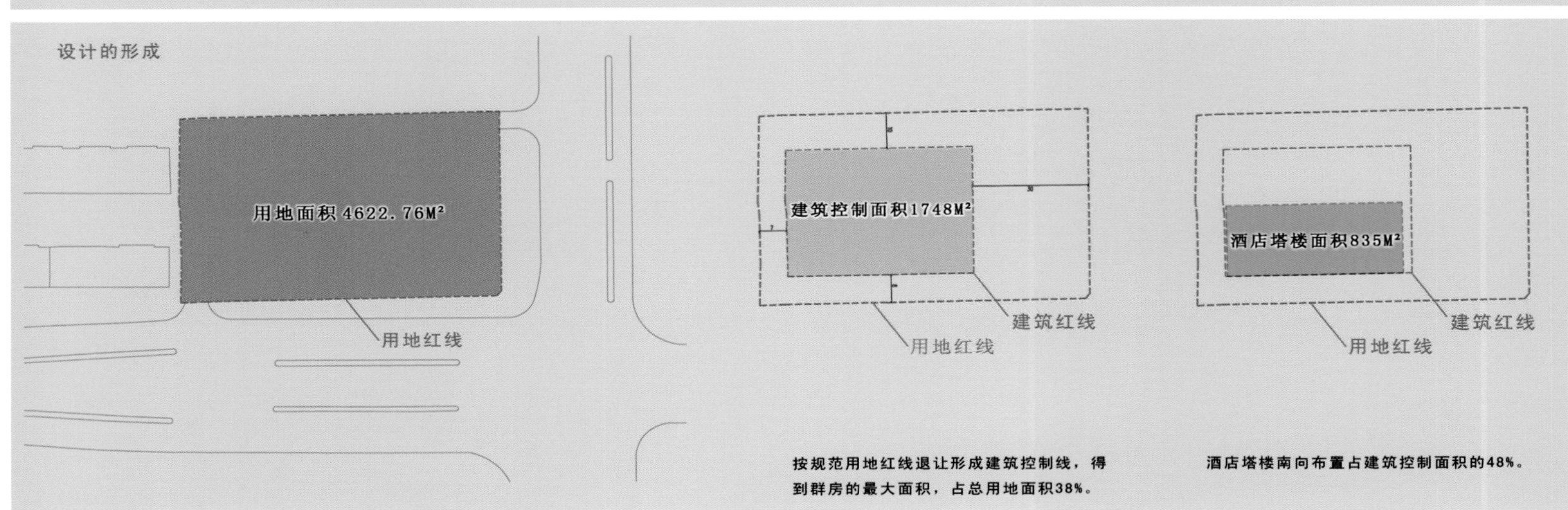

按规范用地红线退让形成建筑控制线，得到群房的最大面积，占总用地面积38%。

酒店塔楼南向布置占建筑控制面积的48%。

酒店塔楼设计的形成1

塔楼形式

长方型orL型

- 裙房
- 塔楼

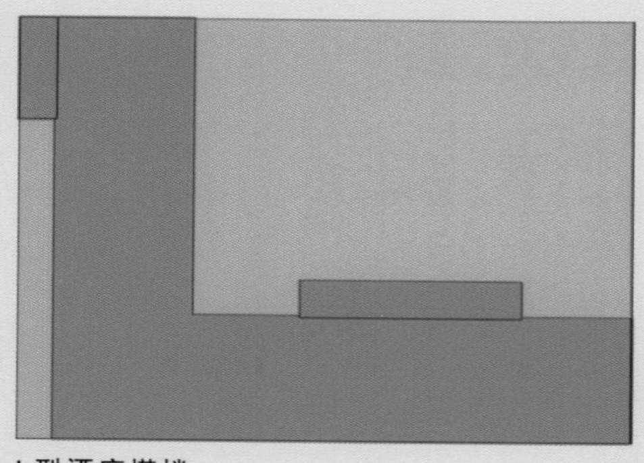

L型酒店塔楼

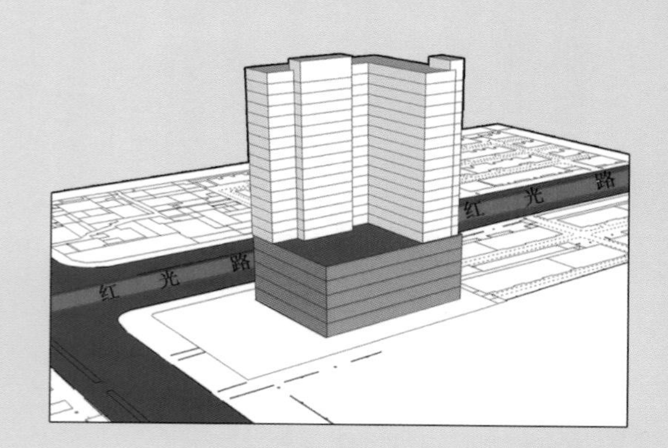

长方型酒店塔楼

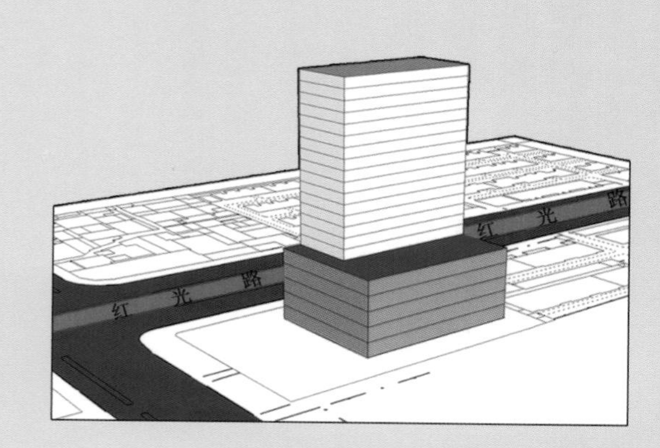

比较结果：经计算对比，L型塔楼因为单面布置房，使用率较底，长条形平面不利于将来改造，而且左侧红光小区现有建筑会对L型塔楼西侧的客房日照产生影响。而长方形板式塔楼的使用率高，层平面方正，对将来的改造有利。综上所述故采用长方型塔楼布置形式。

酒店塔楼设计的形成2

长方型塔楼位置

南向or北向

- 裙房
- 塔楼

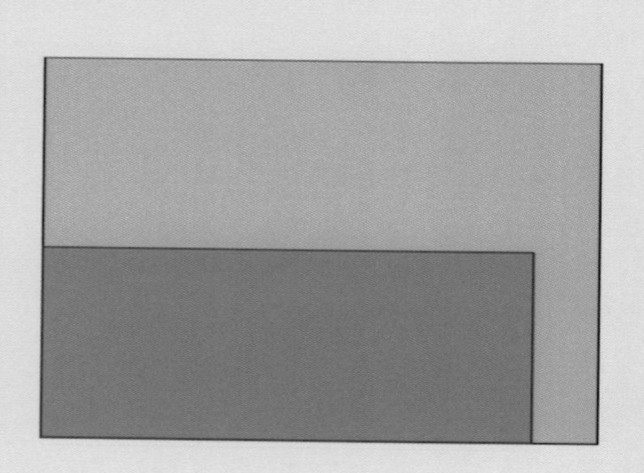

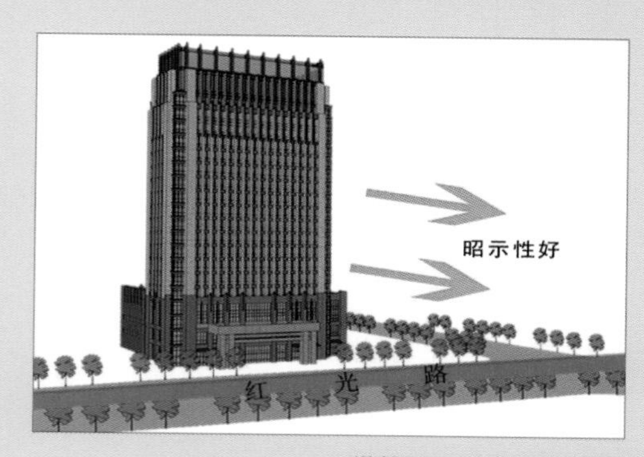

塔楼沿红光路南侧放置

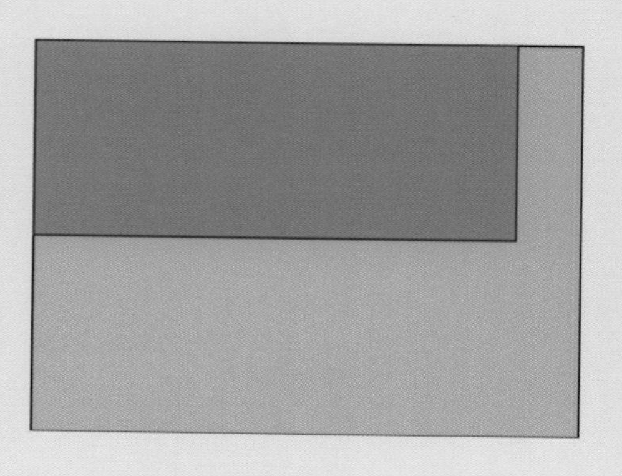

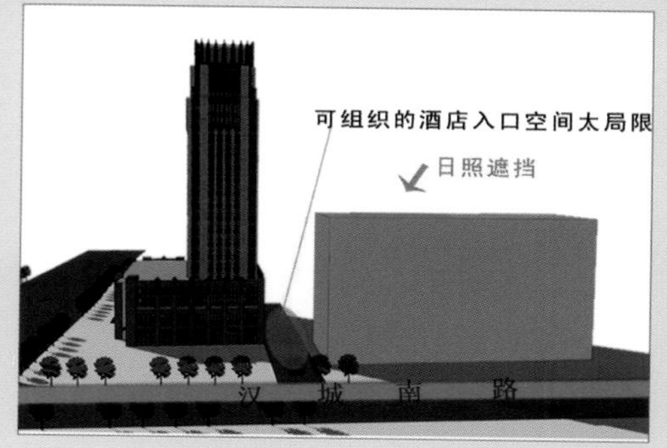

塔楼北侧放置

比较结果：长方型塔楼的位置如果放置在北侧会对北面的建筑产生日照遮挡，塔楼的昭示性差，而且可组织的酒店出入口空间太局促。塔楼南侧临红光路布置，因为临城市干道昭示性好，南侧布置可以在城市界面上形成良好的标识性，而且交通组织也更顺畅。故酒店塔楼临红光路南侧布置。

图例
城市道路
商业人行流线
汉城南路
用地红线
地库范围线
建筑退线

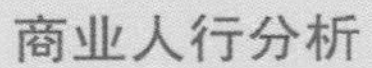

商业人行分析

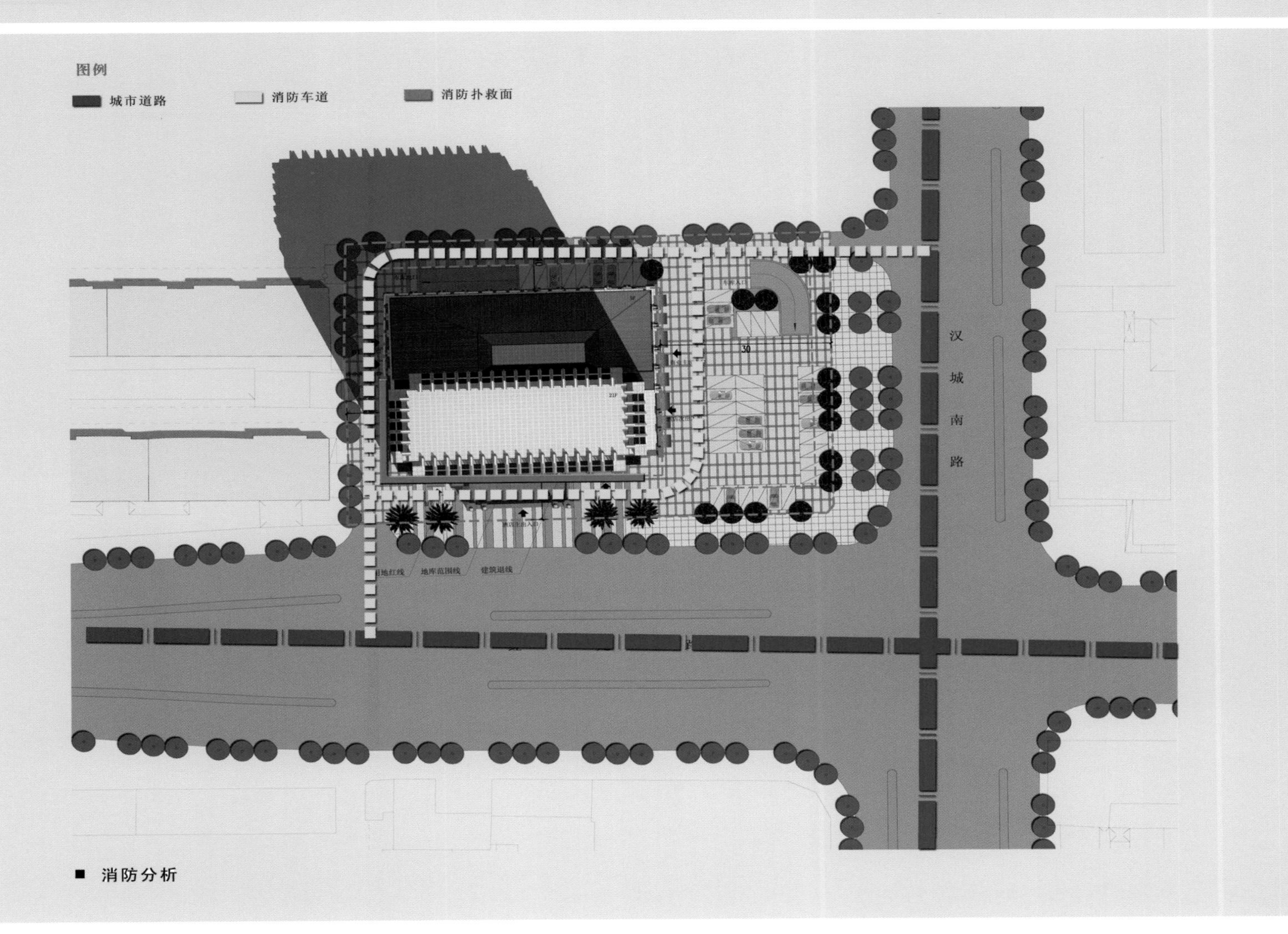
图例
城市道路
消防车道
消防扑救面
汉城南路
用地红线
地库范围线
建筑退线

■ 消防分析

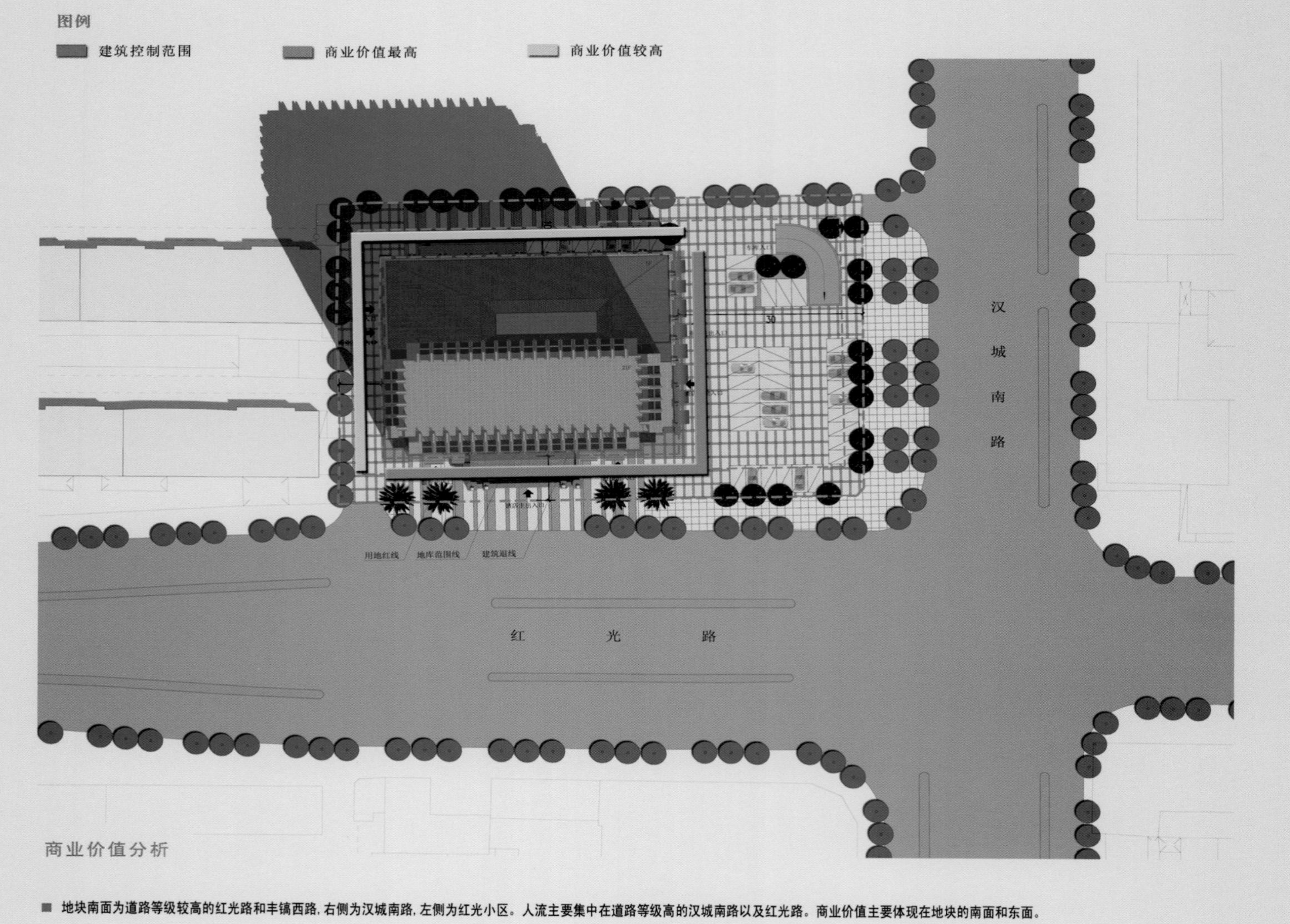

■ 地块南面为道路等级较高的红光路和丰镐西路，右侧为汉城南路，左侧为红光小区。人流主要集中在道路等级高的汉城南路以及红光路。商业价值主要体现在地块的南面和东面。

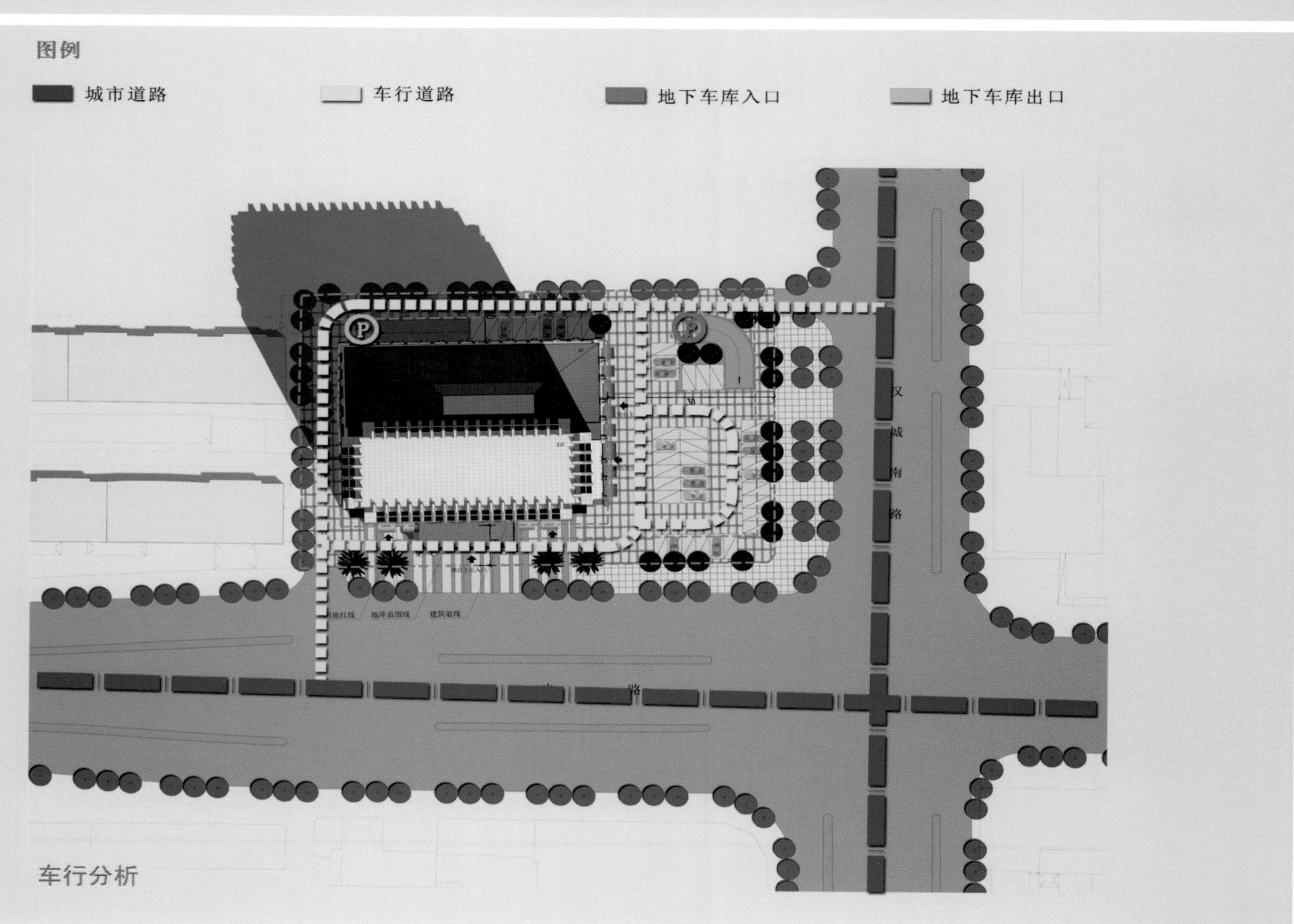

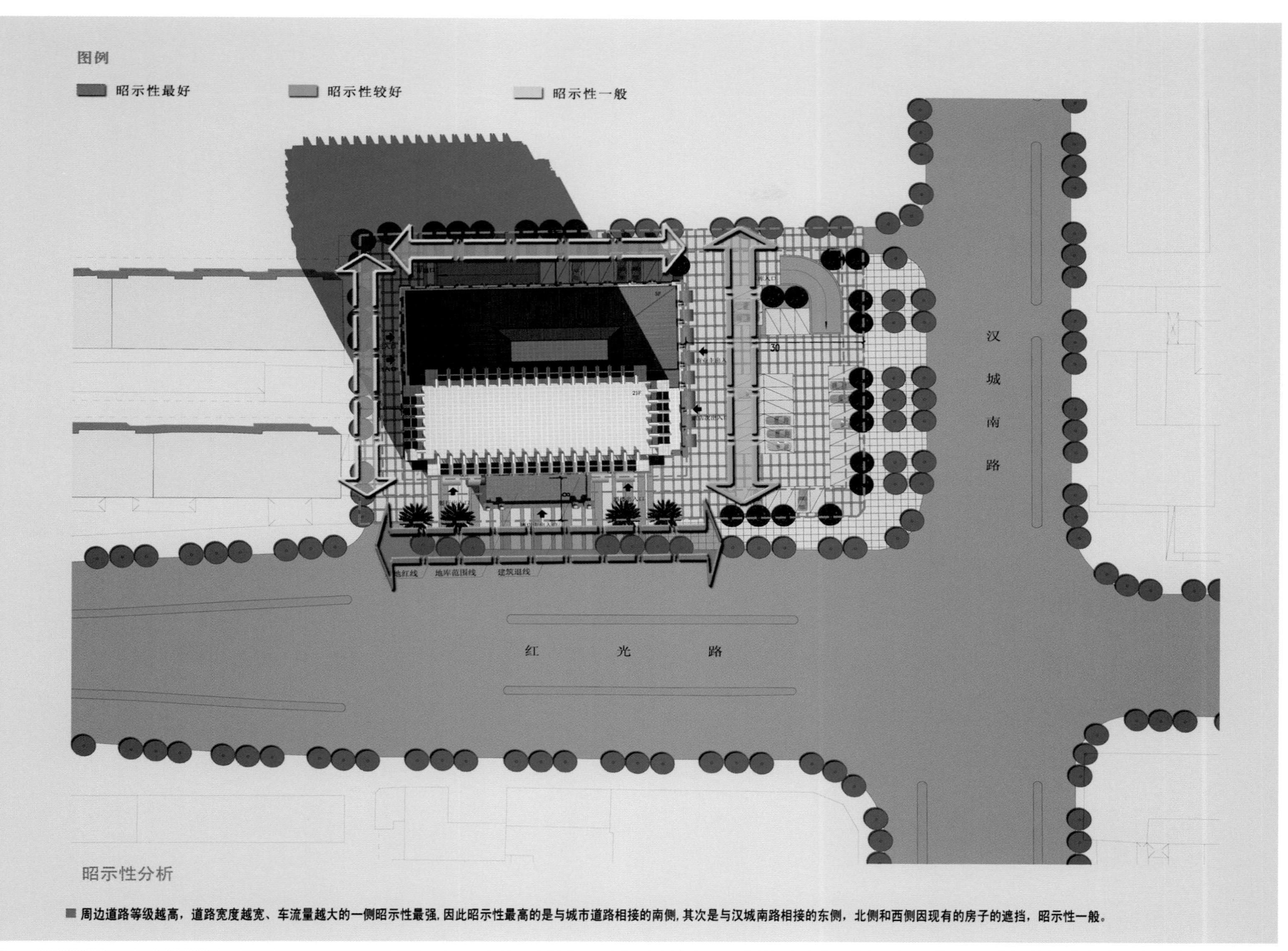

昭示性分析

■ 周边道路等级越高，道路宽度越宽、车流量越大的一侧昭示性最强，因此昭示性最高的是与城市道路相接的南侧，其次是与汉城南路相接的东侧，北侧和西侧因现有的房子的遮挡，昭示性一般。

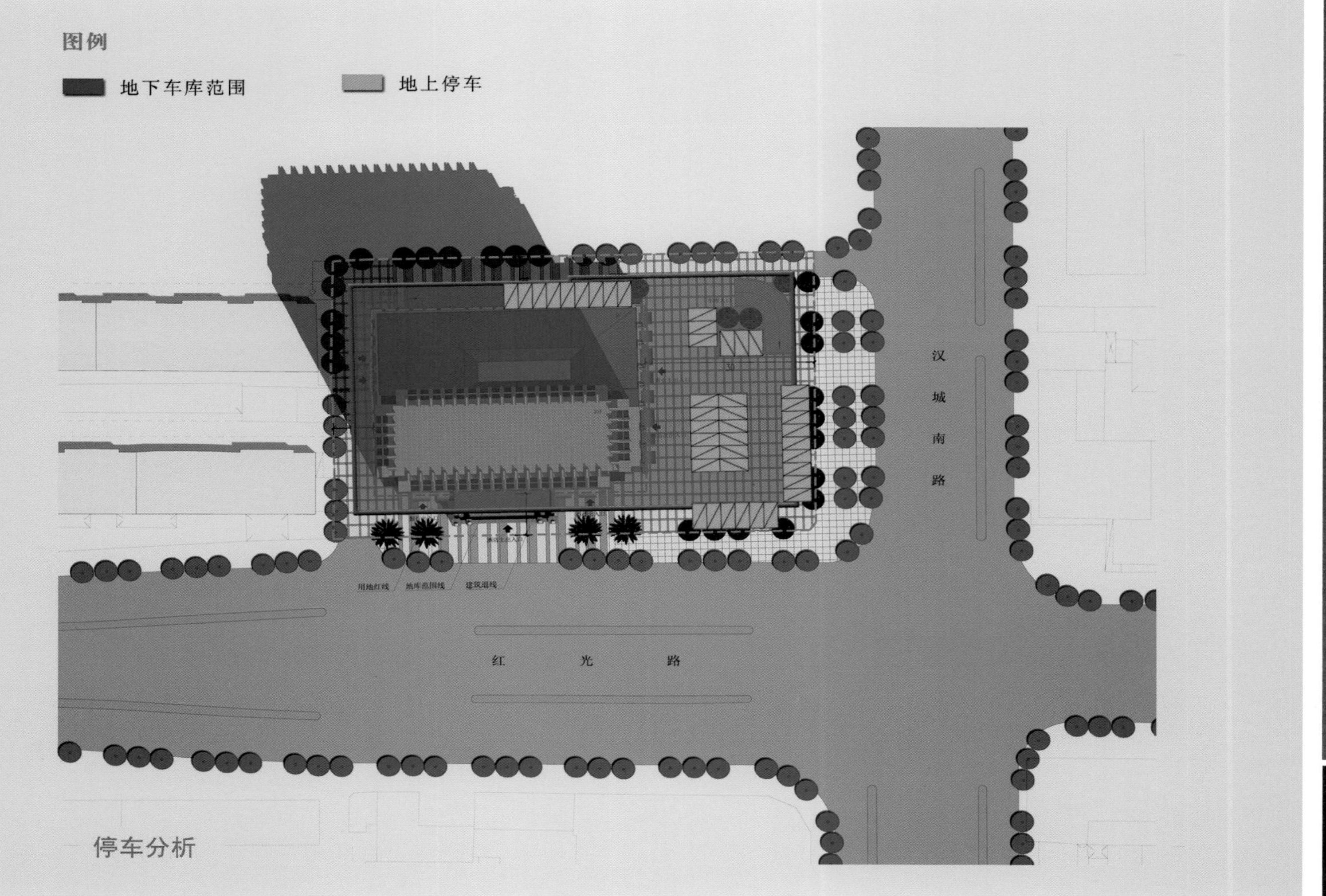

停车分析

马来西亚吉隆坡 KL Gateway 商住综合发展项目

The KL Gateway

设计单位：10 DESIGN（拾稼设计）

开发商：Suez Domain Sdn Bhd

项目地址：马来西亚吉隆坡

建筑面积：250 000 m²

设计团队：Ted Givens　Mohamad Ghamlouch

Emre Icdem　Magic Kwan

Abraham Fung　Maciej Setniewski

Designed by: 10 DESIGN

Client: Suez Domain Sdn Bhd

Location: Kuala Lumpur, Malaysia

Gross Floor Area: 250,000 m²

Design Team: Ted Givens, Mohamad Ghamlouch, Emre Icdem, Magic Kwan, Abraham Fung, Maciej Setniewski

项目概况

大量的公共空间使 KL Gateway 这个综合发展项目成为了一个颇具特色的城市社区，它使居民能在中央庭院、购物商场、会所及私家花园中享受空间的乐趣及高品质的生活环境。

建筑设计

项目的中心是由系列花园围绕而成的露天庭院，其公共空间式的设计主要为吸引城市与周边社区的人流。由大型浮法玻璃组成的环形结构将中央庭院空间定义为整个项目的焦点，并与周边的购物商场相呼应，成为整个综合体的聚集地。

中央庭院周围的塔楼引入了系列花园的设计元素，为住户营造出私密的花园空间，每栋塔楼的屋顶是居民专享的户外花园，压印在塔楼外墙上的弧形肋条与波浪形阳台意喻着花园的肌理，在突出花园区的同时，也进一步表现了项目与自然之间互融的关系。

庭院四周的七座塔楼分别根据水、金、火、土四种基本自然元素来设计。面向干线公路的办公楼，被定义为“水”元素。它的幕墙主要由玻璃和一系列非常具有流动感的金属肋条构成，以刻画出水的感觉。综合体中的住宅双塔楼是以“金”元素为定义，波动起伏的外墙形成了建筑的系列阳台，进一步明确了项目的有机性。

商业裙楼作为住宅与办公楼的实体基座，同样以“金”元素为其设计要点。购物商场被设计分隔成更多较小的、围绕时尚主题的零售空间以支持当地时装工业的发展。“火”元素由回迁房构成，一系列高对比度的灰色石材定义出外墙生动的有机形态。采用“土”元素为设计定义的则是酒店式公寓。

项目还采用了最新的低碳绿色生态技术。两座主办公楼外墙将使用二氧化钛纳米涂料，这是一种光催化涂料，能有效去除尘垢、细菌和空气污染。这种清洁反应由光触发，夜间办公楼外墙的照明装饰灯可使这种空气清洁反应持续 24 小时。

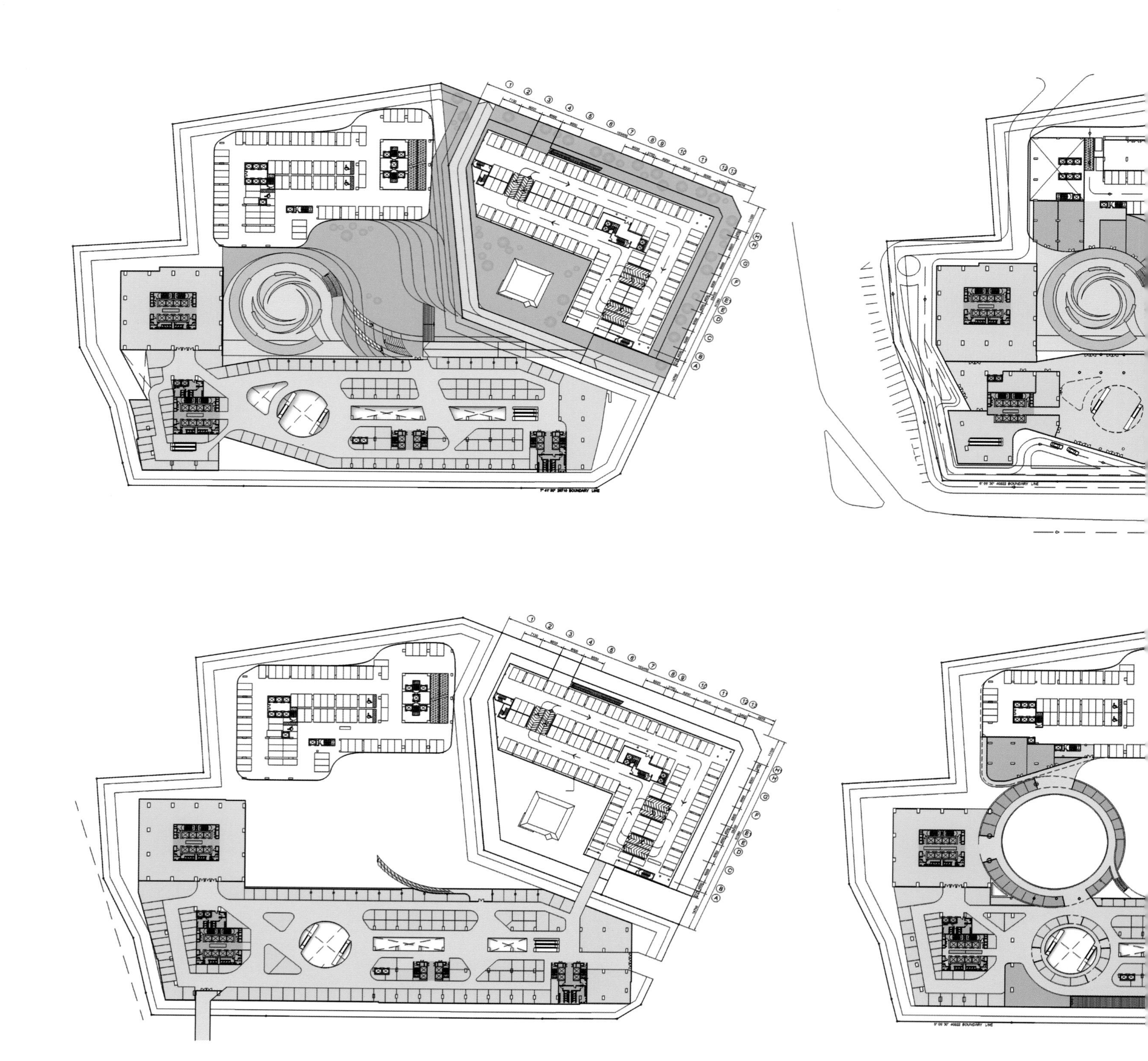

Profile

The extensive public spaces and gardens of the KL Gateway development make it a very special community. Residents can enjoy spaces ranging from the central courtyard, to the retail mall, club houses, and private gardens.

Architectural Design

The heart of the project is an outdoor courtyard surrounded by a series of garden spaces. The courtyard is designed to be a public room attracting people from the surrounding neighborhood and city. A large floating glass ring defines the central courtyard space becoming the focal point for the entire development.

A series of gardens are pulled up into the surrounding towers creating private gardens for the residents. Each roof top area becomes an outdoor garden space for the residents to enjoy. The garden textures are imprinted on the facades of the towers in the form of curving fins and undulating lanais to highlight the lush garden areas contained within the development and to further express the connection to nature.

The 7 buildings wrapping the courtyard are defined by the four basic elements of water, metal, fire, and earth. The office towers facing the main highway are defined as the water element. The facades consist primarily of glass and a series of very fluid metal fins to highlight the feeling of water. The twin residential towers form

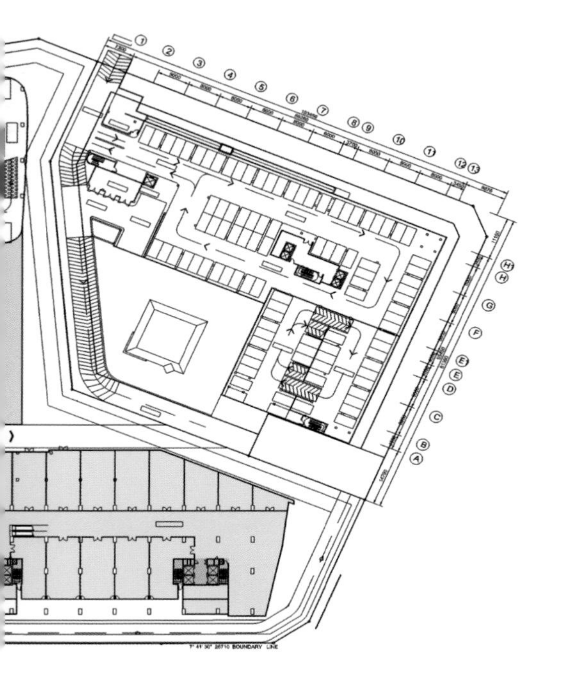

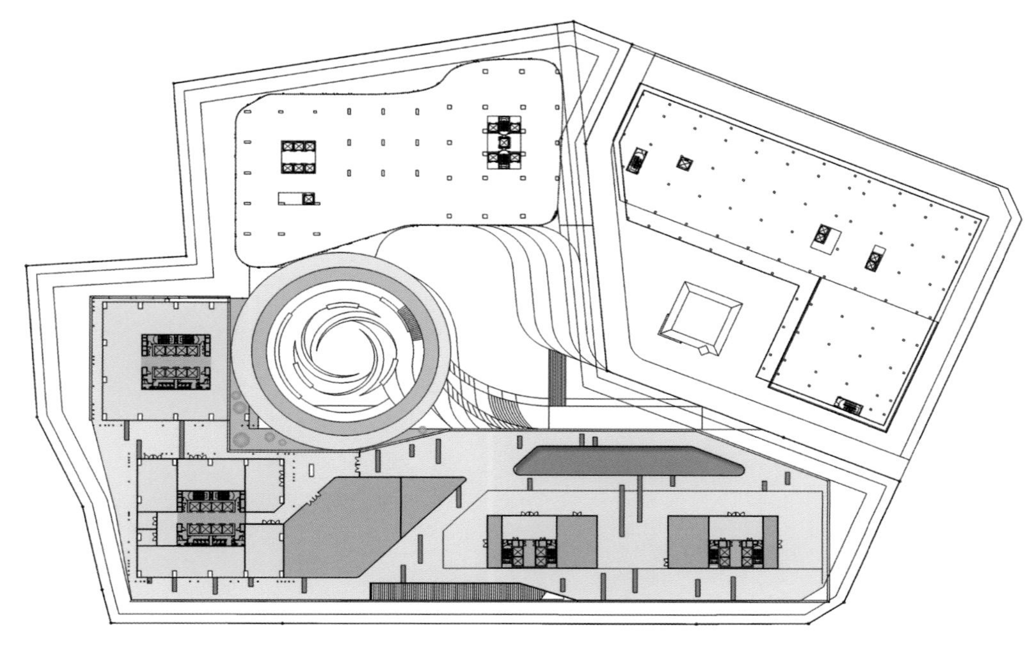

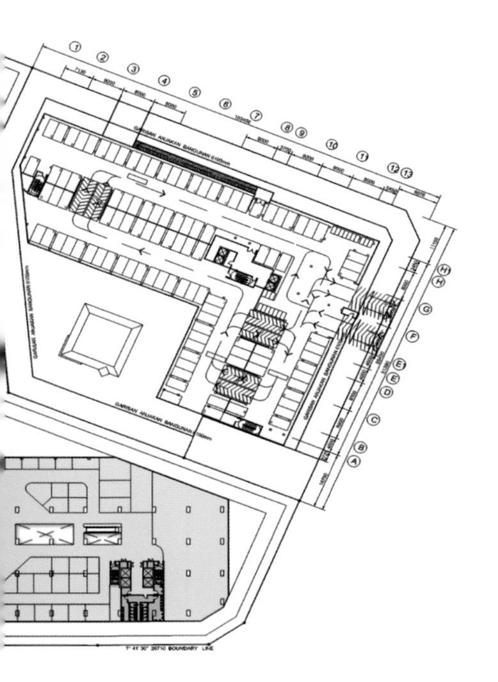

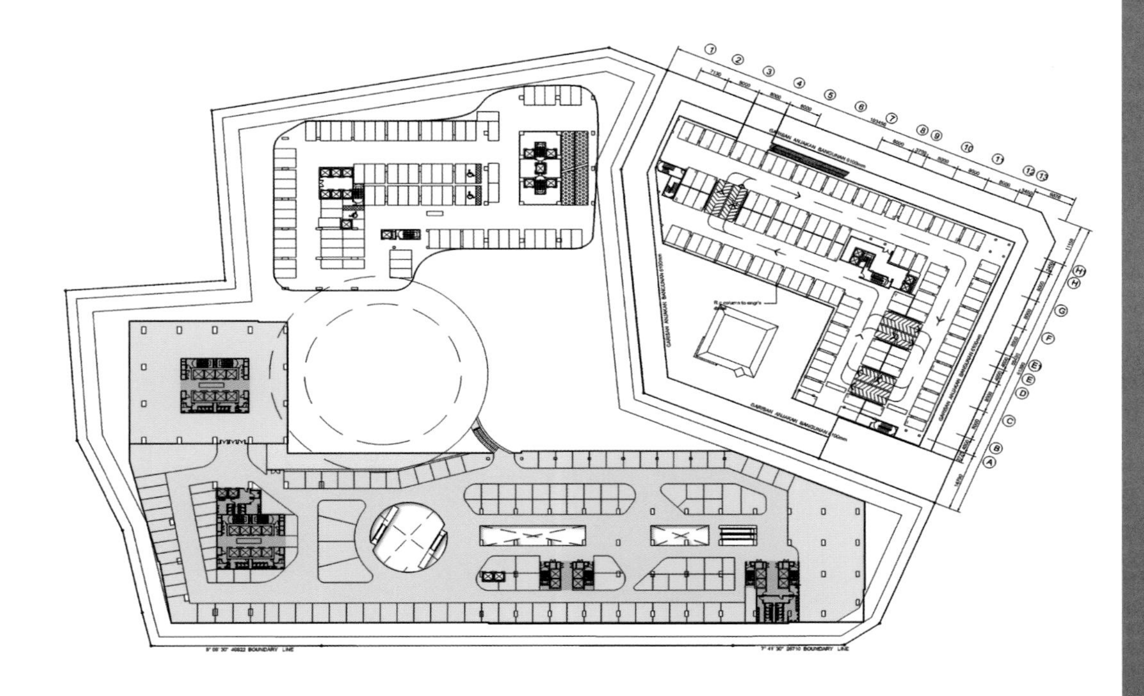

the element of metal. The undulate facades of the residential towers form a series of lanais that further define the organic nature of the project.

The retail podium is also defined by the element of metal. The retail podium is a solid base upon which the residential office towers rest. The retail mall is designed to house smaller localized stores. The mall is themed around the idea of fashion and supporting the local fashion industry. The relocation housing block forms the fire element. The skin is composed of dramatic organic pattern defined by a series of high contrast grey tones. The service apartment tower is defined by the element of earth.

The development will also make use of the latest developments in low-carbon sustainable technology. The two main office towers will be covered in a nano-coating of Titanium Dioxide. This is a photocatalytic coating that will remove dirt, bacteria, and pollution from the air. The cleaning reaction can be triggered by light. At night a series of lights will illuminate the office tower facades to make the air cleaning lasting twenty four hours a day.

瑞典斯德哥尔摩大陆塔

Stockholm Continental Tower

设计单位：C.F.Møller Architects
开发商：FOLKSAM
Citybanan
Scandic Hotels
Stockholms Stadsbyggnadskontor
Stockholms Exploateringskontor
项目地址：瑞典斯德哥尔摩
项目面积：20 000 ㎡

Designed by: C.F. Møller
Client: FOLKSAM; Citybanan; Scandic Hotels;
Stockholms Stadsbyggnadskontor;
Stockholms Exploateringskontor
Location: Stockholm, Sweden
Area: 20,000 m²

项目概况

项目位于赛格尔广场与中央车站的中间地带，紧临克拉拉教堂，横跨了位于 Klarabergsviadukten 地区的一个新地铁站，优越的地理位置将使之成为该地区最具发展潜力的项目之一。

设计理念

设计的灵感来源于斯德哥尔摩城，其原型来自位于邻近区域的斯德哥尔摩最大的教堂——克拉拉教堂。设计以多功能理念为基础，构建一栋能同时容纳不同使用功能的建筑。

建筑设计

具有反光面的复合塔楼宛若一个雕塑般的体量，在不同的方面都清晰可见。这个具备城市认同感的建筑受当地传统的历史建筑的尖屋顶和高水平的细节设计的启发，展现了一个朝向城市的生动外立面，这个立面有着多变的开合度。

建筑的横向稳定性由外部钢架结构所支撑，这个外部钢架结构呈三角形状，相对较坚固。在建筑的上部，钢架与核心墙壁共同作用，这个核心墙壁将由钢筋建造，核心结构与每个楼层通过钢梁和楼板里的钢筋混凝土黏合在一起。

为了支撑位于地下自动扶梯之上的塔楼，设计采用了钢材转化结构，它将支撑核心部分和一部分楼层的荷载，并将这些荷载转移到延伸至建筑边缘的外部结构上。

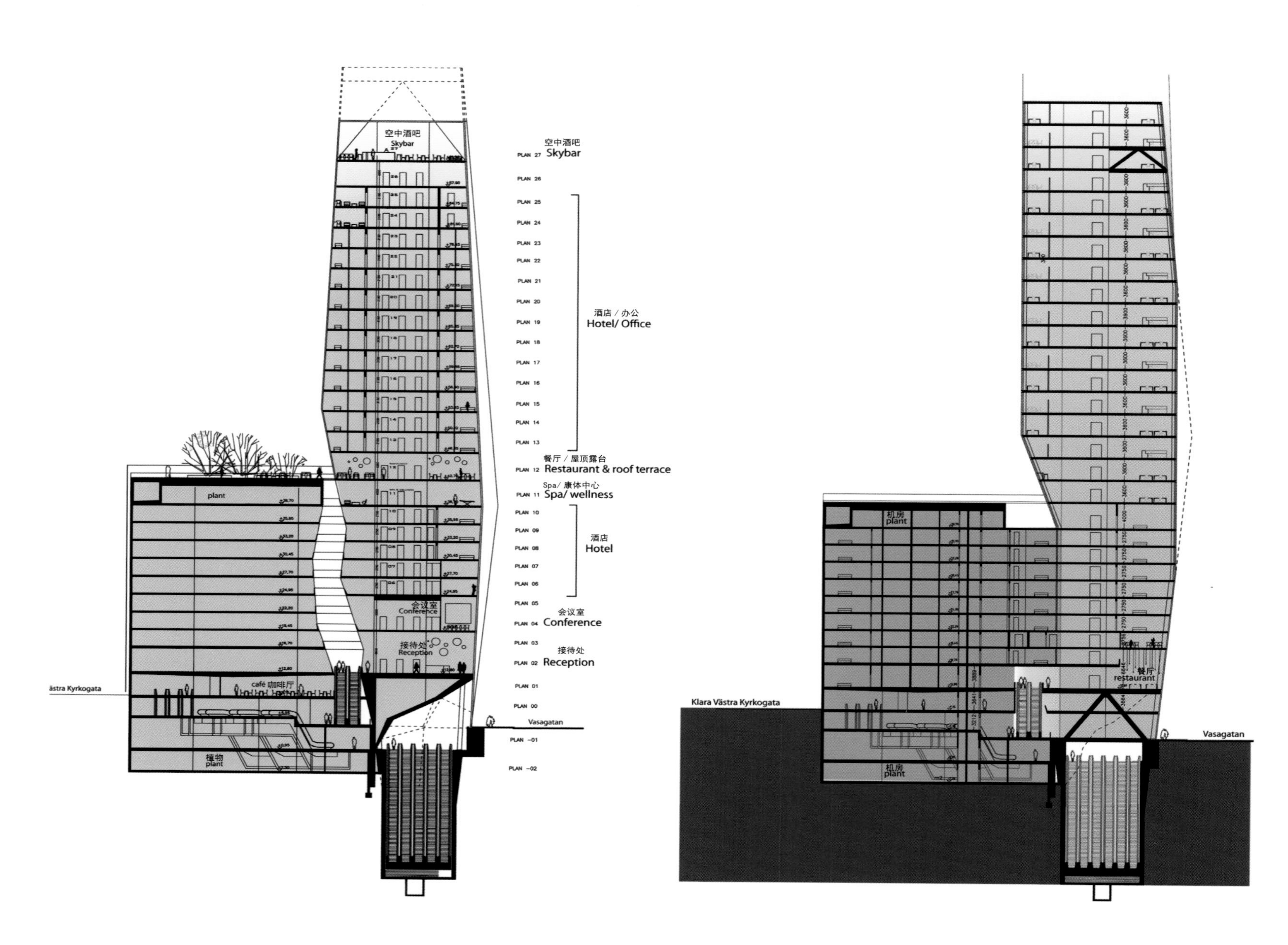

空中酒吧
Skybar
PLAN 27
PLAN 26
PLAN 25
PLAN 24
PLAN 23
PLAN 22
PLAN 21
PLAN 20
PLAN 19
PLAN 18
PLAN 17
PLAN 16
PLAN 15
PLAN 14
PLAN 13
酒店 / 办公
Hotel/ Office
餐厅 / 屋顶露台
PLAN 12 Restaurant & roof terrace
Spa/ 康体中心
PLAN 11 Spa/ wellness
PLAN 10
PLAN 09
PLAN 08
PLAN 07
PLAN 06
酒店
Hotel
PLAN 05
会议室
PLAN 04 Conference
PLAN 03
接待处
PLAN 02 Reception
PLAN 01
PLAN 00
PLAN −01
PLAN −02
会议室
Conference
接待处
Reception
plant
café 咖啡厅
植物
plant
ästra Kyrkogata
Vasagatan
机房
plant
餐厅
restaurant
Klara Västra Kyrkogata
Vasagatan

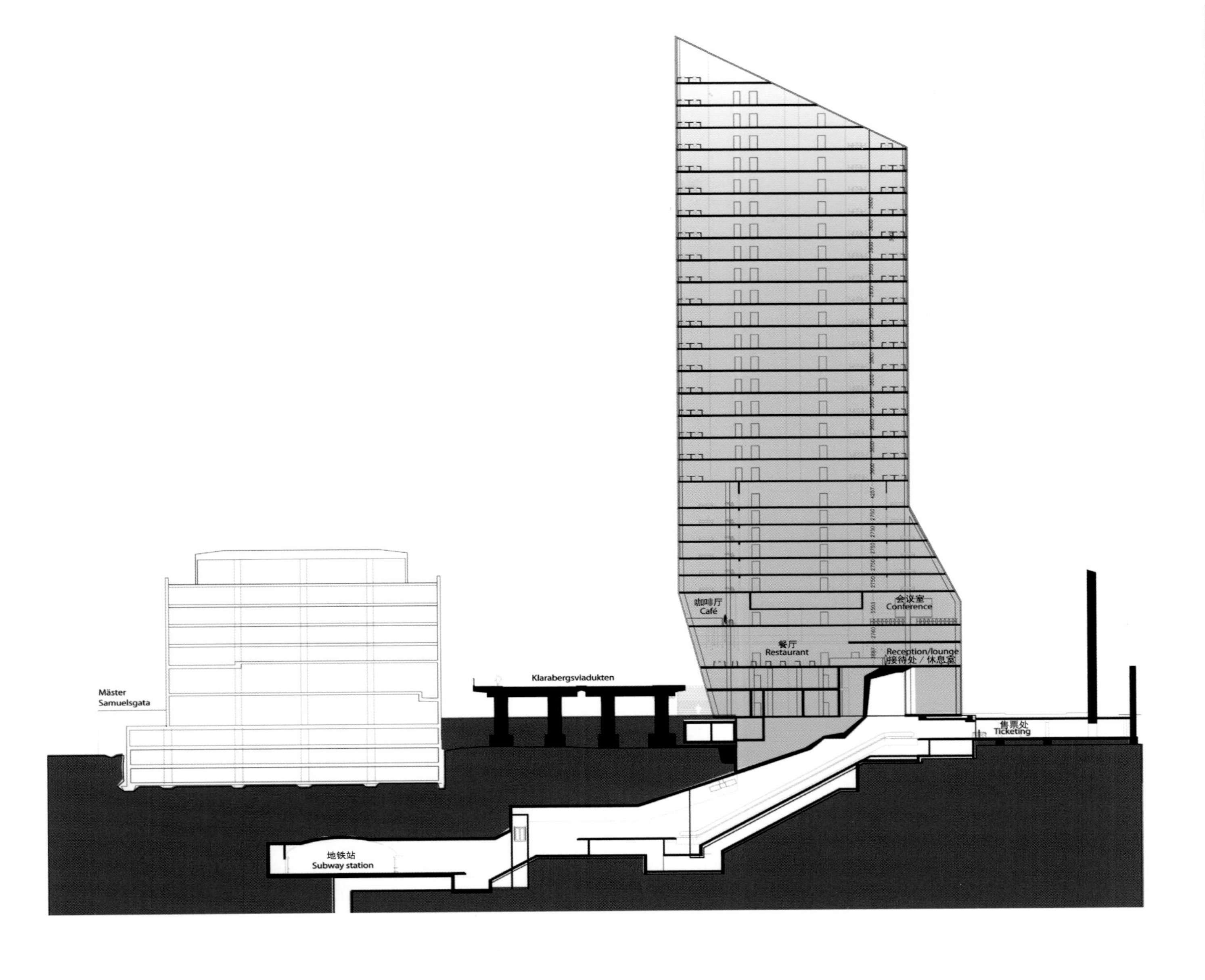
咖啡厅
Café
会议室
Conference
餐厅
Restaurant
Reception/lounge
接待处／休息室
Klarabergsviadukten
Mäster
Samuelsgata
售票处
Ticketing
地铁站
Subway station

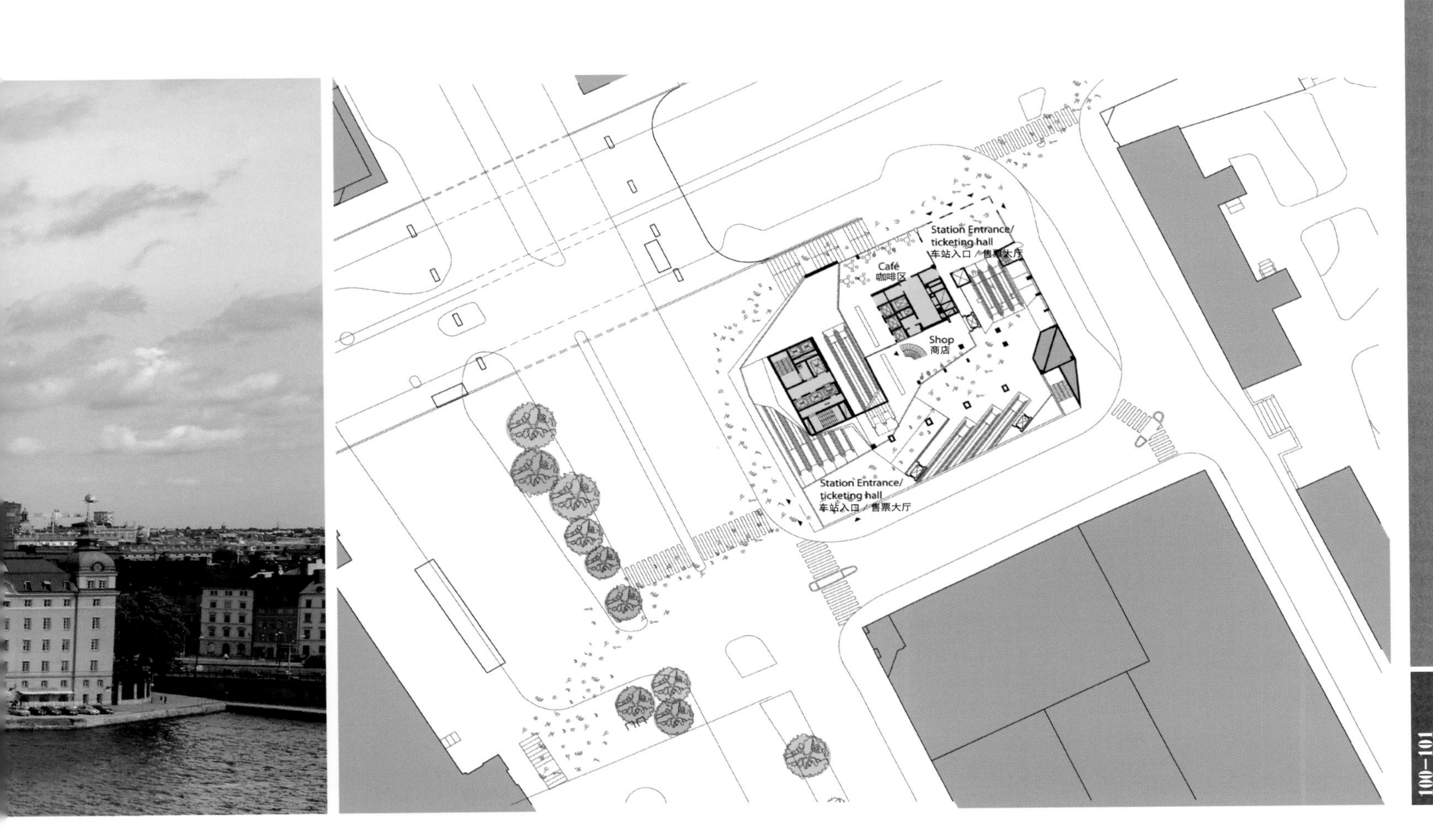
Station Entrance/
ticketing hall
车站入口／售票大厅
Café
咖啡区
Shop
商店
Station Entrance/
ticketing hall
车站入口／售票大厅

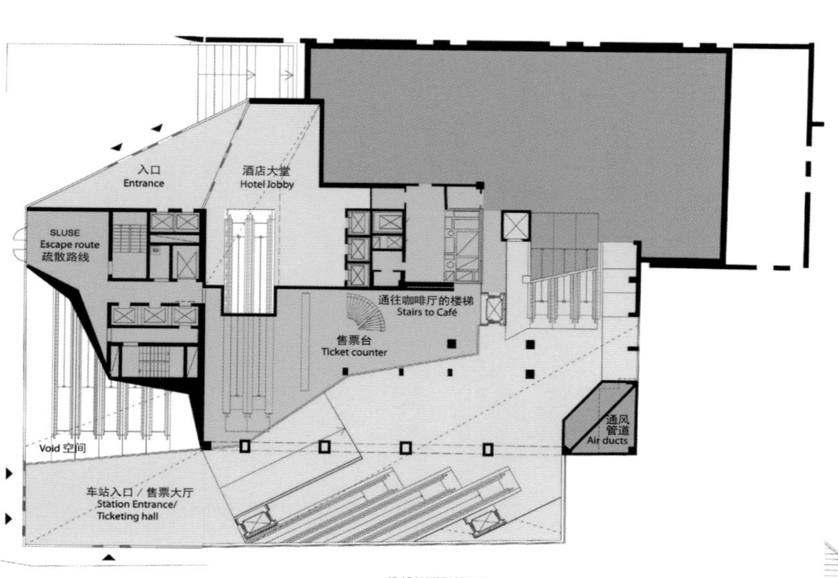
入口
Entrance
酒店大堂
Hotel lobby
SLUSE
Escape route
疏散路线
通往咖啡厅的楼梯
Stairs to Café
售票台
Ticket counter
通风
管道
Air ducts
Void 空间
车站入口 / 售票大厅
Station Entrance/
Ticketing hall
KLARA VATTUGRUND

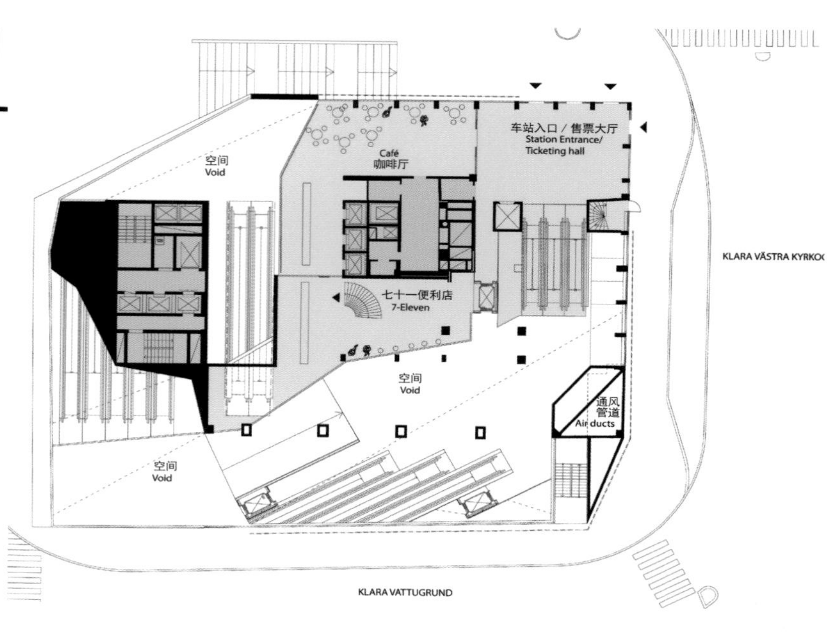
空间
Void
Café
咖啡厅
车站入口 / 售票大厅
Station Entrance/
Ticketing hall
七十一便利店
7-Eleven
空间
Void
通风
管道
Air ducts
空间
Void
KLARA VÄSTRA KYRKO
KLARA VATTUGRUND

Profile

The tower block construction in Stockholm has been planned in connection with the extension of the City Line rail link.

Design Concept

The tower draws inspiration from the city of Stockholm, an example of which is Stockholm's largest church Klara Kyrka situated right next to the site. The design is based on a concept of multi-function, a building where many varying uses can take place simultaneously.

Architectural Design

The complex of 100 meters, the equivalent of approximately 30 stories, will include a hotel, offices and possibly housing. The building will be equipped with an entrance to a new station at street level directly opposite the city's old main railway station and will therefore be a landmark for Stockholm's new metro.

The project reinterprets the city's beautiful town hall and classic church steeples which are combined to make up the city's historic landmarks and create Stockholm's historic skyline, but at the same time integrates the unique urban feeling defining the area around Centralstationen.

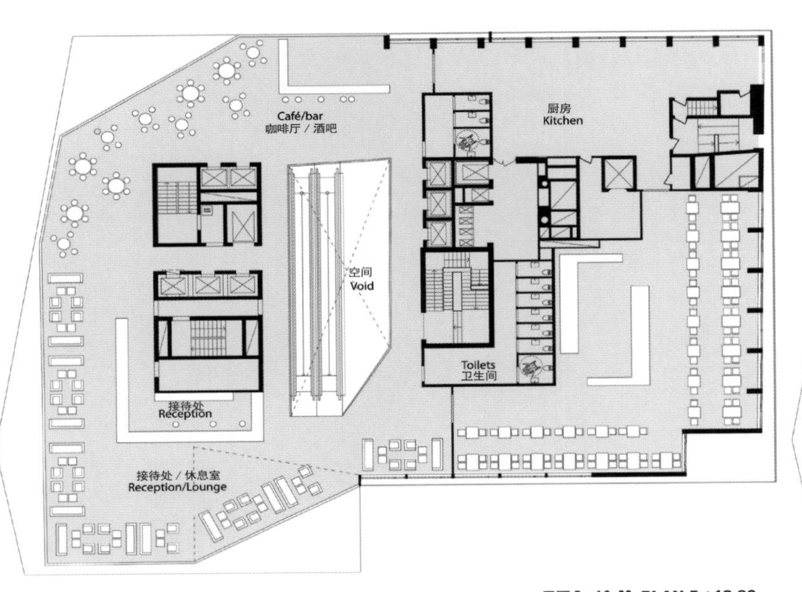

平面 5 +12.80 PLAN 5 +12.80
比例尺 1/200 SCALE 1/200

平面 6 +16.70 PLAN 6 +16.70
会议中心 Conference centre

平面 11+30.45　**PLAN11+30.45**

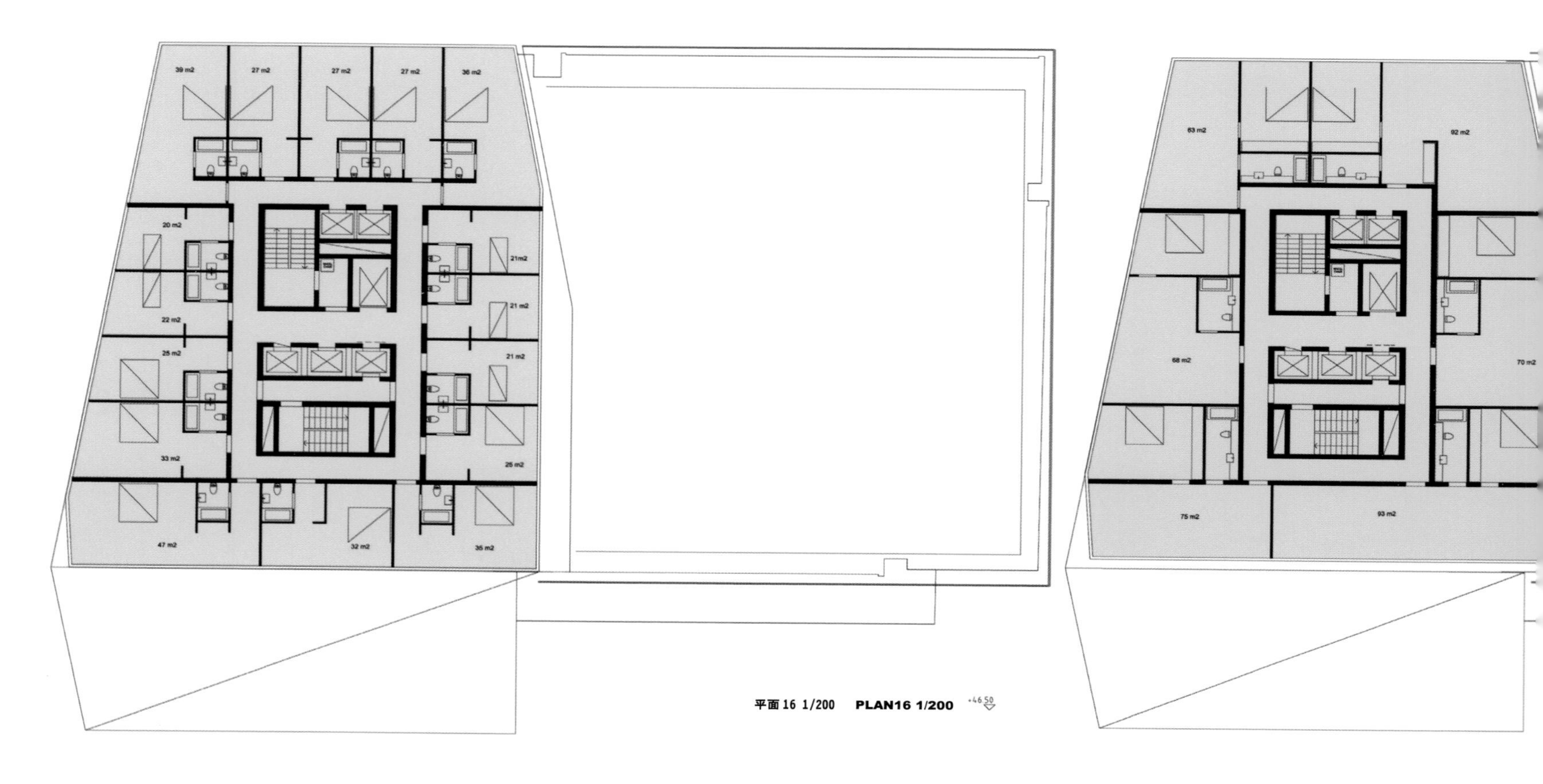

平面 16 1/200　**PLAN16 1/200**

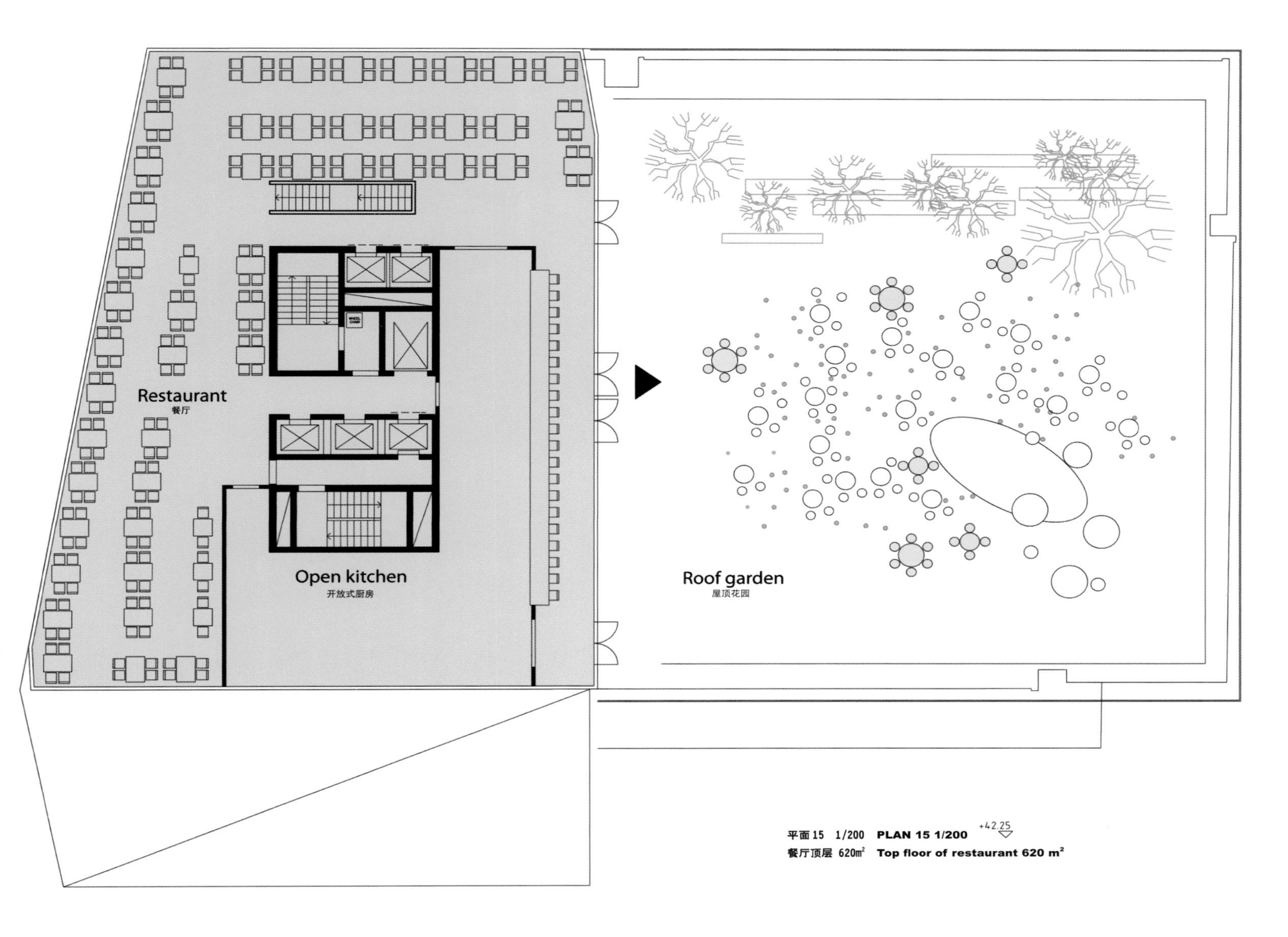

平面 15　1/200　**PLAN 15 1/200** +42.25
餐厅顶层 620m²　**Top floor of restaurant 620 m²**

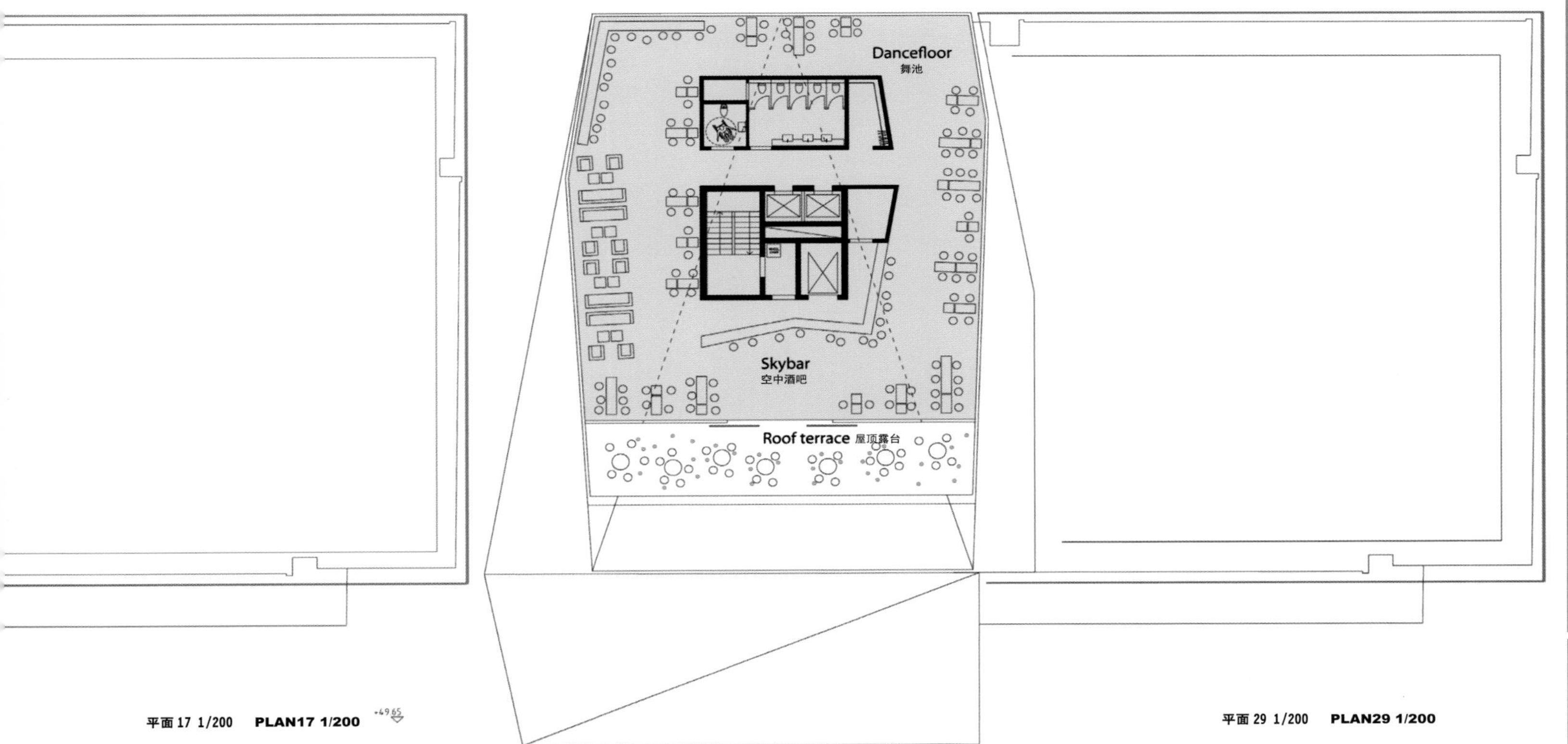

平面 17 1/200　**PLAN17 1/200** +49.65

平面 29 1/200　**PLAN29 1/200**

福建莆田正荣·金融财富中心

Fujian Putian Zhenro Financial Wealth Center

设计单位：日本 M.A.O. 一级建筑士事务所
开发商：正荣集团
项目地址：中国福建省莆田市
占地面积：199 998.6 m²
总建筑面积：719 465.8 m²
绿化率：30%
容积率：3.6

Designed by: M.A.O.
Developer: Zhenro Group
Location: Putian, Fujian, China
Land Area: 199,998.6 m²
Gross Floor Area: 719,465.8 m²
Greening Ratio: 30%
Plot Ratio: 3.6

项目概况

项目以国际大都会为标准，综合金融商务、民俗风情、主题商业、国际化花园社区等元素，将打造成为一个拥有大型国际级购物中心、5A级写字楼、民俗博物馆、国际化学校和顶级滨水宜居住宅等物业形态的，集文化、餐饮、购物、娱乐、休闲、办公及商住等为一体的大型滨水城市综合体。

设计理念

设计旨在超越层面上的符号化设计，不但在城市格局上整合完善了莆田新城商务中心，并且顺应时代需求，立体复合地延续了生态空间，更新了城市形象与活力。同时，设计创造性地发挥了业态组合的经济效力，对城市生态、文化、经济等功能进行重组，通过提供多样的业态组合，形成一个以人为本的群体地标。

分区设计

商业中心

通过对当地商业开发和各种业态的了解，项目设计了一条商业动线，这将是一条充满了主题性的商业街，将乐活、时尚、品位、优享、安逸五大元素融于一体，给莆田市民一种休闲、时尚、精致的全新购物体验。

设计采用铝板群与玻璃作为主要的造型元素进行立面设计，这两种材质组合成张弛有度的线条机理，使整个建筑群形成丰富的立面层次和光影变化，同时也给夜景照明设计创造了很好的条件。

酒店

建筑造型稳重大方，且同时具备鲜明的特色。外立面及建筑造型的设计灵感来源于莆田市市花——月季，设计师以简洁流畅的抽象化花蕾造型，将莆田这个城市的气质完全展现出来。

住宅区

单体立面采用新古典建筑风格，突出建筑的色彩及体量，同时强调建筑细节，体现典雅高贵的特色和风格。住宅均采用平屋顶的形式，造型多变而富有特色。幼儿园结合其建筑特性，采取了活泼现代的处理手法，使其有别于住宅建筑，又保持与整体环境的和谐统一。

Profile

The project takes international metropolis as its standard. Integrating financial business, native custom, theme business, international garden community and other elements, the project is envisaged to be a large waterfront urban complex that containing large international shopping center, 5A office building, folk custom museum, international school and top waterfront livable residences and integrating culture, F&B, mall, entertainment, leisure, office and etc.

Design Concept

The design aims to surpass symbolic design. It has not only integrated and completed Putian New Urban Business Center, but also achieved stereoscopic and integrated continuity of ecological space complying with the times. Meanwhile, it has managed to renew city image and energy. The design creatively exerts economic benefits of industries. It strives to reorganize urban ecology, culture and economy, and to create a people-oriented landmark by providing diversified combination of industries.

Subarea Design

Business Center

Based on understanding of local business development and types of industries, the project design has envisaged an active commercial line — a thematic commercial street to integrate living, fashion, taste, enjoyment and leisure as well as provide leisure, fashionable and exquisite shopping experience.

Aluminum panel and glass are main materials for building facades. Their flexible lines texture creates rich façade levels and shadow changes. Also, it is a fine condition for nightscape lighting.

Hotel

The shape of building is dignified and of good taste, with its own distinctive features. The design inspiration of facades and architectural image comes from the city flower of Putian —rosa chinensis. Simple, smooth and abstract shape has fully exerted the temperament of this city.

Residential Area

The façade of residential buildings has a neoclassical architectural design, highlighting the buildings' colors and volumes, emphasizing architectural details and reflecting elegant, dignified feature and style. These residential buildings mainly adopts flat roof. It is diversified and unique. Active and modern buildings of the kindergarten are distinguished from other residential buildings while harmoniously unified with the surroundings.

BASIC HOUSE

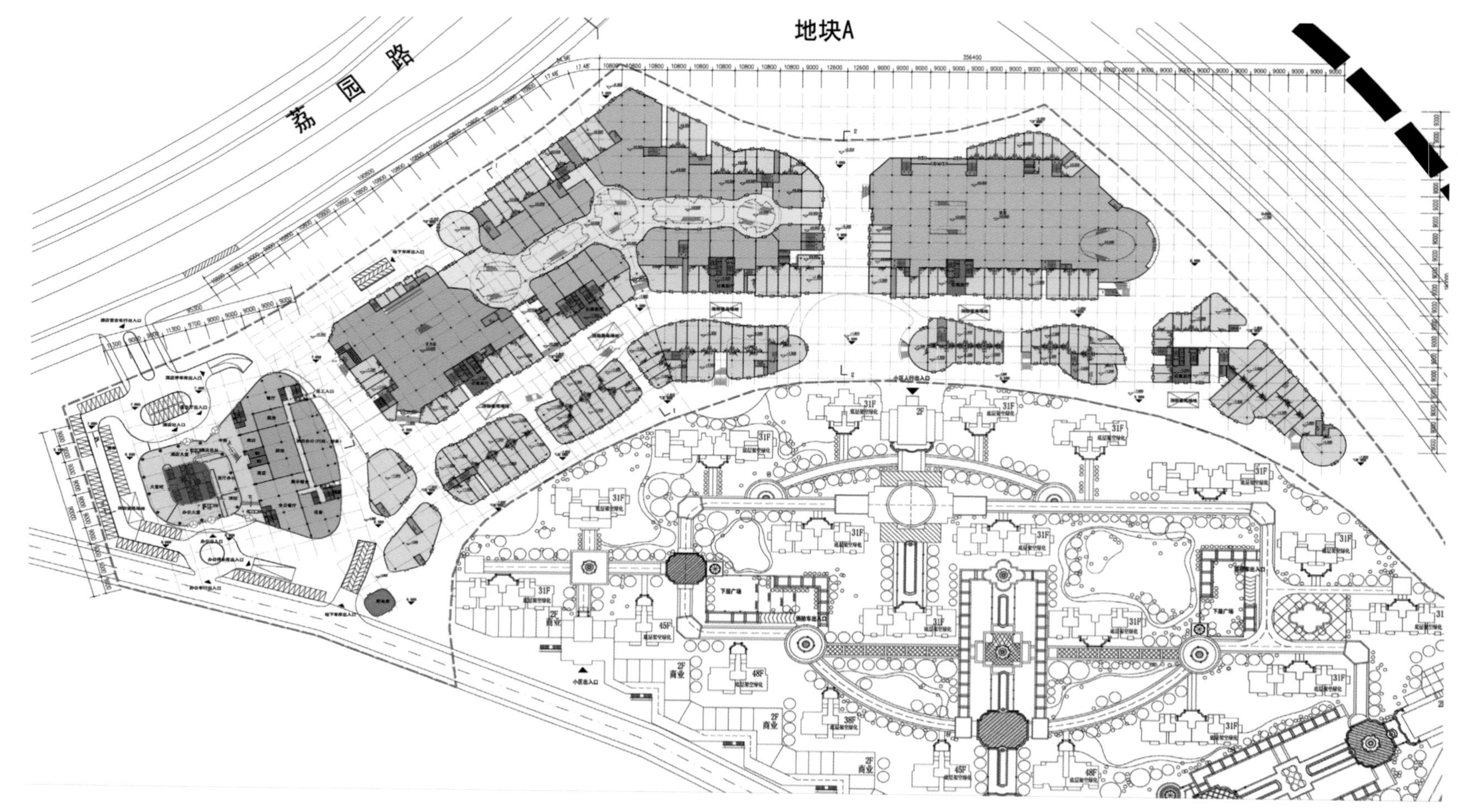

地块A
荔园路

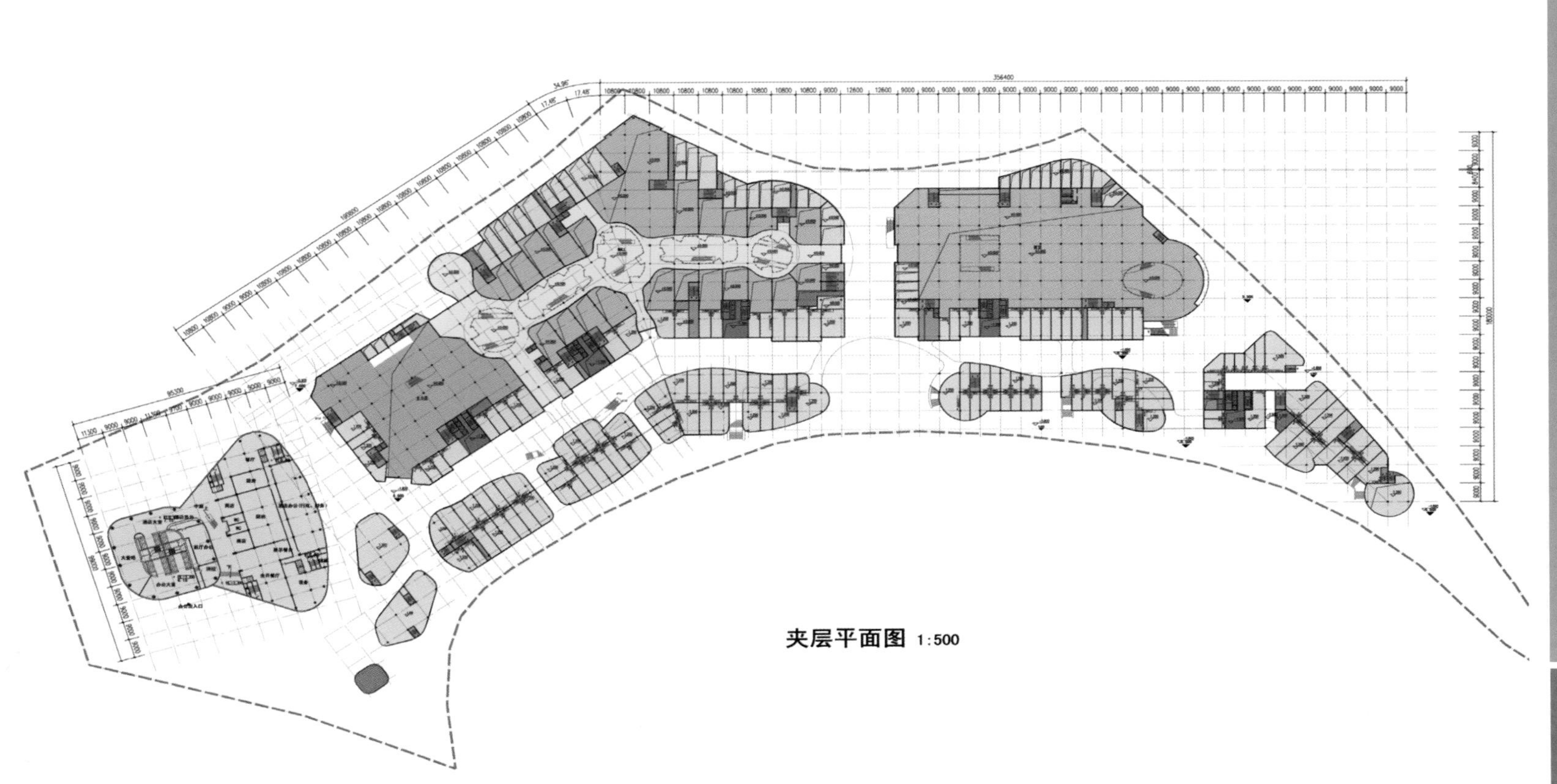

夹层平面图 1:500

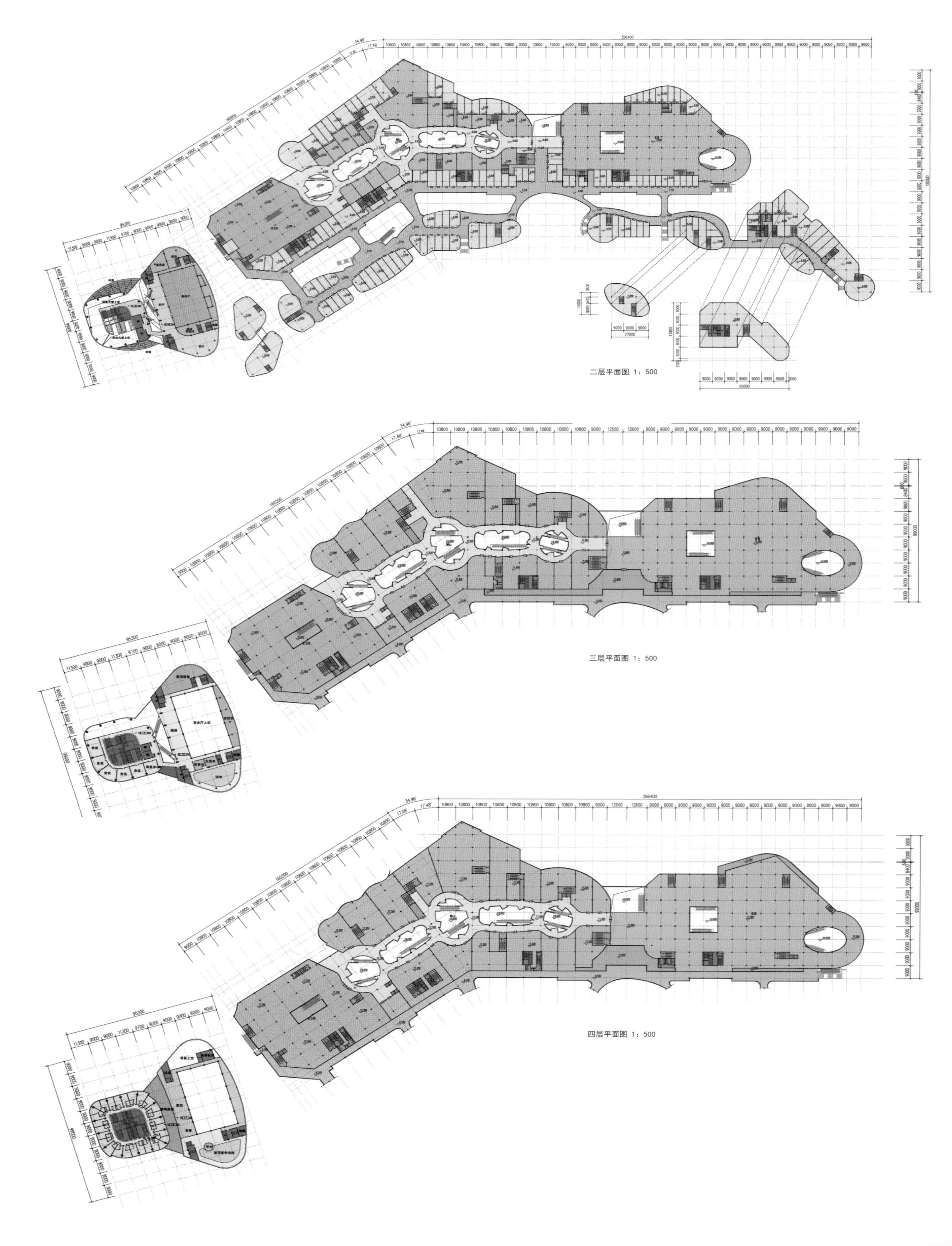

二层平面图 1：500

三层平面图 1：500

四层平面图 1：500

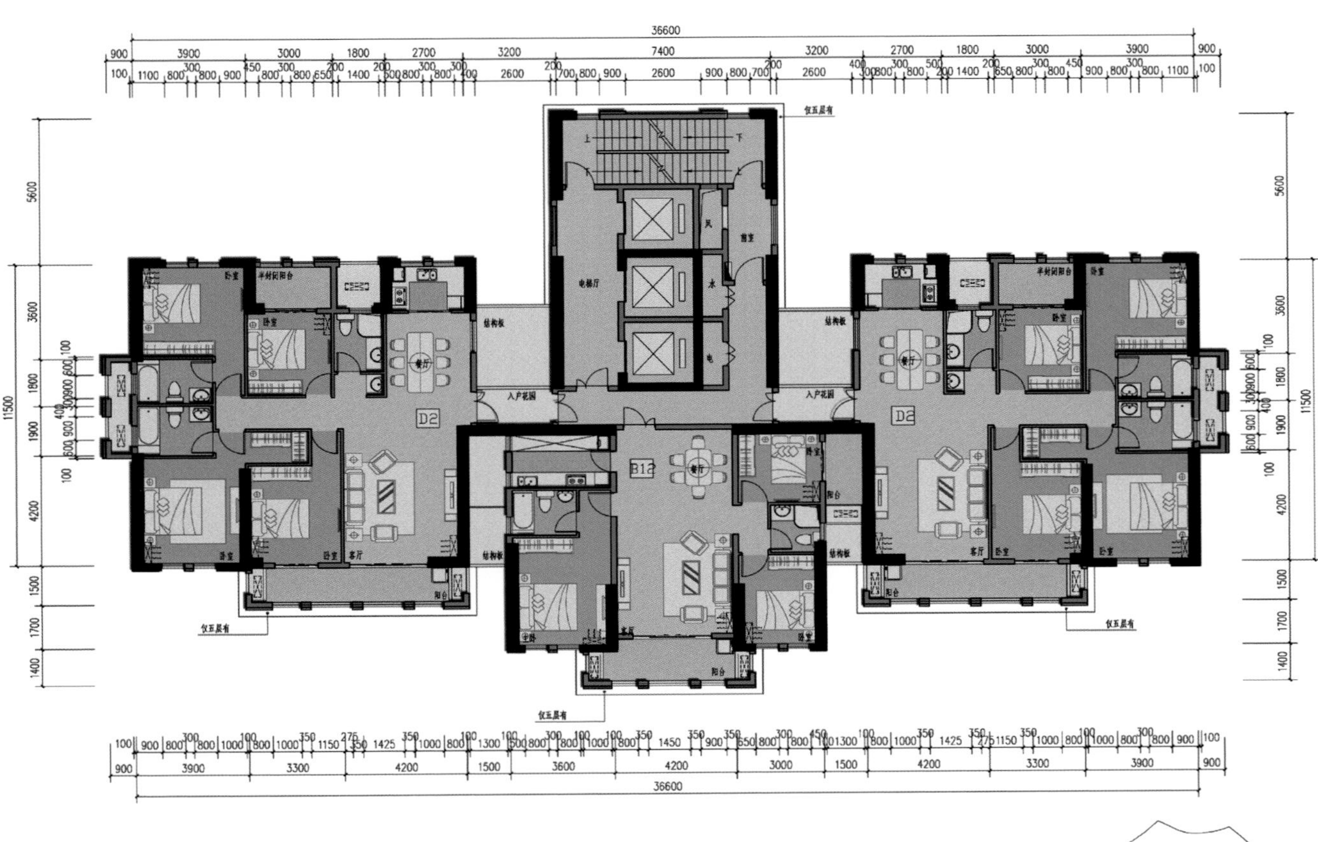

标准层平面图

注：本层建筑面积 476.70m²
其中：计容建筑面积 476.70m²
不计容建筑面积：0
阳台建筑面积（一半计）：21.59m²

户型	房型	套内建筑面积	公摊面积	1/2阳台建筑面积	套型建筑面积	户型建筑面积	阳台投影面积占套型建筑建筑面积的比例
B12	三房二厅两卫	89.04m²	30.47m²	5.87m²	119.51m²	125.38m²	9.82%
D2	四房二厅三卫	124.35m²	42.28m²	7.36m²	166.63m²	173.99m²	8.83%

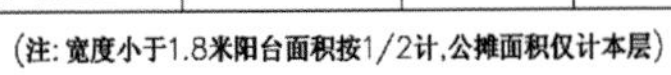
（注：宽度小于1.8米阳台面积按1/2计，公摊面积仅计本层）

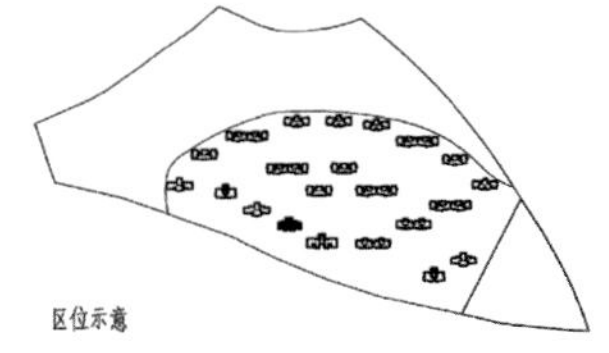

区位示意

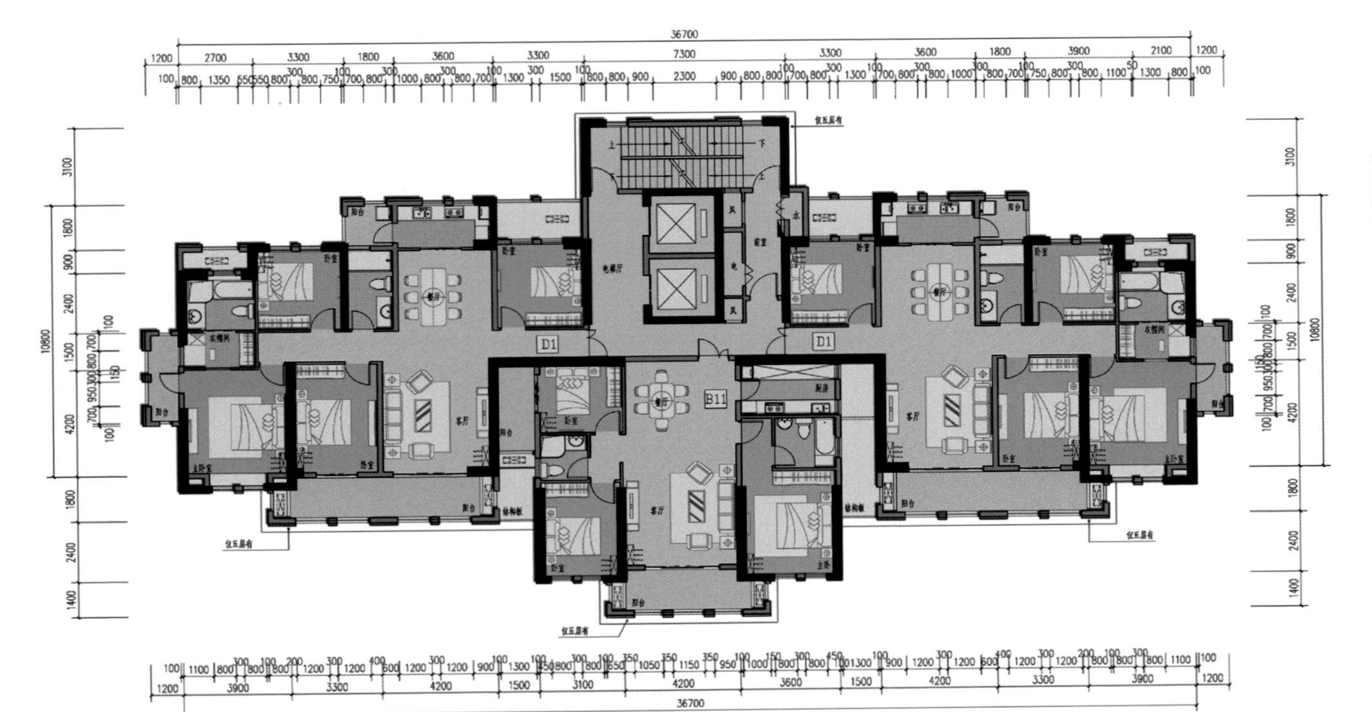

标准层平面图

注：本层建筑面积 451.11m²
其中：计容建筑面积 451.11m²
不计容建筑面积：0
阳台建筑面积（一半计）：27.80m²

户型	房型	套内建筑面积	公摊面积	1/2阳台建筑面积	套型建筑面积	户型建筑面积	阳台投影面积占套型建筑建筑面积的比例
B11	三房二厅两卫	93.15m²	23.25m²	6.10m²	116.40m²	122.50m²	10.48%
D1	四房二厅两卫	122.49m²	30.10m²	10.28m²	153.59m²	163.87m²	13.39%

（注：宽度小于1.8米阳台面积按1/2计，公摊面积仅计本层）

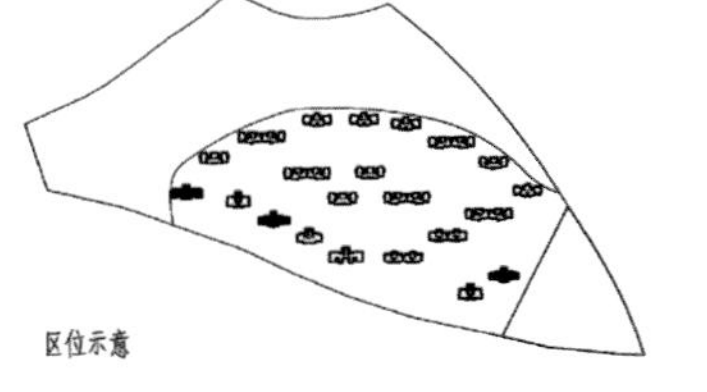

区位示意

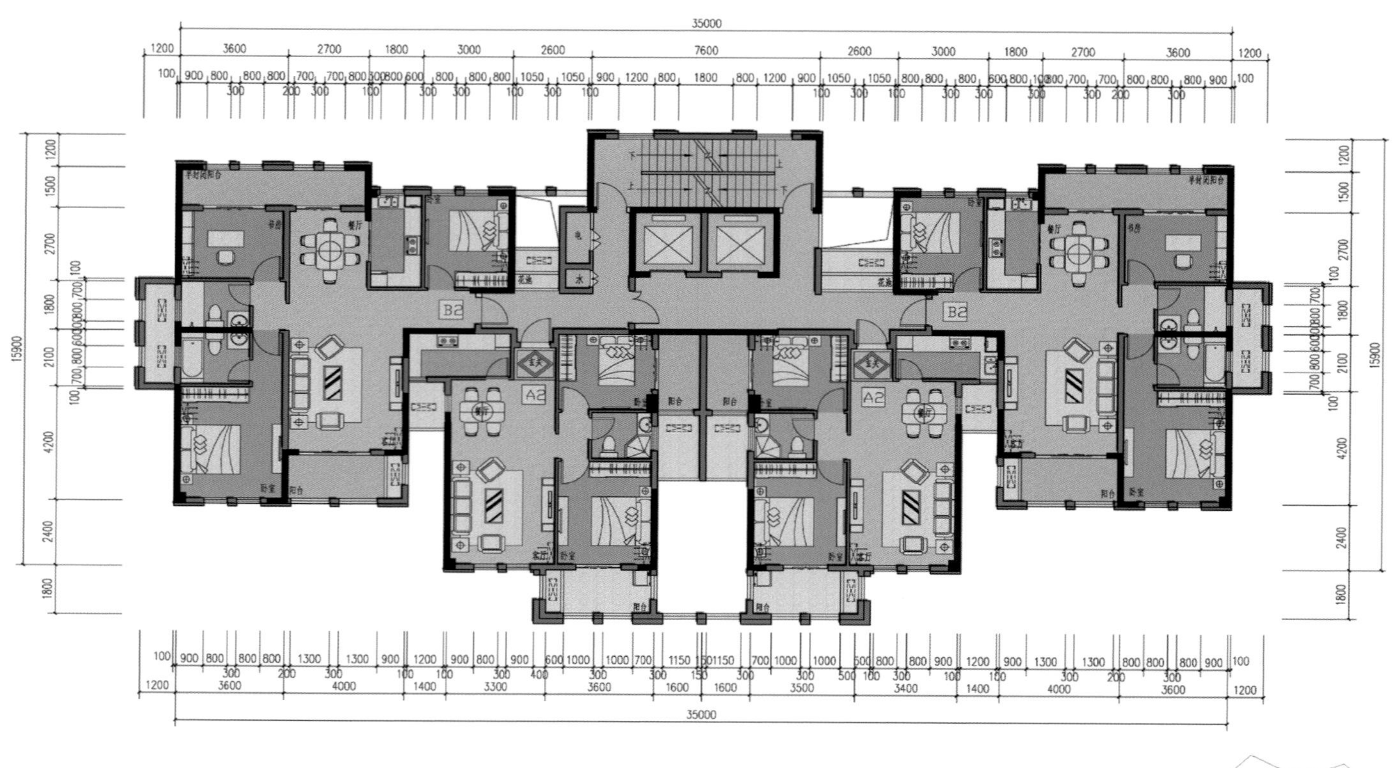

标准层平面图

注：本层建筑面积 418.97m²
其中：计容建筑面积 418.97m²
不计容建筑面积：0
阳台建筑面积（一半计）：25.82m²

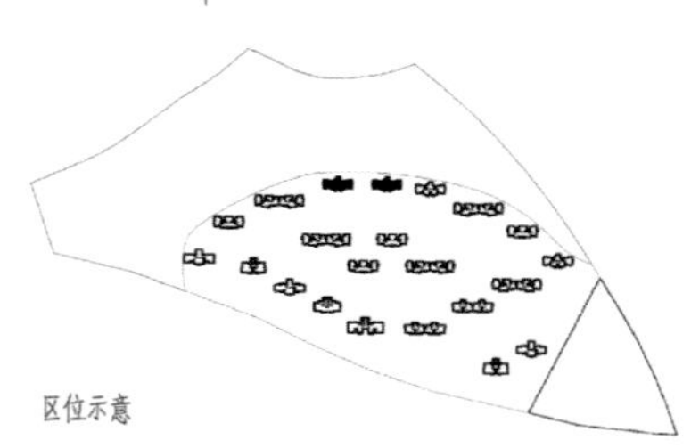

区位示意

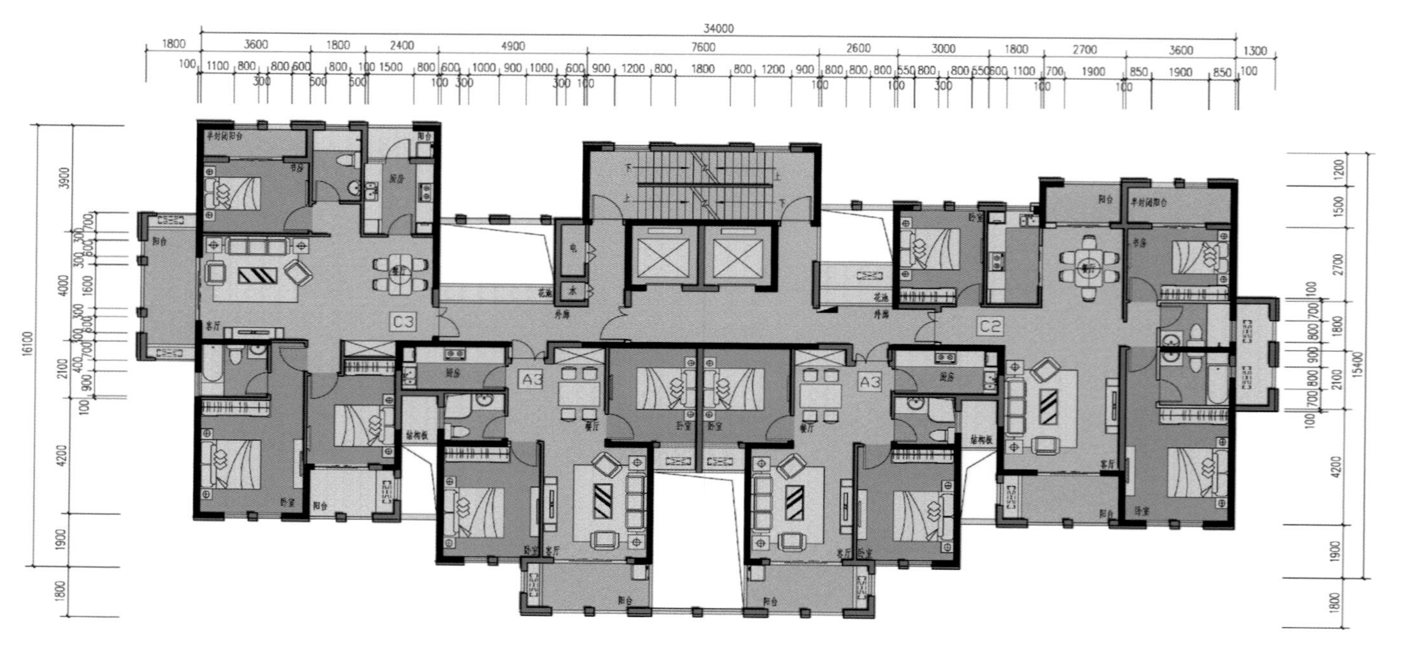

标准层平面图

注：本层建筑面积 407.41m²
其中：计容建筑面积 407.41m²
不计容建筑面积：0
阳台建筑面积（一半计）：20.11m²

户型	房型	套内建筑面积	公摊面积	1/2阳台建筑面积	套型建筑面积	户型建筑面积	阳台投影面积占套型建筑建筑面积的比例
C2	三房二厅两卫	93.51m²	25.04m²	7.78m²	110.77m²	126.34m²	14.06%
C3	三房二厅两卫	90.60m²	23.80m²	5.67m²	108.73m²	120.07m²	10.43%
A3	两房二厅一卫	61.21m²	15.96m²	3.33m²	73.84m²	80.50m²	9.02%

（注：宽度小于1.8米阳台面积按1/2计，公摊面积仅计本层）

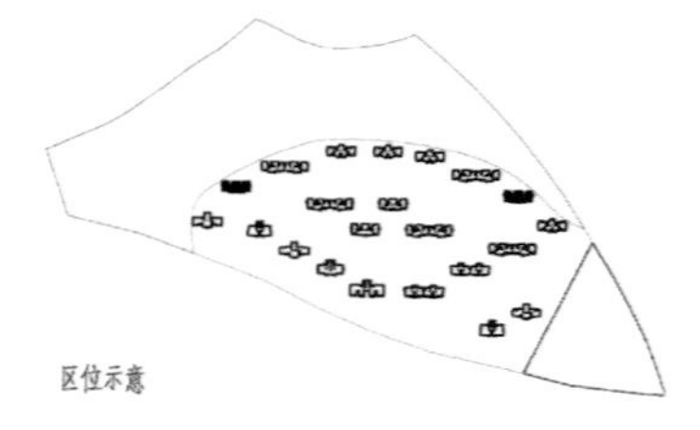

区位示意

KOSE

HUM
BASIC HOUSE
Honeys

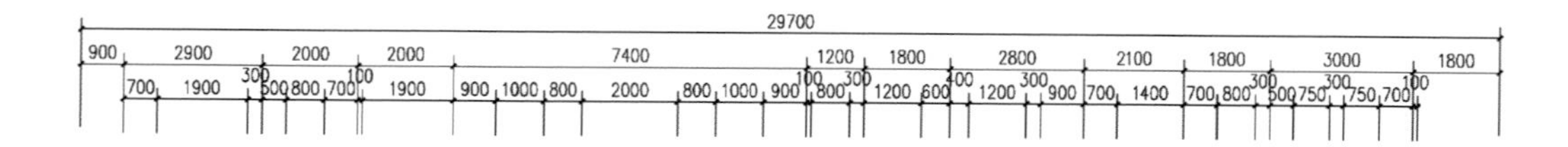

标准层平面图

注：本层建筑面积 340.13m²
其中：计容建筑面积 340.13m²
不计容建筑面积：0
阳台建筑面积（一半计）：20.76m²

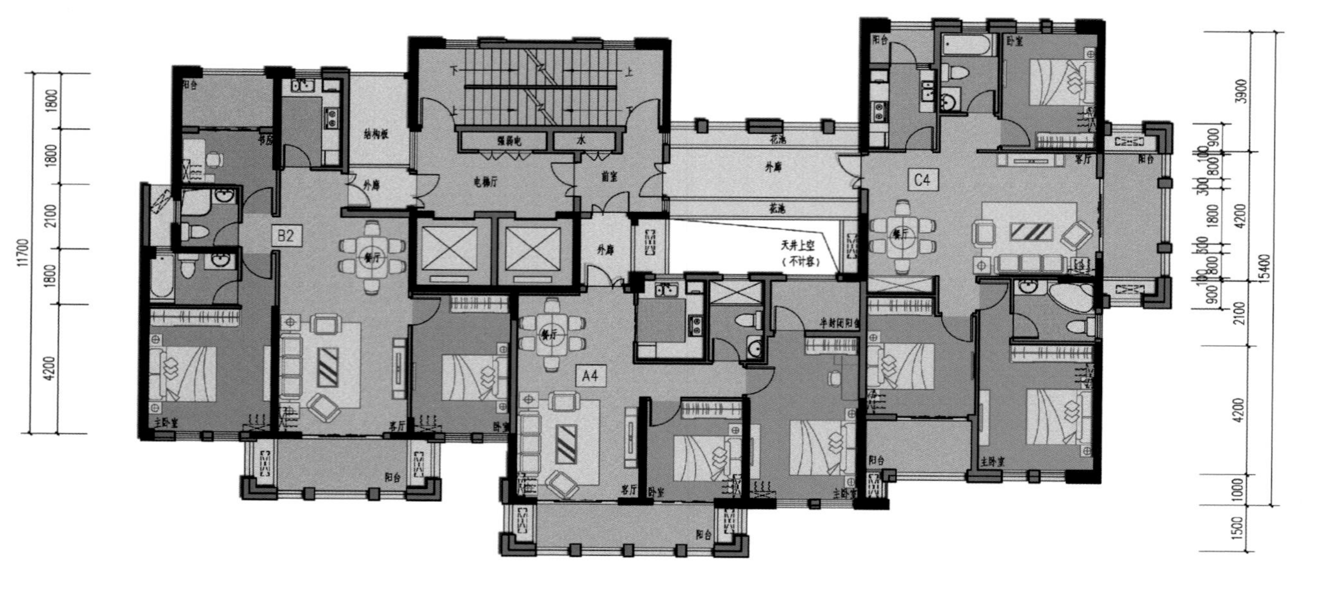

户型	房型	套内建筑面积	公摊面积	1/2阳台建筑面积	套型建筑面积	户型建筑面积	阳台投影面积占套型建筑建筑面积的比例
C4	三房二厅两卫	90.90m²	27.57m²	8.01m²	118.47m²	126.48m²	11.08%
A4	二房二厅一卫	68.02m²	20.78m²	6.54m²	88.80m²	95.34m²	14.73%
B2	两房半二厅两卫	86.31m²	25.79m²	6.21m²	112.10m²	118.31m²	13.52%

（注：宽度小于1.8米阳台面积按1/2计，公摊面积仅计本层）

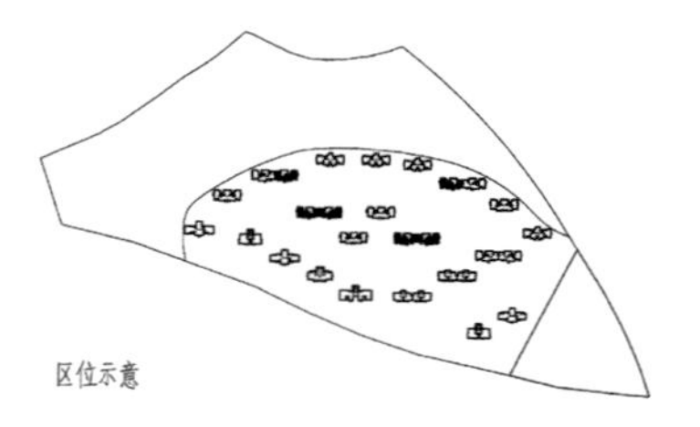

区位示意

标准层平面图

注：本层建筑面积 439.14m²
其中：计容建筑面积 439.14m²
不计容建筑面积：0
阳台建筑面积（一半计）：26.94m²

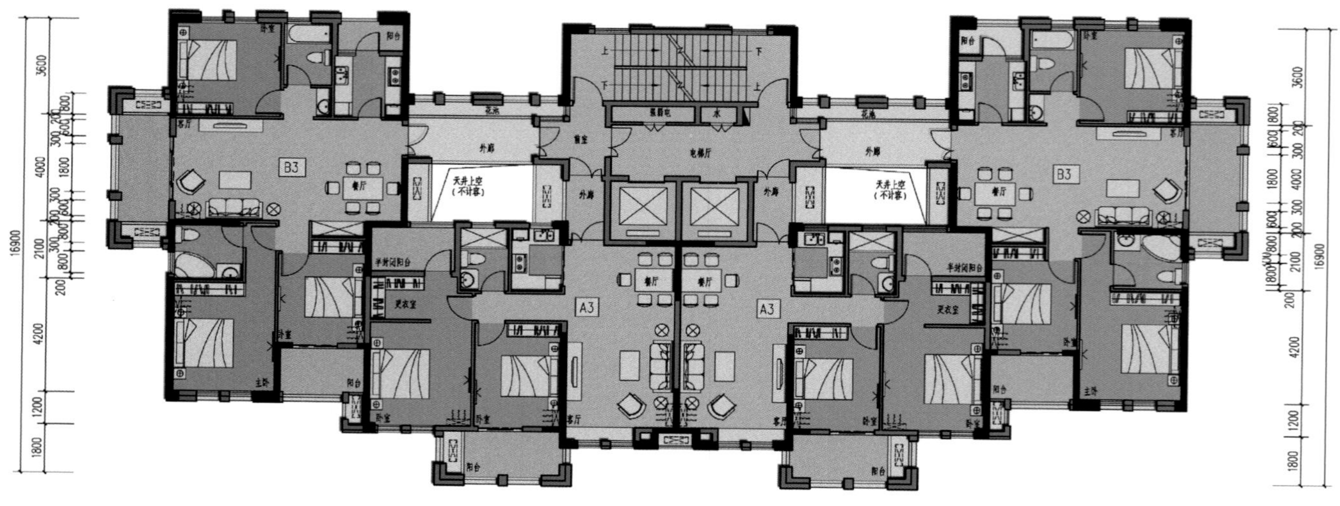

户型	房型	套内建筑面积	公摊面积	1/2阳台建筑面积	套型建筑面积	户型建筑面积	阳台投影面积占套型建筑建筑面积的比例
B3	三房二厅两卫	92.58m²	24.57m²	7.74m²	117.15m²	124.89m²	13.21%
A3	两房二厅一卫	70.32m²	18.63m²	5.73m²	88.95m²	94.68m²	12.88%

（注：宽度小于1.8米阳台面积按1/2计，公摊面积仅计本层）

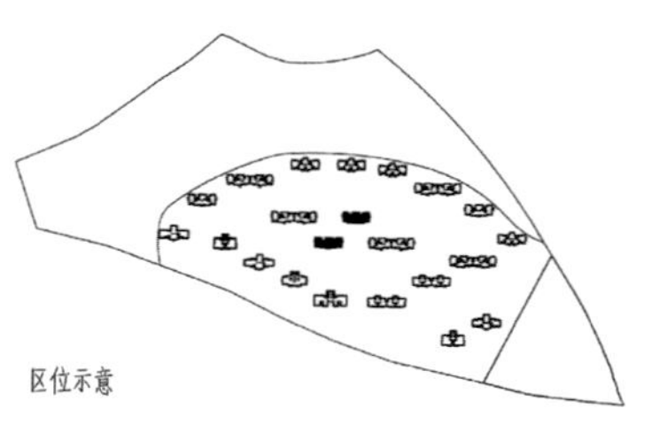

区位示意

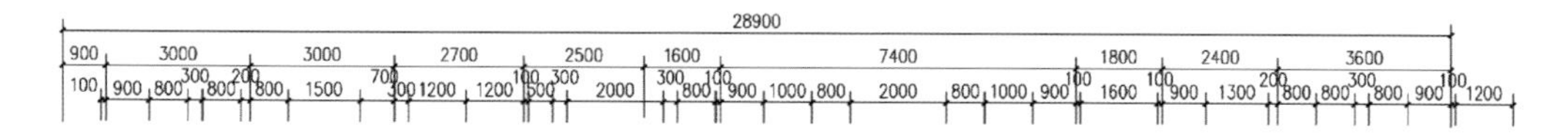

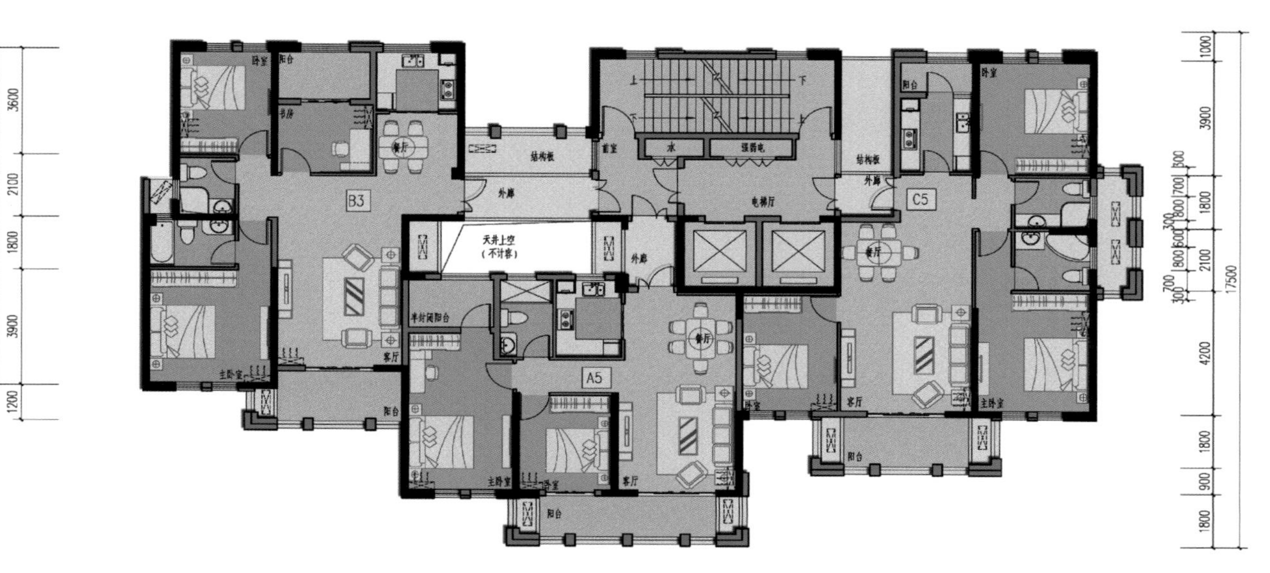

标准层平面图

注：本层建筑面积 338.96m²
其中：计容建筑面积 337.74m²
不计容建筑面积：0
阳台建筑面积（一半计）：18.30m²

户型	房型	套内建筑面积	公摊面积	1/2阳台建筑面积	套型建筑面积	户型建筑面积	阳台投影面积占套型建筑建筑面积的比例
C5	三房二厅两卫	93.60m²	26.96m²	5.04m²	120.65m²	125.69m²	8.36%
A5	二房二厅一卫	68.02m²	20.38m²	6.54m²	88.46m²	95.00m²	14.79%
B3	两房半二厅两卫	86.82m²	26.96m²	6.00m²	112.27m²	118.27m²	10.69%

(注：宽度小于1.8米阳台面积按1/2计，公摊面积仅计本层)

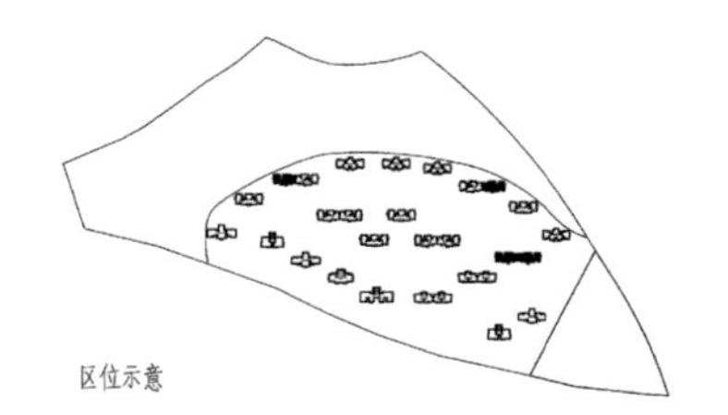

区位示意

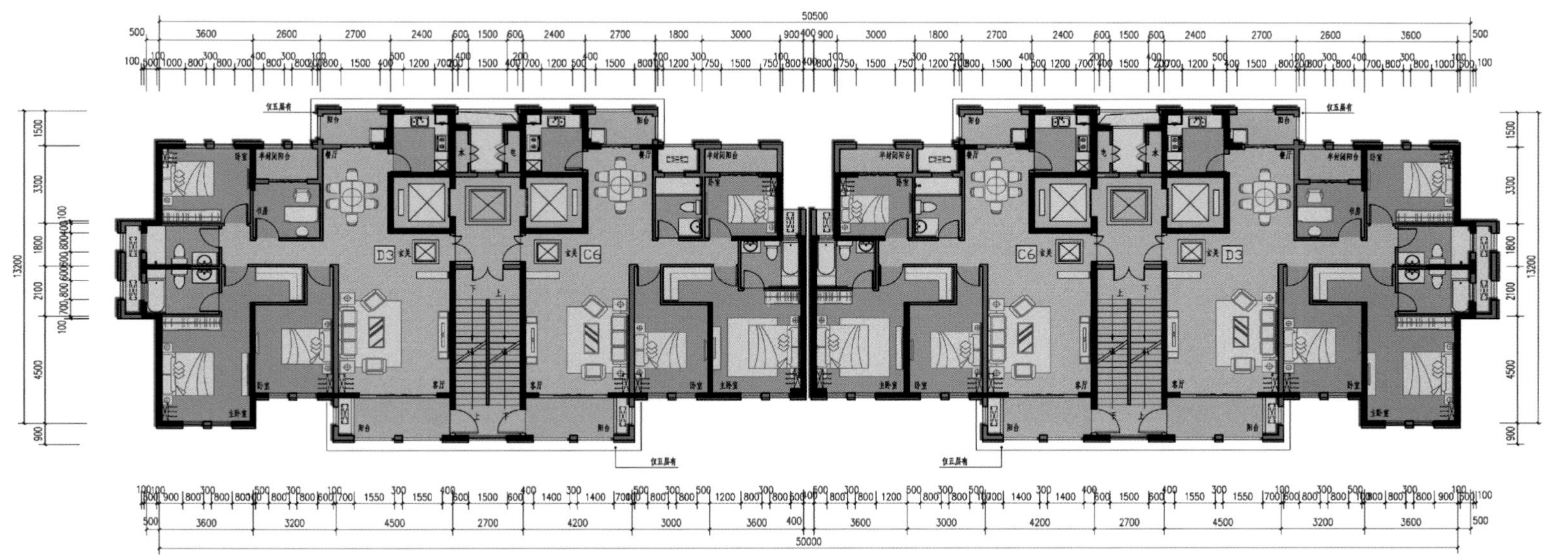

标准层平面图

注：本层建筑面积 582.98m²
其中：计容建筑面积 582.98m²
不计容建筑面积：0
阳台建筑面积（一半计）：28.68m²

户型	房型	套内建筑面积	公摊面积	1/2阳台建筑面积	套型建筑面积	户型建筑面积	阳台投影面积占套型建筑建筑面积的比例
C6	三房二厅两卫	101.97m²	25.89m²	7.09m²	127.86m²	134.95m²	11.09%
D3	三房半二厅两卫	119.25m²	30.04m²	7.25m²	149.29m²	156.54m²	9.71%

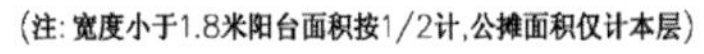

(注：宽度小于1.8米阳台面积按1/2计，公摊面积仅计本层)

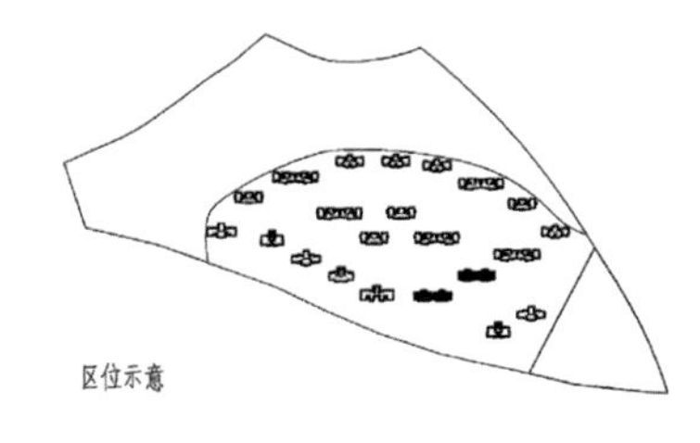

区位示意

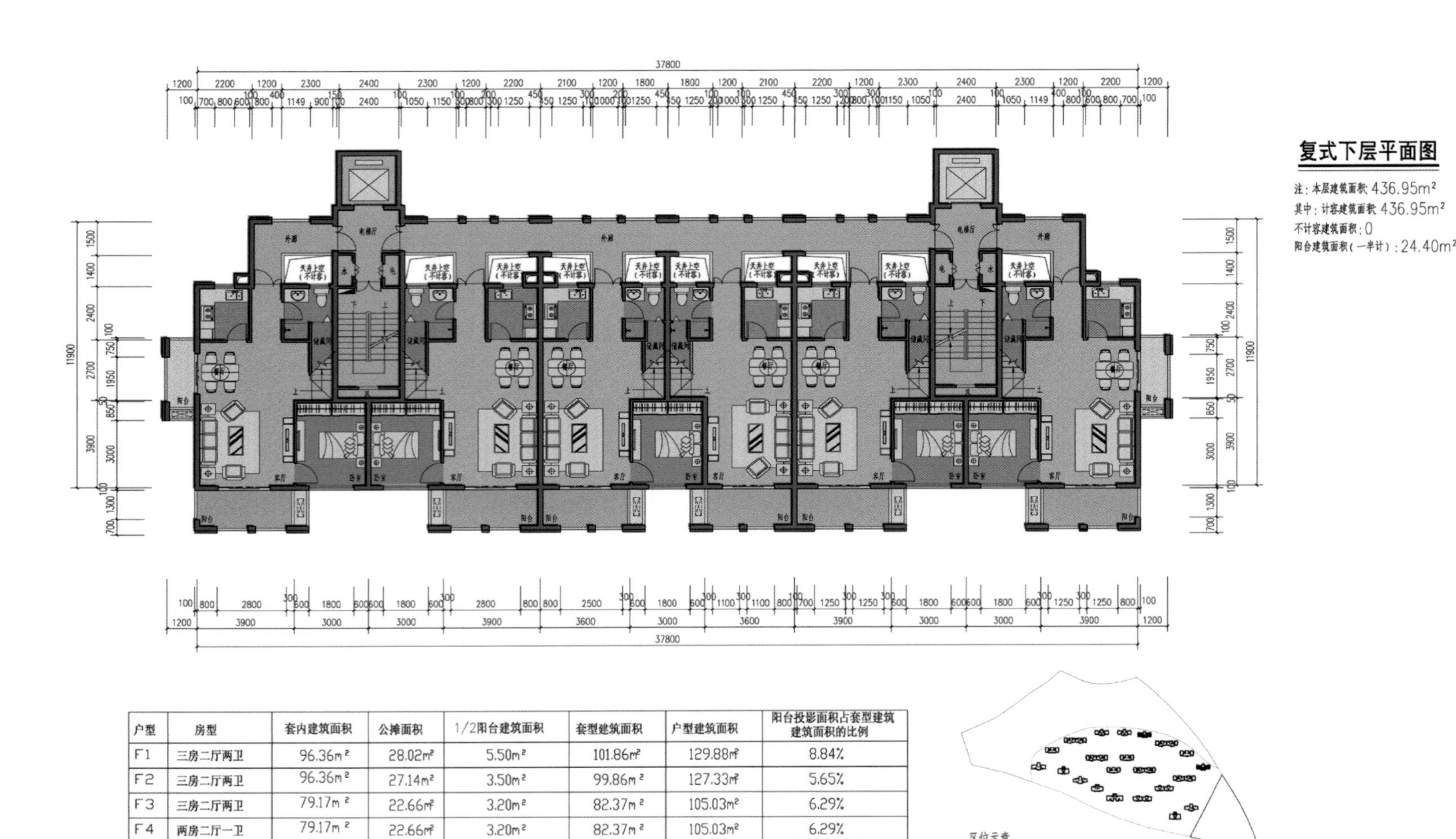

复式下层平面图

注：本层建筑面积 436.95m²
其中：计容建筑面积 436.95m²
不计容建筑面积：0
阳台建筑面积（一半计）：24.40m²

户型	房型	套内建筑面积	公摊面积	1/2阳台建筑面积	套型建筑面积	户型建筑面积	阳台投影面积占套型建筑建筑面积的比例
F1	三房二厅两卫	96.36m²	28.02m²	5.50m²	101.86m²	129.88m²	8.84%
F2	三房二厅两卫	96.36m²	27.14m²	3.50m²	99.86m²	127.33m²	5.65%
F3	三房二厅两卫	79.17m²	22.66m²	3.20m²	82.37m²	105.03m²	6.29%
F4	两房二厅一卫	79.17m²	22.66m²	3.20m²	82.37m²	105.03m²	6.29%

（注：宽度小于1.8米阳台面积按1/2计，公摊面积仅计本层）

区位示意

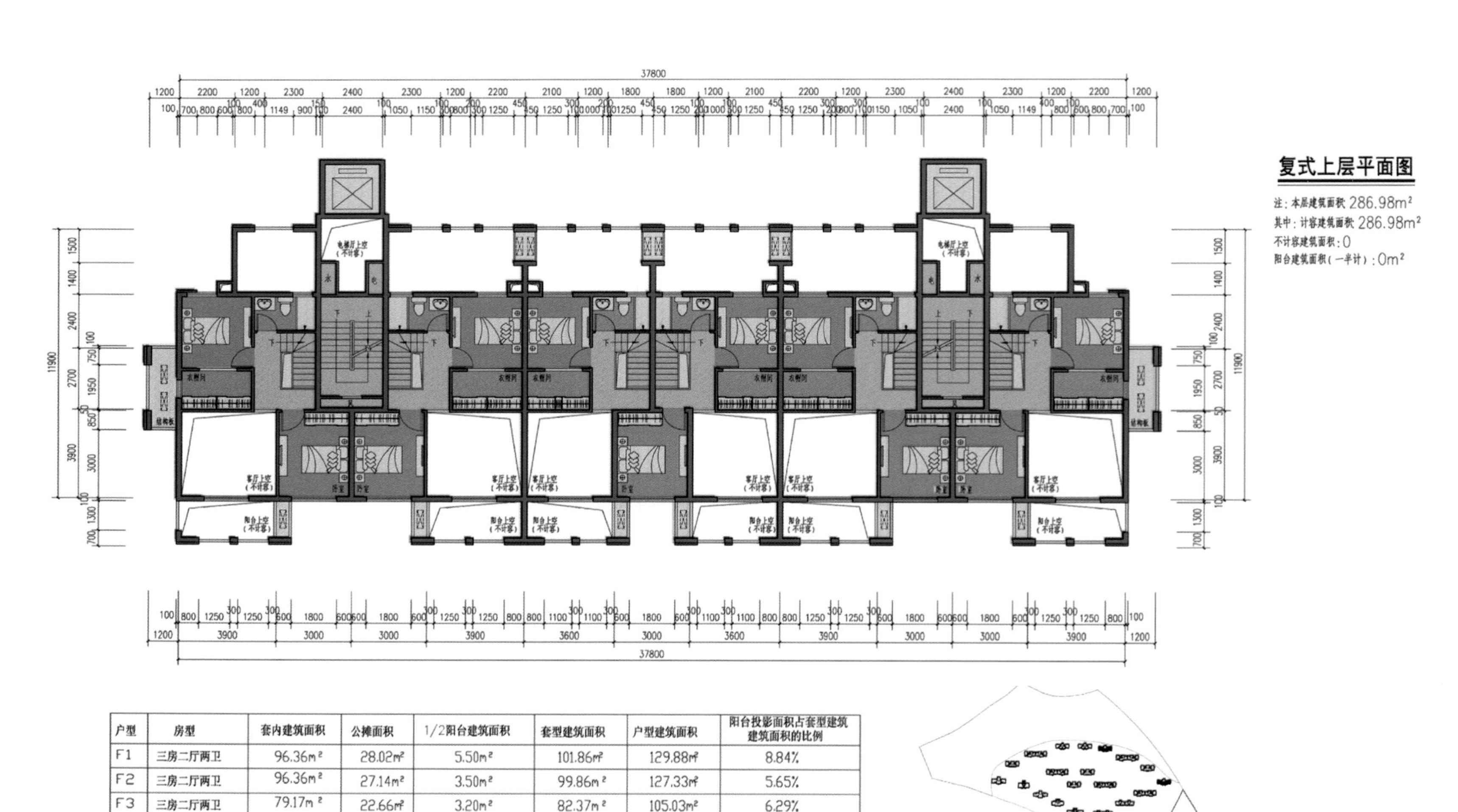

复式上层平面图

注：本层建筑面积 286.98m²
其中：计容建筑面积 286.98m²
不计容建筑面积：0
阳台建筑面积（一半计）：0m²

户型	房型	套内建筑面积	公摊面积	1/2阳台建筑面积	套型建筑面积	户型建筑面积	阳台投影面积占套型建筑建筑面积的比例
F1	三房二厅两卫	96.36m²	28.02m²	5.50m²	101.86m²	129.88m²	8.84%
F2	三房二厅两卫	96.36m²	27.14m²	3.50m²	99.86m²	127.33m²	5.65%
F3	三房二厅两卫	79.17m²	22.66m²	3.20m²	82.37m²	105.03m²	6.29%
F4	两房二厅一卫	79.17m²	22.66m²	3.20m²	82.37m²	105.03m²	6.29%

（注：宽度小于1.8米阳台面积按1/2计，公摊面积仅计本层）

区位示意

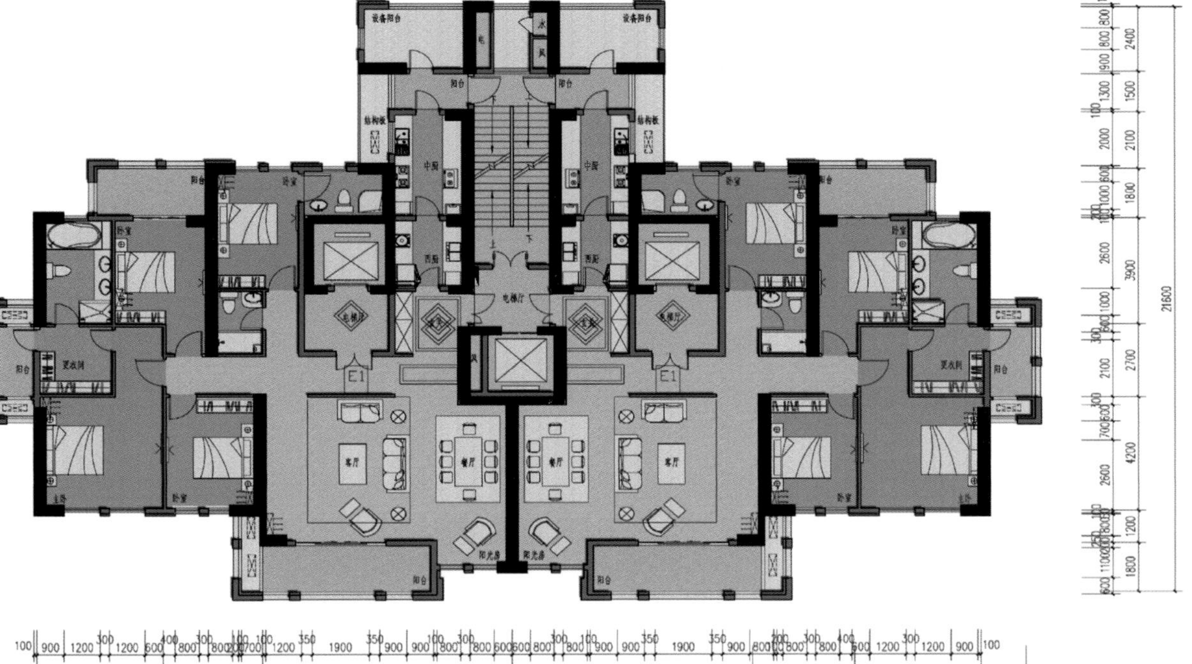

标准层平面图

注：本层建筑面积 463.60m²
其中：计容建筑面积 463.60m²
不计容建筑面积：0
阳台建筑面积（一半计）：24.90m²

户型	房型	套内建筑面积	公摊面积	1/2阳台建筑面积	套型建筑面积	户型建筑面积	阳台投影面积占套型建筑建筑面积的比例
E1	四房二厅三卫	179.76㎡	39.50㎡	12.45㎡	222.59㎡	235.98㎡	11.19%

（注：宽度小于1.8米阳台面积按1/2计，公摊面积仅计本层）

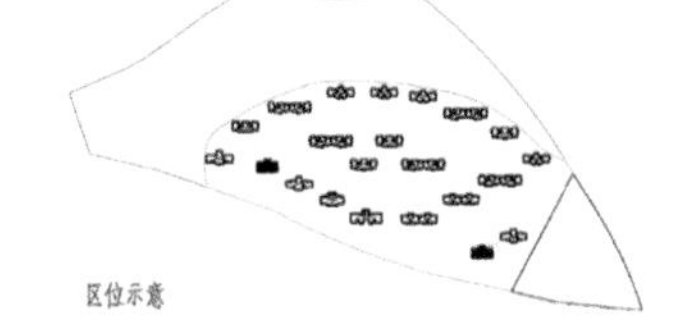

区位示意

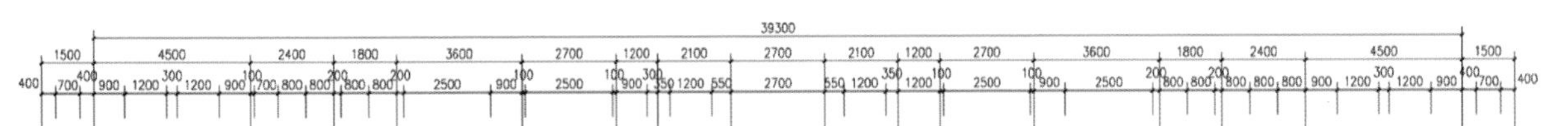

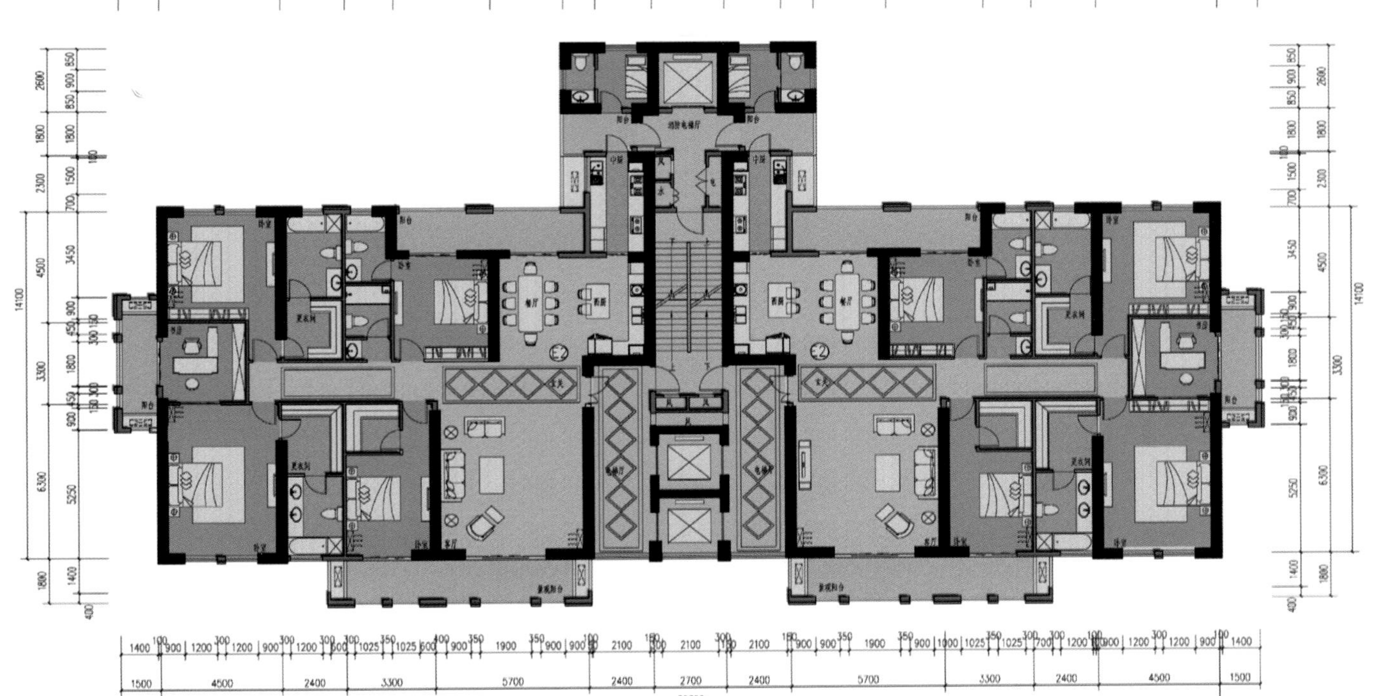

标准层平面图

户型	房型	套内建筑面积	公摊面积	1/2阳台建筑面积	套型建筑面积	户型建筑面积	阳台投影面积占套型建筑建筑面积的比例
E2	四房二厅四卫	239.86㎡	56.58㎡	19.40㎡	296.44㎡	315.84㎡	13.05%

（注：宽度小于1.8米阳台面积按1/2计，公摊面积仅计本层）

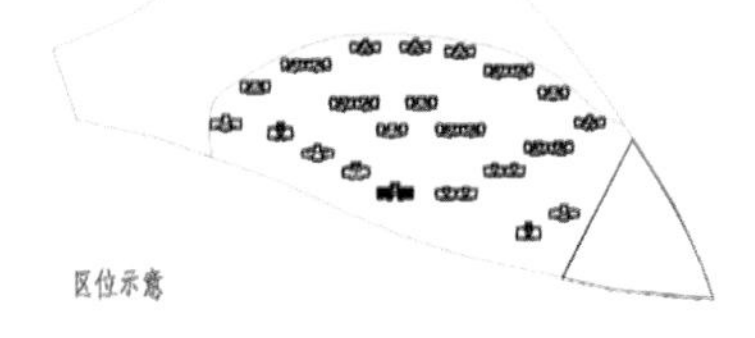

区位示意

Hum

DECLEOR

Honeys
BASIC HOUSE
CELINE
ST JOHN

法国波尔多 Tour Saint-Jean

Tour Saint-Jean

设计单位：ECDM Architectes

开发商：Crédit Agricole Immobilier

项目地址：法国波尔多

Designed by: ECDM Architectes

Client: Crédit Agricole Immobilier

Location: Bordeaux, France

设计理念

一个新兴的元素，在一定程度上会改变地区结构和景观格局，其形态、规模以及蕴含的文化内涵，都有可能使之成为当地独特的存在。项目位于法国西南部的波尔多，设计考虑了后现代社会的发展需求，使高度复杂的城市发展策略成为可能。

设计思路

波尔多是法国西南部的一个港口城市，其平坦的地形，形成了当地密集、平缓的城市肌理。设计旨在打破对建筑高度的限制，将建筑融入城市和景观中，打破原有的城市天际线，强化由此而形成的特定的视觉效果。

项目紧凑又轻盈，理性又复杂，它给予居民身居高处、栖息在自然景观之中的体验，建立起与环境之间的密切关系。在这里，人们能获得同时居住在城市和乡间的双重体验。

App T4
89.70m²
App T1
36.30m²
R+7
+21.00
Accès
Local FO
Accès
Local poubelles
Local vélos
30.50m²
Local
Fibre optique
14.93m²
Local poubelles
21.03m²
RdC
+00.00
DK287
Hall
17.40m²
Loge
13.06m²
Surface commerciale
81.87m²
Accès
Logements
Accès
Commerce
R+13
+38.40
App T1
37.80m²
R+3
+09.40
DK287
App T2
75.13m²
App T1 bis
42.15m²
App T1
36.30m²

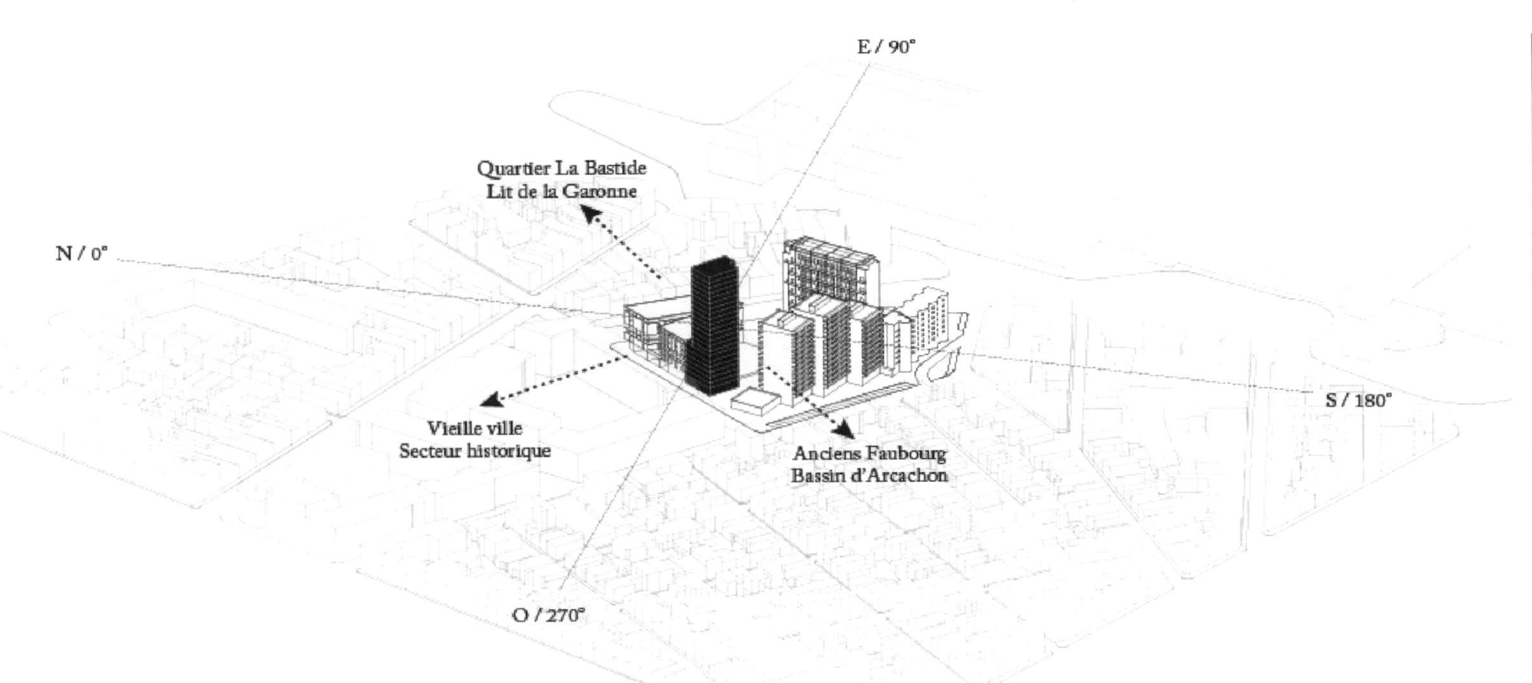

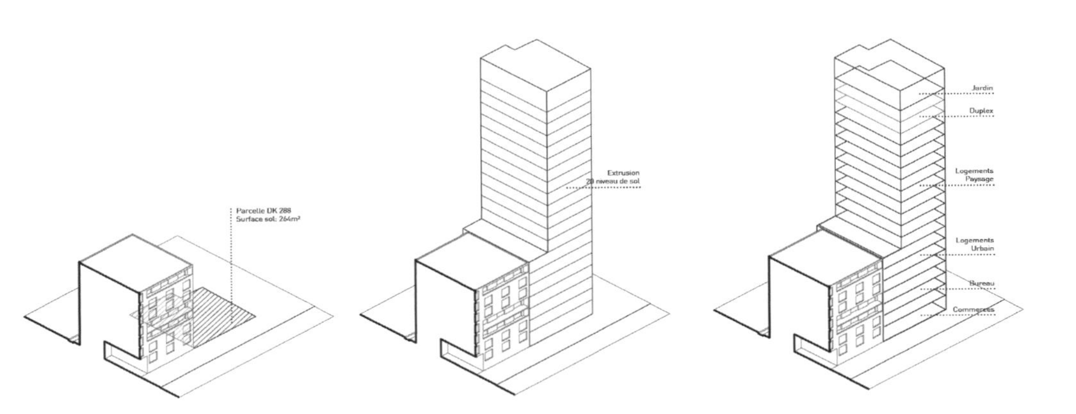

Design Concept

An emergent element shall to some extent change the relationship between the new neighborhood and the surrounding landscape. Its status, size and cultural connotation will possibly make it a unique existence of local district. The project is located in Bordeaux, the south of France. Its design has taken the development demands of a postmodern society into account, enabling the implementation of a highly complex urban strategy.

Design Idea

Bordeaux is a harbor city in the southwest of France. Its geography is characterized by flatness upon which is inscribed a dense and low urban fabric. The project aims to break limits related to the height, to bring buildings into the city and urban landscape, to pierce the sky to permit a broad yet organized reading of the skyline, and to enhance specific visual effect.

The project is compact, light, dense and slender, rational and complex. For residents living above the Bordeaux skyline, living up high is to inhabit the landscape, to have a privileged relationship with one's environment, and to live at once in the city and the countryside.

土耳其马尼萨市政大楼

Municipality Building, Manisa, Turkey

设计单位：RTA-Office 建筑事务所
合作单位：DOME Partners
开发单位：马尼萨政府
项目地址：土耳其马尼萨
项目面积：81 700 ㎡
设计团队：Santiago Parramón　Simona Assiero Brá
Mariana Rapela　Miguel Vilacha
Rodrigo Schiavoni　Agnieszka Zagorska
Kelly Tuink　Sujing Zhang
Feifei Zhang　Murat Yilmaz
Rami Idlby　Lucas Garcia de Oteyza
摄影：RTA-Office 建筑事务所

Designed by: RTA-Office
Collaboration: DOME Partners
Client: The Municipality of Manisa
Location: Manisa, Turkey
Area: 81,700 m^2
Design Team: Santiago Parramón, Simona Assiero Brá, Mariana Rapela, Miguel Vilacha, Rodrigo Schiavoni, Agnieszka Zagorska, Kelly Tuink, Sujing Zhang, Feifei Zhang, Murat Yilmaz, Rami Idlby, Lucas Garcia de Oteyza
Photography: RTA-Office

项目概况

设计方案旨在构建一个绿色环保、经济效益高的建筑，为人们提供一个高质量的生活和工作环境。为实现这一目标，设计师采用了跨学科合作的方式，通过综合考虑当地的社会、经济、文化因素，提出实际可行的设计方案。

设计理念

设计师试图通过创造一个具备举办大型活动的功能、具有强大吸引力的城市中心来重新定位城市地位和形象。设计方案融入了美学理念，以构建一个外观独特、形象鲜明的建筑，同时，设计充分考虑到了市民的需求，以简单的方式将适宜的人体尺度融入城市活动中。

建筑设计

设计师将生物气候学应用到建筑设计中。建筑外观、表皮、透明的大型窗户和涂抹了不同绝缘度的覆盖层，使得建筑具备了高水平的防晒功能。建筑具有独特的热调节和光调节功能。庭院的屋顶是一个网状的通风罩，太阳光透过大型的玻璃表面照射到建筑内部，既弱化了太阳光的强度，也营造了一个明亮的室内环境。建筑的自然通风通过露台来实现，露台的气温可自动监控和调节，为维持宜人的室内小气候提供了条件，每个模块都设有露台，减少了建筑对传统散热系统的依赖。

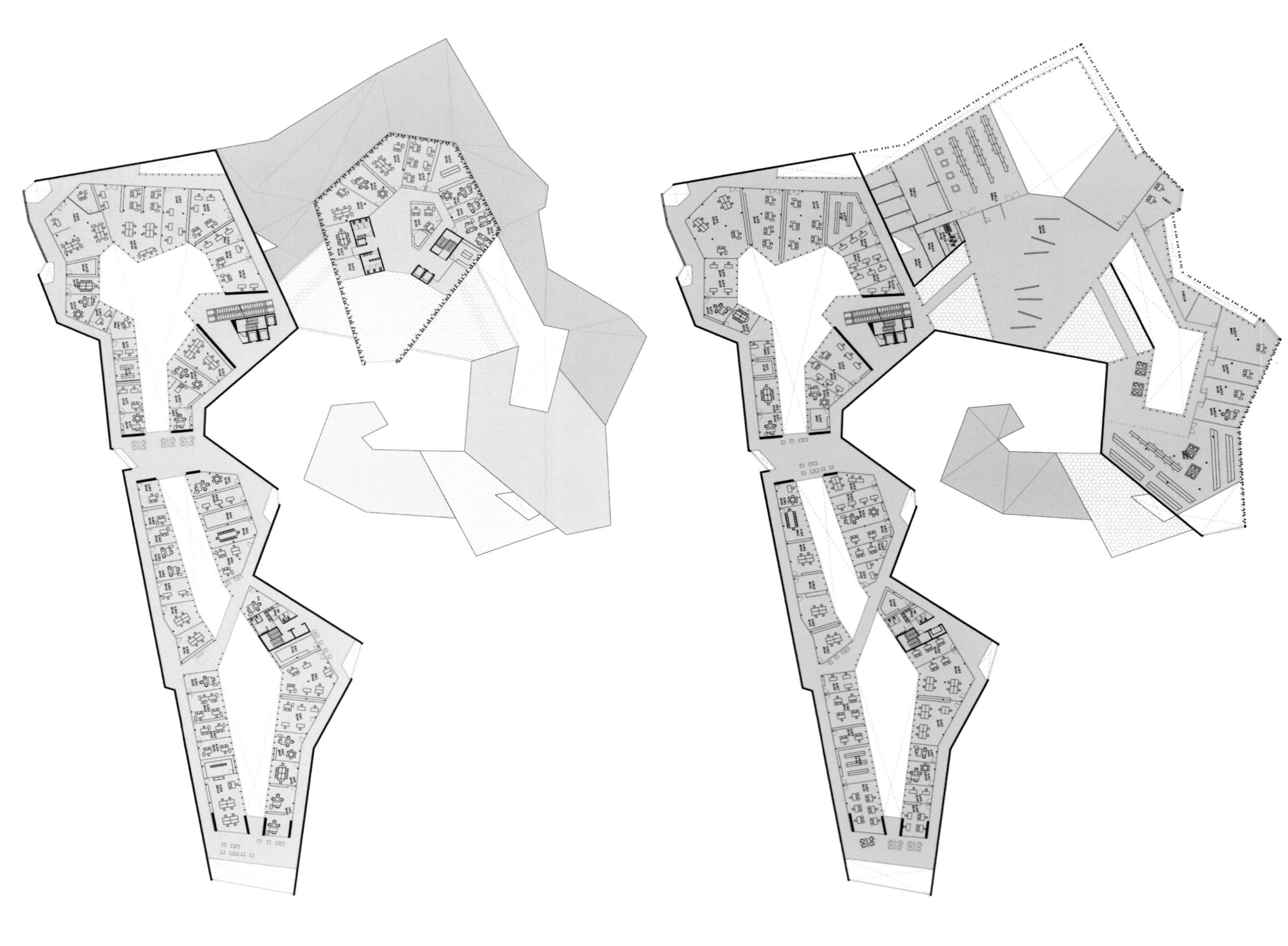

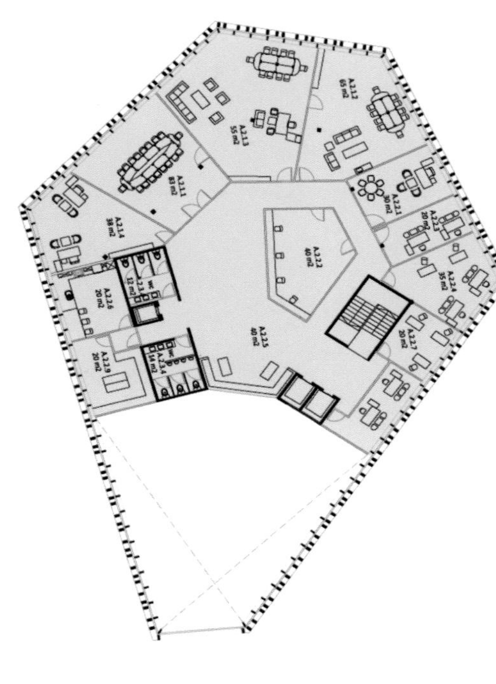

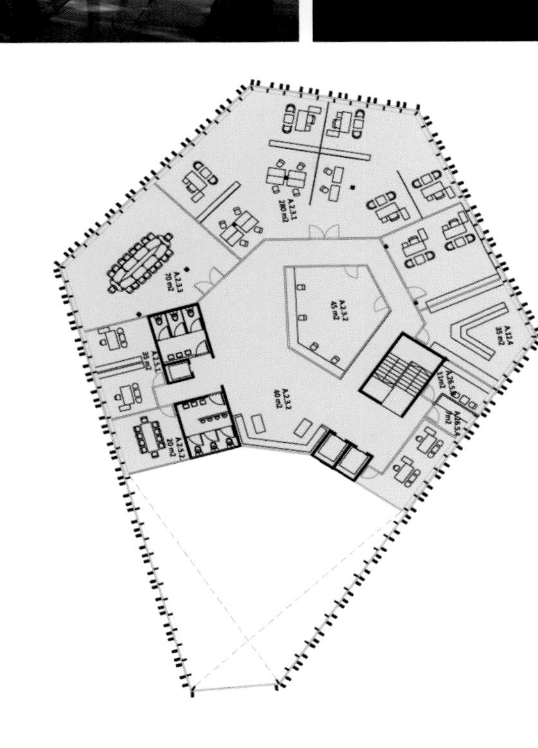

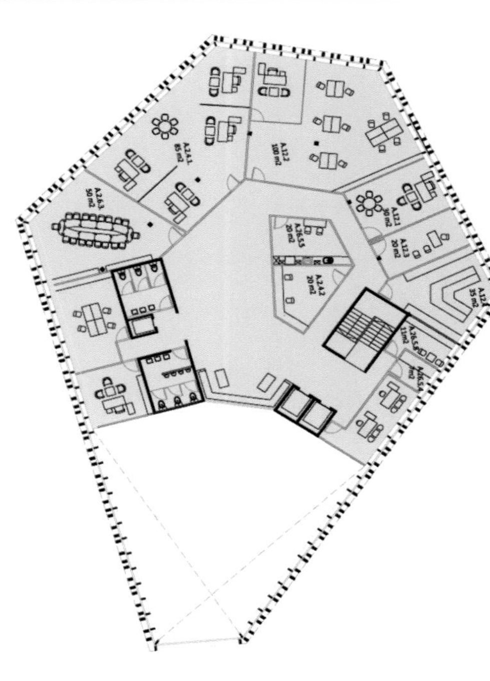

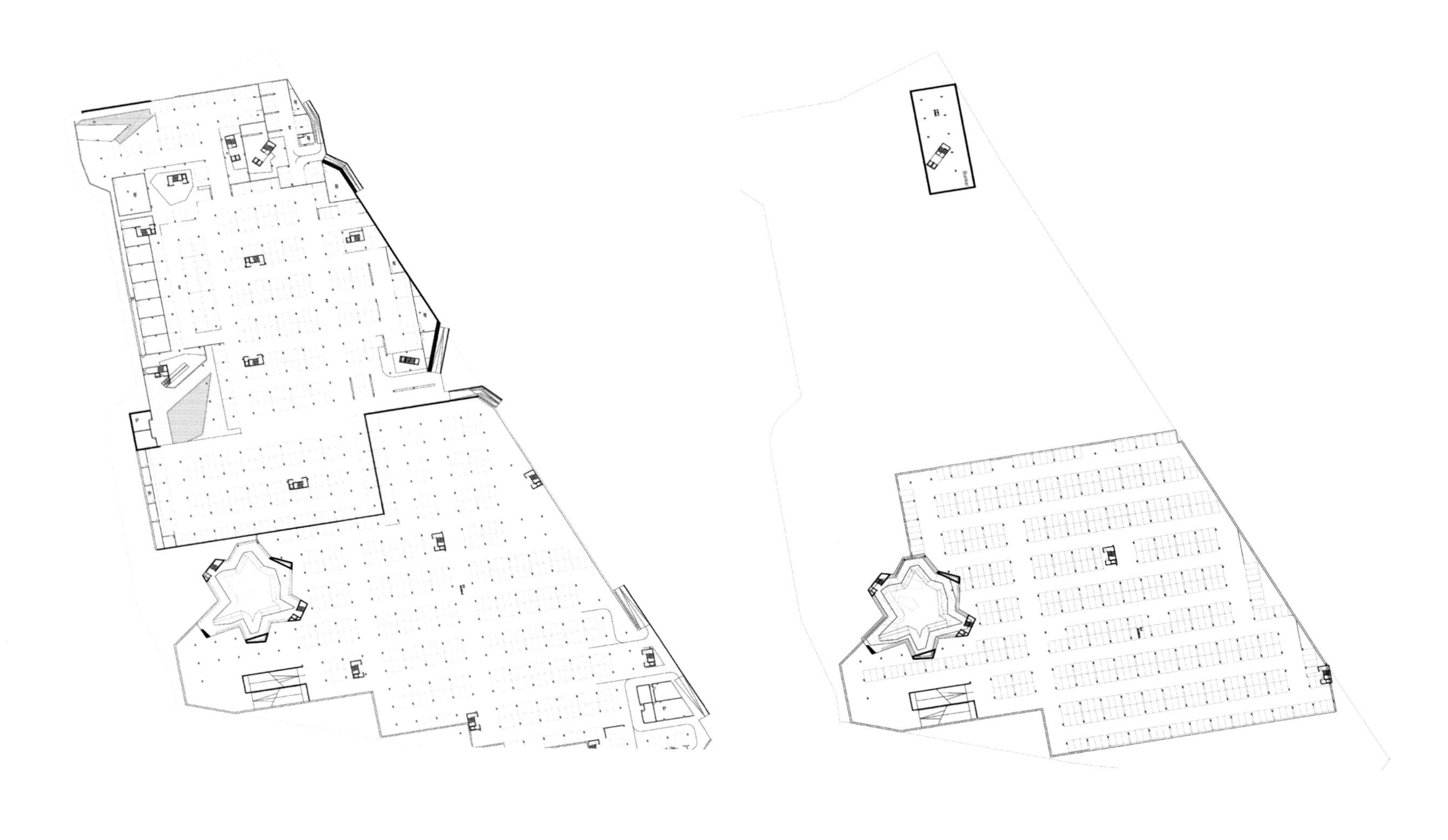

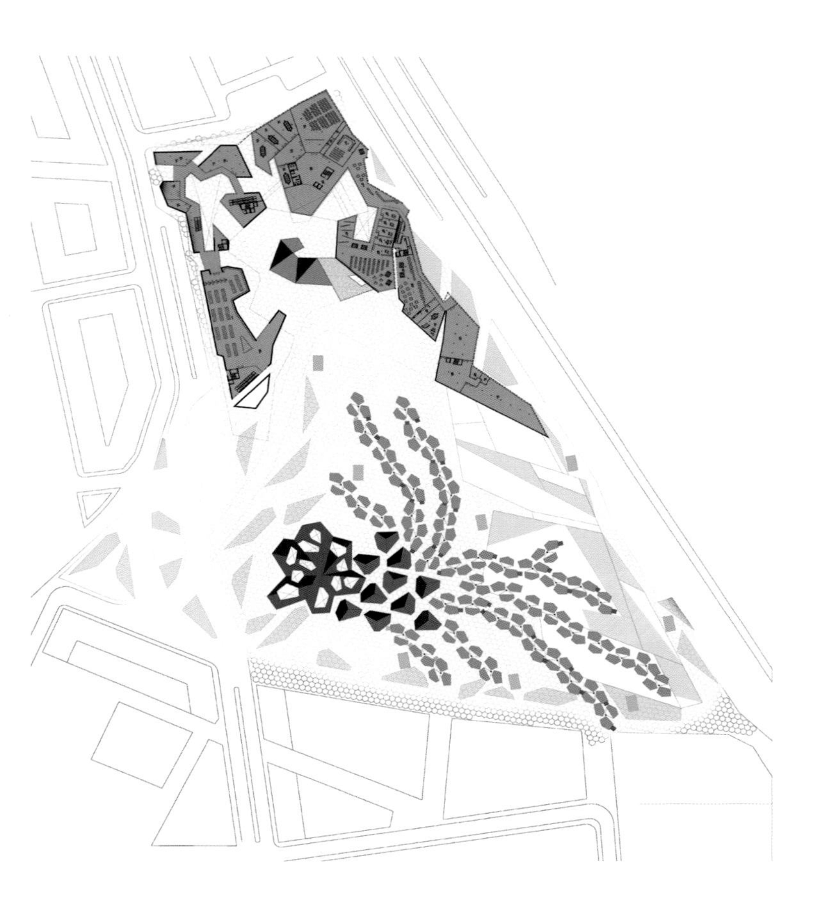

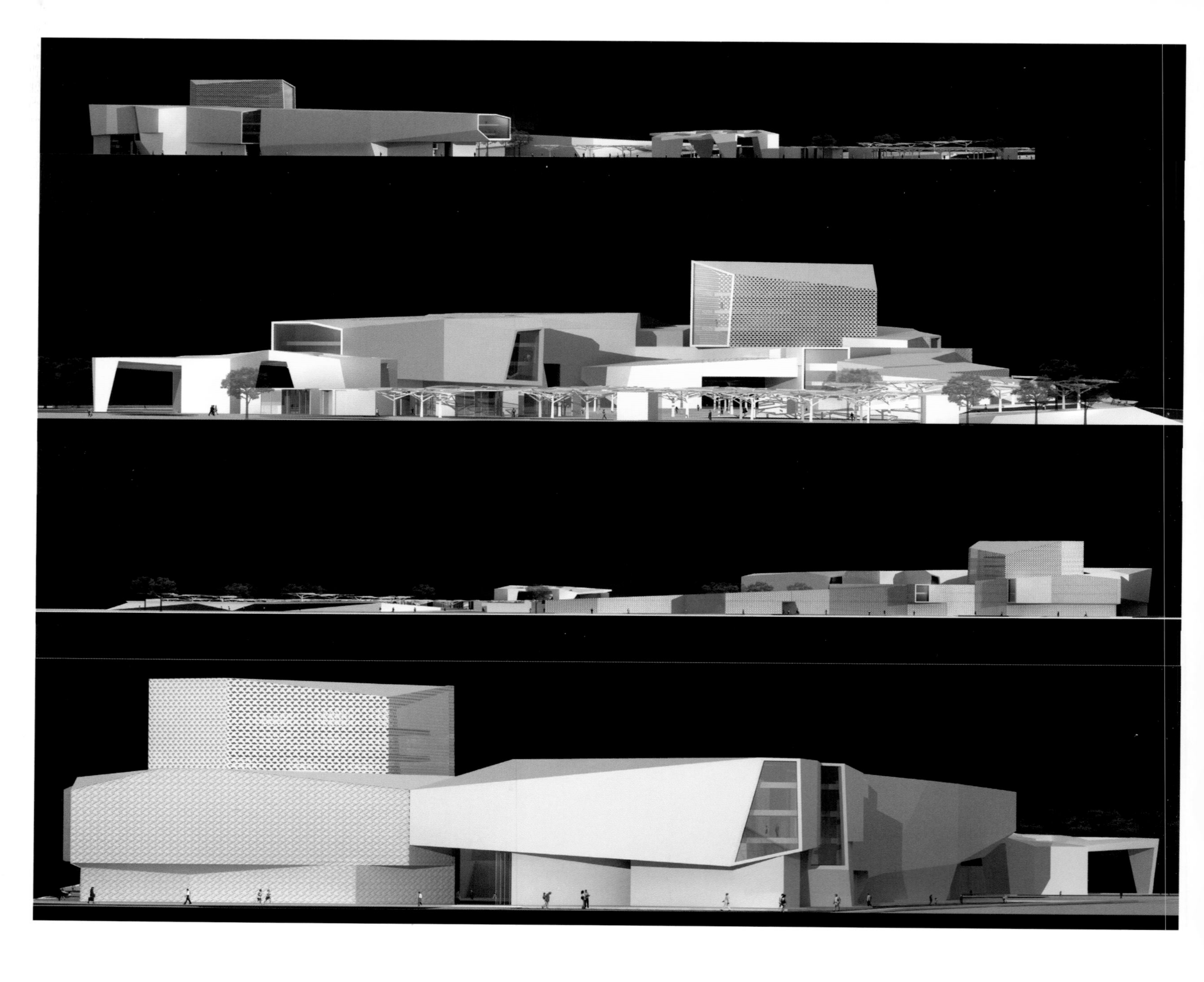

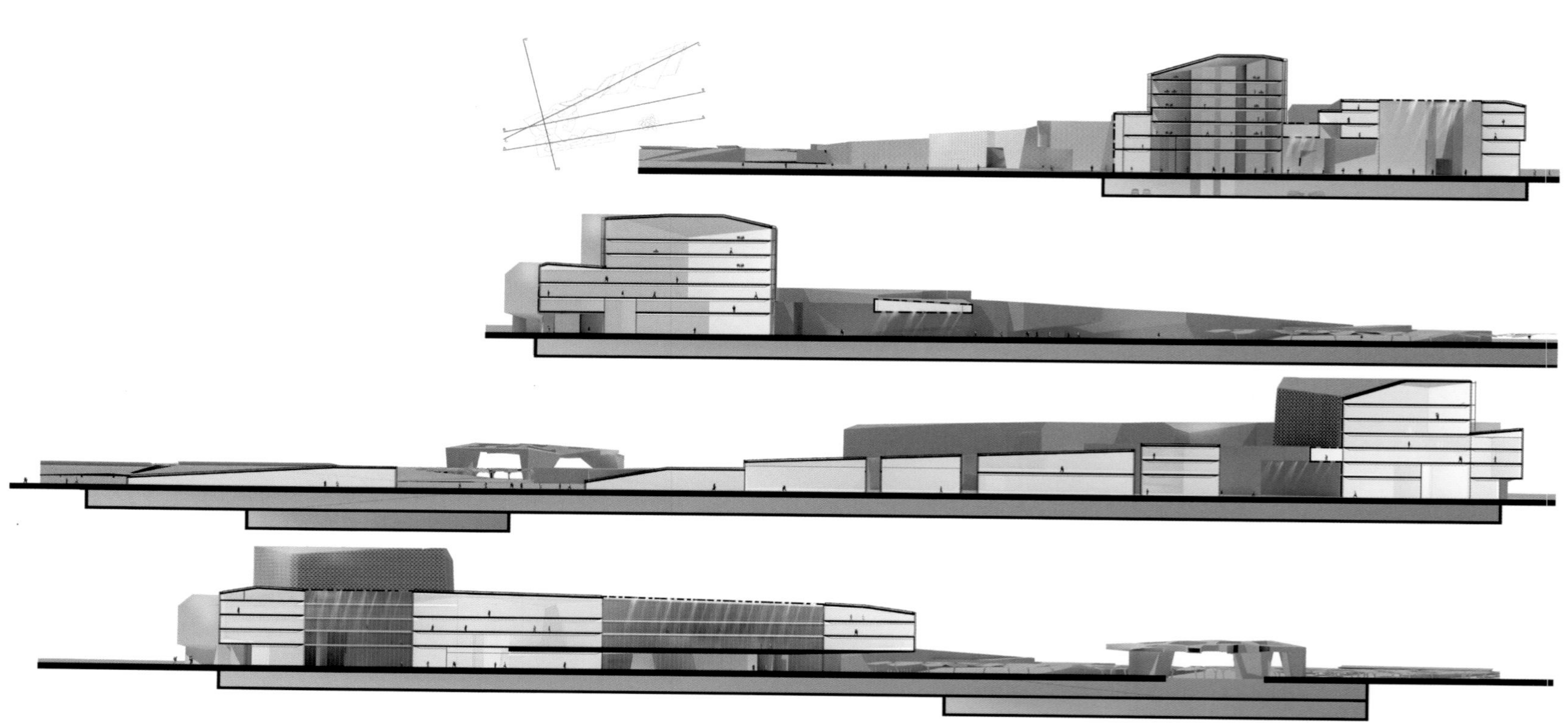

Profile

The design scheme is to build an environmentally responsible, profitable building, where people can live and work in a high-quality environment. In order to achieve this, designers have used synergies between different disciplines and proposed technologies according to the social, economical and cultural reality of the place.

Design Concept

Designers try to create an attractive city center which has the potential to hold large activities to reposition city status and image. Aesthetics integrated in the design contributes to construct such a building with unique appearance and distinctive image. For the consideration of public demands, the project is integrated with urban activities in a simple way based on proper human scale.

Architectural Design

Bioclimatology is applied in architectural design of the project. Building façade, skin and large transparent windows together with different insulation degrees of coating provide a high level of sun protection. The building is characterized by exceptional thermal and lighting conditions. The roof of the patio is a structural netting perforated with an organic formalization. Sunlight permeating large glass surface gets into the interior in a softened manner. Thanks to the patios, natural ventilation has achieved in this building. The air temperature in the patios will be automatically monitored and regulated to maintain enjoyable indoor microclimate. Patios set in each module reduce dependence on traditional cooling system.

斯洛文尼亚卢布尔雅那“Shopping Pillow Terraces”商住大楼

Shopping Pillow Terraces

设计单位： OFIS 建筑事务所
开发单位： LE Nepremicnine
项目地址： 斯洛文尼亚卢布尔雅那
用地面积： 1 900 m²
总建筑面积： 8 696 m²
设计团队： Rok Oman　Spela Videcnik　Andrej Gregoric　Janez Martincic　Janja Del Linz　Katja Aljaz　Verena Smahel

Designed by: *OFIS Arhitekti*
Client: *LE Nepremicnine*
Location: *Ljubljana, Slovenia*
Site Area: *1,900 m²*
Gross Floor Area: *8,696 m²*
Project Team: *Rok Oman, Spela Videcnik, Andrej Gregoric, Janez Martincic, Janja Del Linz, Katja Aljaz, Verena Smahel*

项目概况

OFIS 建筑事务所为斯洛文尼亚卢布尔雅那设计的这座名为“Shopping Pillow Terraces”的小型城市综合体，是一个商业和住宅的综合项目，已获得卢布尔雅那市议会的批准。

建筑设计

项目位于卢布尔雅那市中心的主步行街上，周边被多层的老房子环绕，故设计师将建筑高度规划为 7 层。建筑以“梯田”的形态出现，这一结构使建筑与旧城和城堡之间形成了最佳的视觉联系。

建筑的底下 4 层为沿街商铺和商场，上面 3 层则是公寓。为了提高其开敞性，在底层商场设置为可穿透的步行通道，4 层的顶楼为开敞的大露台，形成了一个露天咖啡厅，是一个绝佳的喝咖啡和欣赏城市景观的场所。

设计特色

Shopping Pillow Terraces 的每一层都不同程度地被一种有机金属网包裹，这是建筑的特色所在，也是其名称的由来。绿色植物被置于金属网内，产生立体花园的效果，结合斯洛文尼亚季节分明的气候特征，建筑在自然的妆点下，呈现出不同的外观：夏季植被茂盛，建筑披上了绿色的外衣；冬季花草枯萎，金属网则将被白雪覆盖。

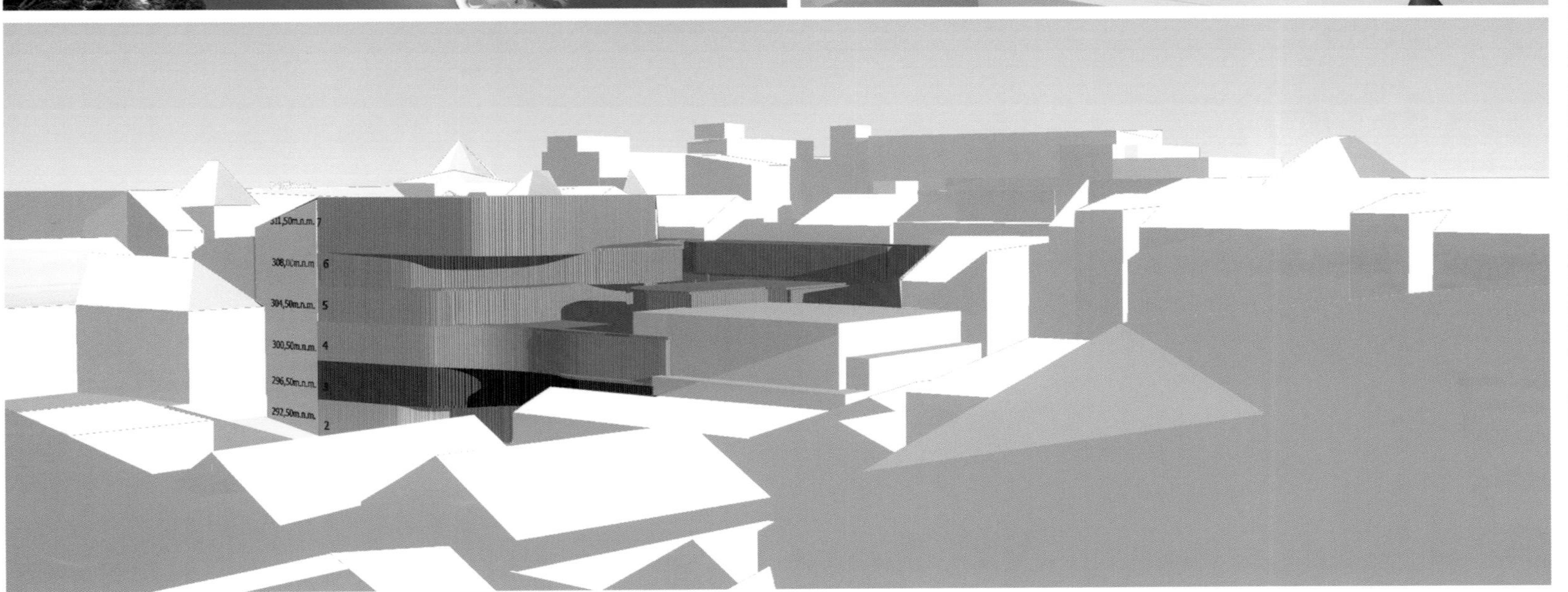
311,50m.n.m. 7
308,00m.n.m 6
304,50m.n.m. 5
300,50m.n.m. 4
296,50m.n.m. 3
292,50m.n.m. 2

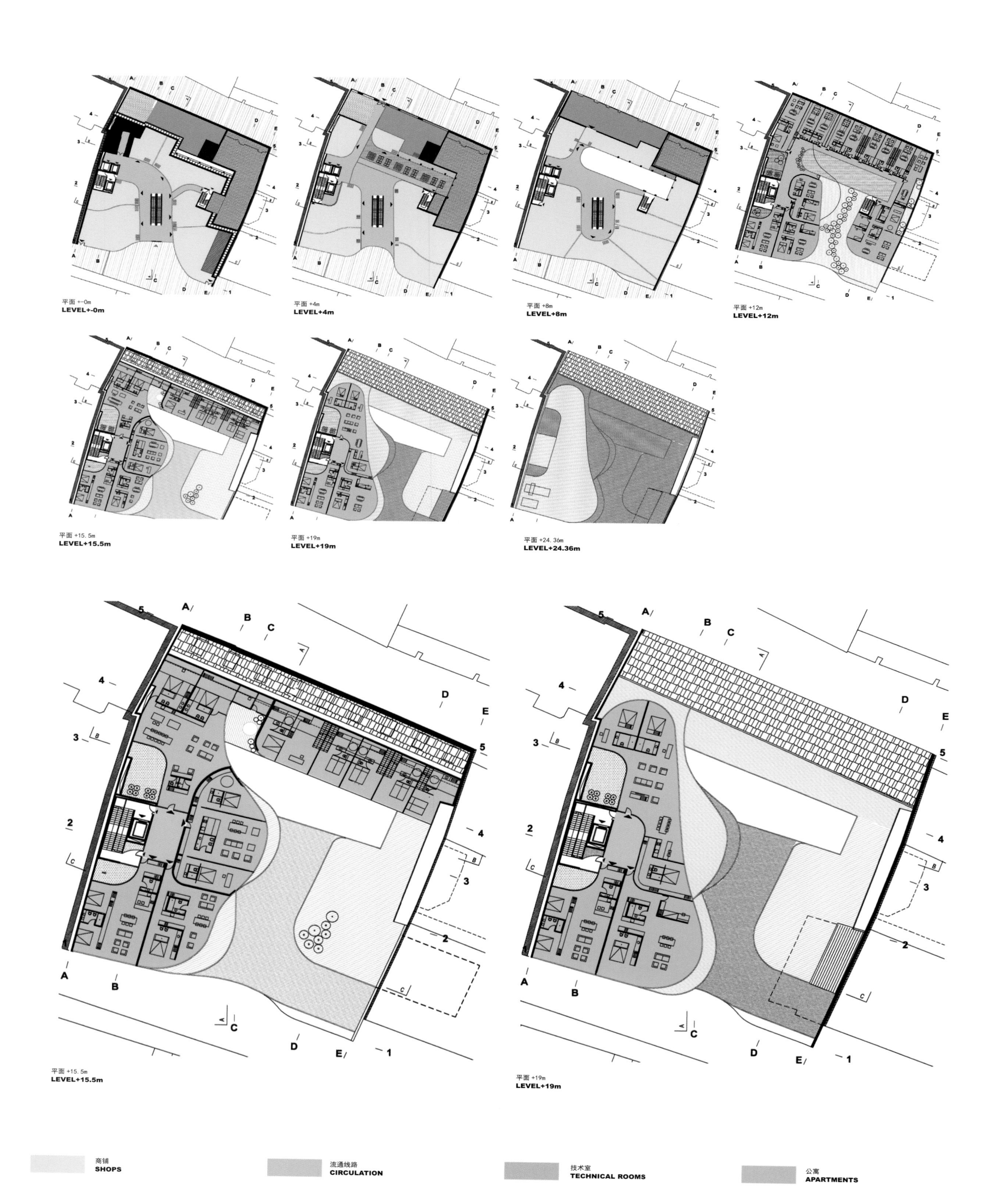
平面 +-0m
LEVEL+-0m
平面 +4m
LEVEL+4m
平面 +8m
LEVEL+8m
平面 +12m
LEVEL+12m
平面 +15.5m
LEVEL+15.5m
平面 +19m
LEVEL+19m
平面 +24.36m
LEVEL+24.36m
平面 +15.5m
LEVEL+15.5m
平面 +19m
LEVEL+19m
商铺
SHOPS
流通线路
CIRCULATION
技术室
TECHNICAL ROOMS
公寓
APARTMENTS

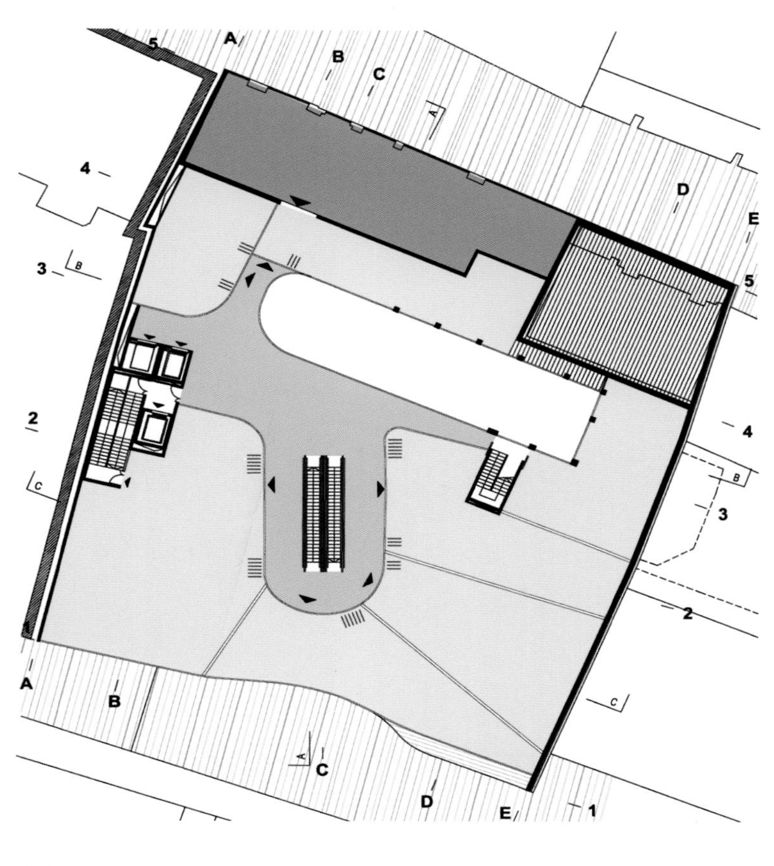

平面 +8m
LEVEL+8m

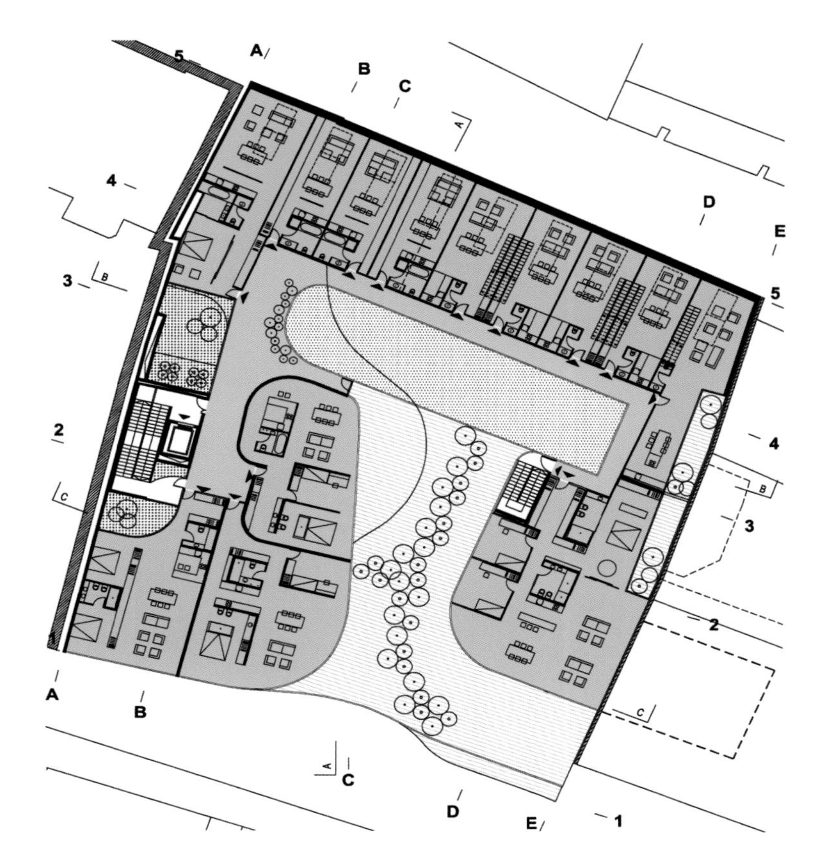

平面 +12m
LEVEL+12m

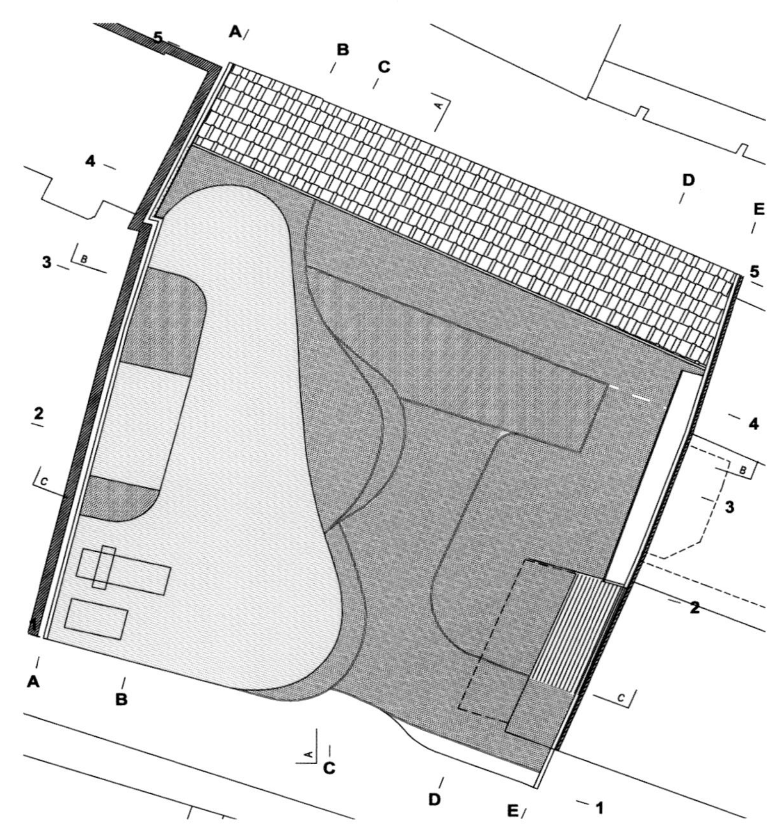

平面 +24. 36m
LEVEL+24.36m

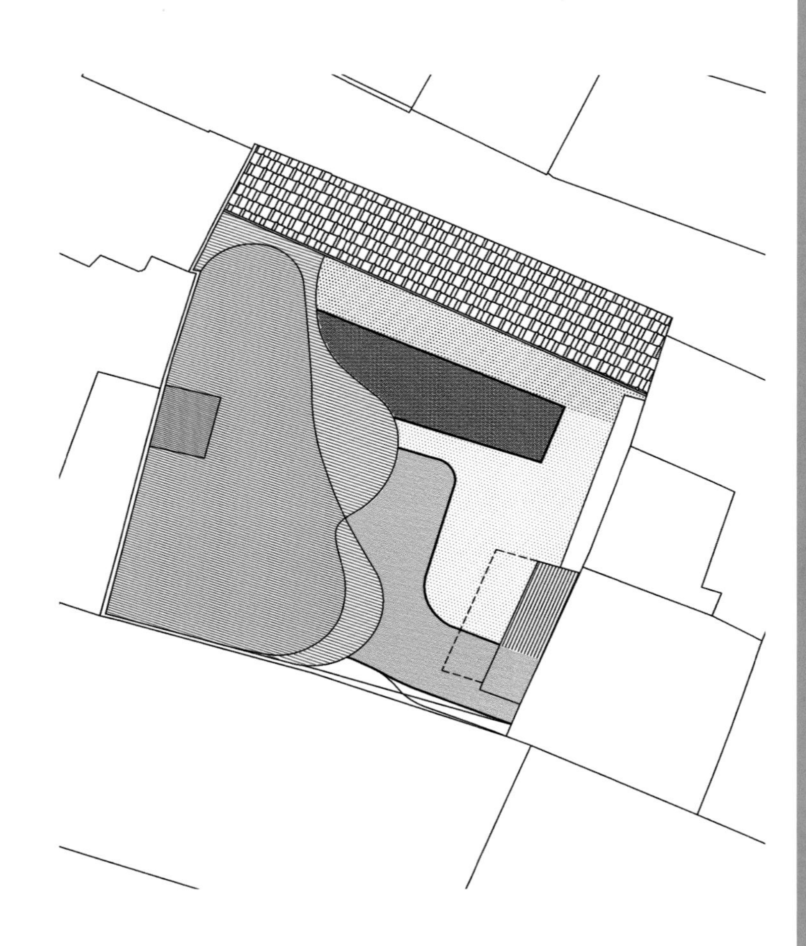

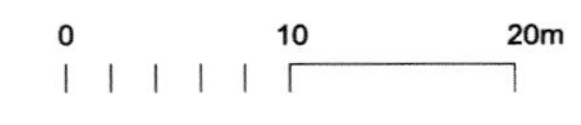

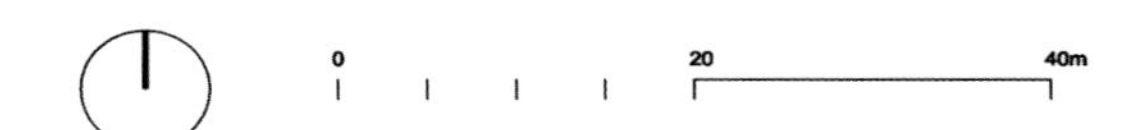

一层
Level 1

商场 Mall
商店 1 Shop 1
商店 2 Shop 2
商店 3 Shop 3
商店 5 Shop 5
技术室 Technical room
卫生间 WC
商店 4 Shop 4
商店 6 Shop 6

二层 — 入口公园
Level 2-entrance park

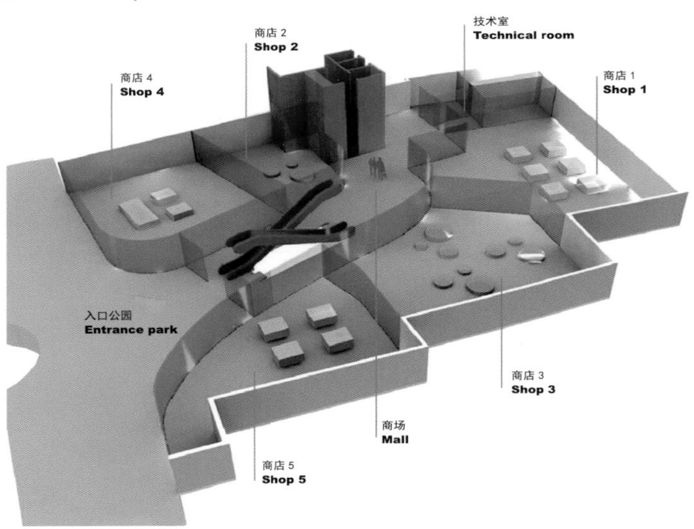

三层入口街道
Level 3-entrance street

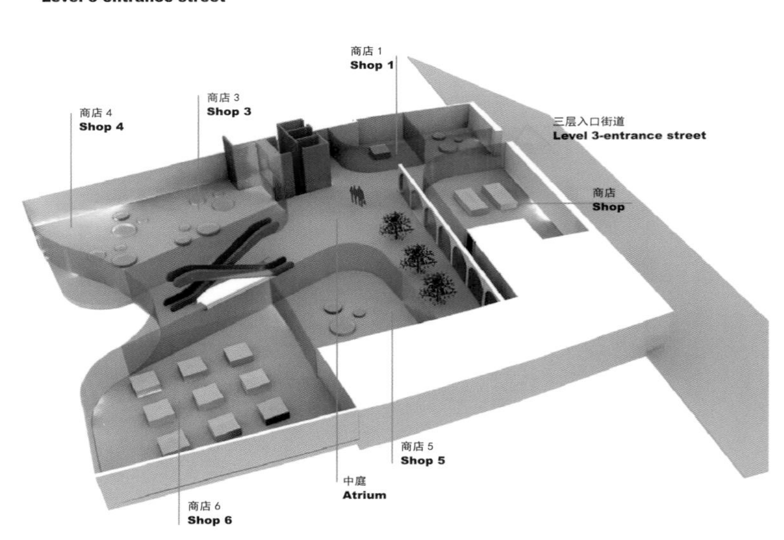

四层
Level 4

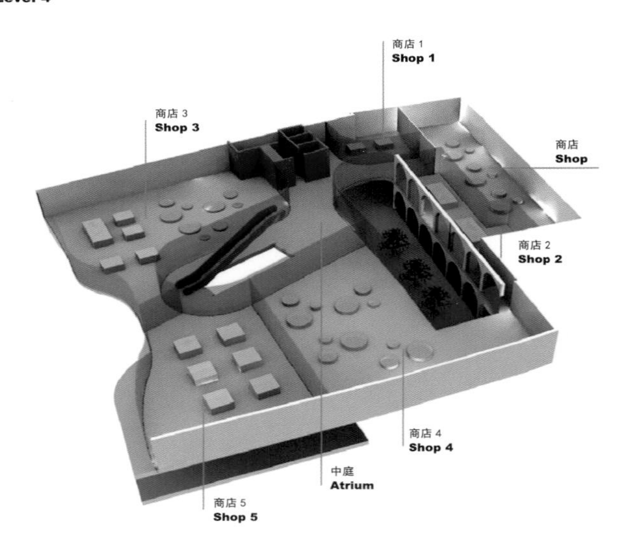

五层
Level 5

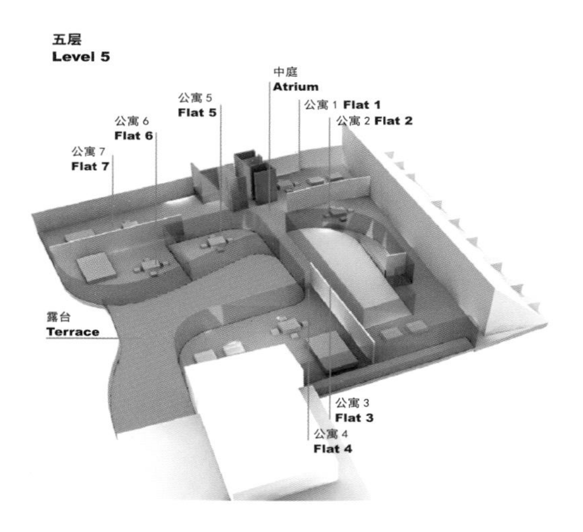

六层
Level 6

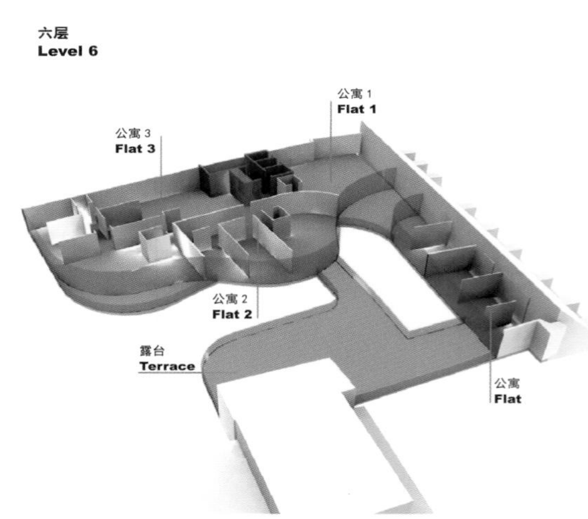

七层
Level 7

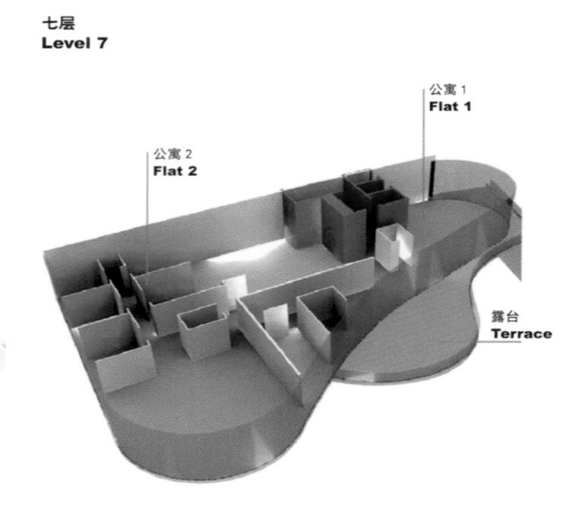

Profile

The small urban complex "Shopping Pillow Terraces" designed by OFIS Arhitekti for Ljubljana of Slovenia is a commercial & residential complex project granted by Ljubljana City Council.

Architectural Design

Located in main pedestrian street of Ljubljana and surrounded by multistory old houses, the project is planned to be seven storeys high. A "Terrace" form offers views towards the old city and the castle.

The lower four storeys are occupied by shops and malls while the upper three floors by apartments. To further open ground floor mall, pedestrian passages are set here. A large open terrace on the top of the fourth floor provides an outdoor café for coffee and urban landscape enjoying as well.

Design Feature

Each floor of Shopping Pillow Terraces is covered by organic wire netting to varying degrees. It's a distinctive feature of the project while more importantly is where the name comes from. Vegetations planted inside the wire netting generate an effect of vertical garden. With distinctive climatic characteristics they present changeful appearance: luxuriant vegetations in summer cover a green coat for the building; when vegetations withered in winter, the wire netting will be covered by snow.

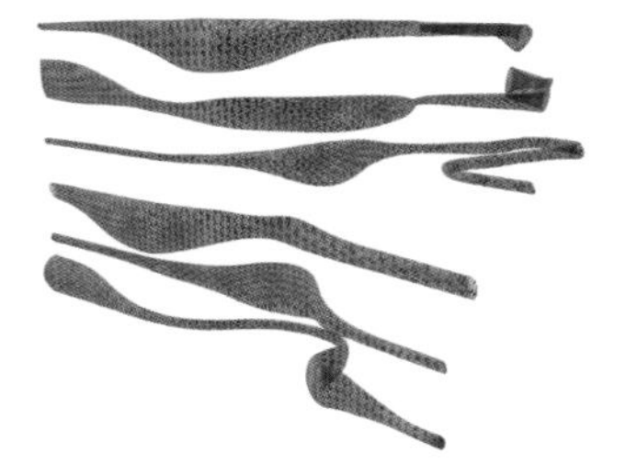

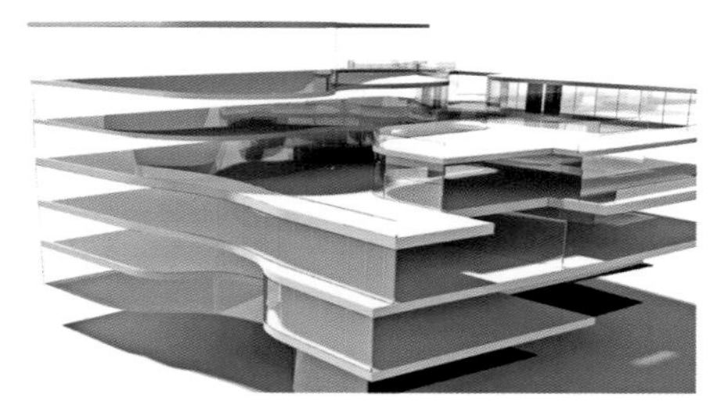

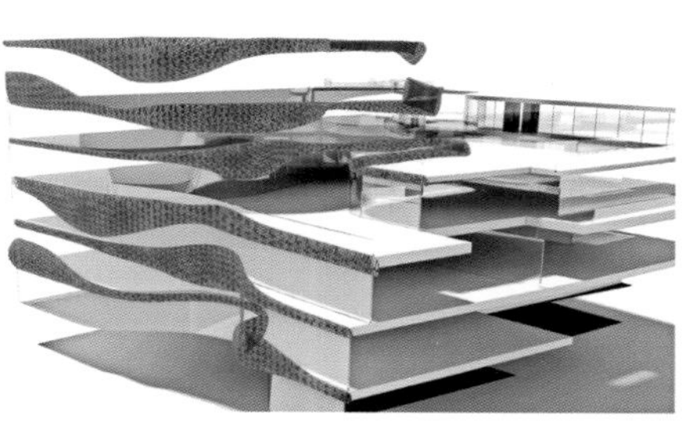

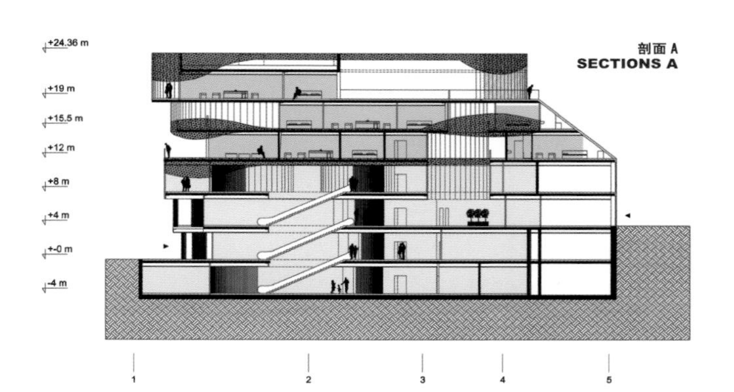
剖面 A
SECTIONS A
+24.36 m
+19 m
+15.5 m
+12 m
+8 m
+4 m
+-0 m
-4 m
1
2
3
4
5

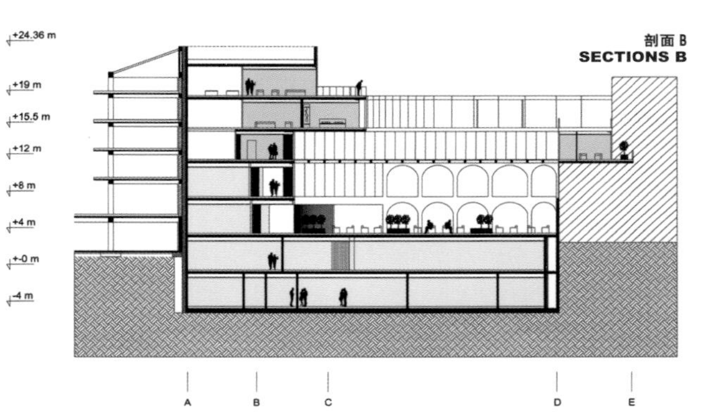
剖面 B
SECTIONS B
+24.36 m
+19 m
+15.5 m
+12 m
+8 m
+4 m
+-0 m
-4 m
A
B
C
D
E

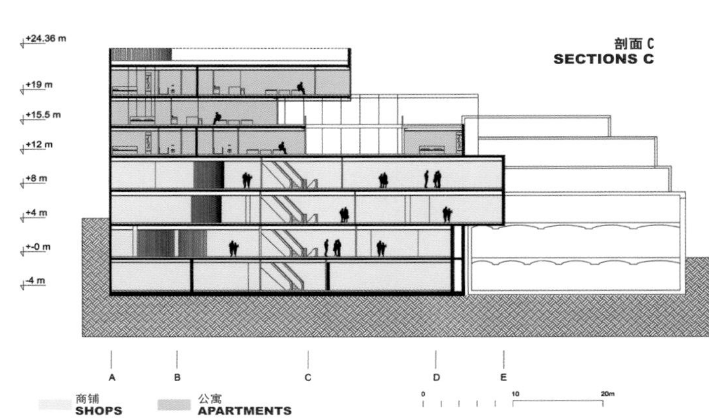
剖面 C
SECTIONS C
+24.36 m
+19 m
+15.5 m
+12 m
+8 m
+4 m
+-0 m
-4 m
A
B
C
D
E
商铺
SHOPS
公寓
APARTMENTS
0
10
20m

墨西哥合众国墨西哥市“摩地楼”

The Earthscraper

设计单位：BNKR Arquitectura
项目地址：墨西哥墨西哥市
项目面积：775 000 ㎡
设计团队：Arief Budiman　Diego Eumir
　　　　　Adrian Aguilar　Guillermo Bastian

Designed by: BNKR Arquitectura
Location: Mexico City, Mexico
Area: 775,000 m^2
Design Team: Arief Budiman, Diego Eumir,
Adrian Aguilar, Guillermo Bastian

项目概况

BNKR Arquitectura 为墨西哥城设计了一座深入地下 300 米的倒金字塔形的“摩地楼”。这座令人惊异的“摩地楼”将建在墨西哥城中部的宪法广场，将为墨西哥城面临的人口增长、建筑用地稀缺、历史建筑物保护以及建筑物限高等一系列难题提供解决方案。

建筑设计

这座倒置的金字塔建筑共 65 层，大楼的最上 10 层是博物馆和游客中心，供游人欣赏墨西哥的阿兹特克古文化；位于中部的是购物中心和公寓；最下层则是 35 层的办公空间。

由于建筑结构转入地下，“摩地楼”也避开了墨西哥城对新建筑的高度限制这一问题。“摩地大楼”的“楼顶”与城市地面平齐，顶部面积为 240 米 ×240 米，将由玻璃板覆盖，自然光线可透过玻璃的表面射入金字塔深处。同时，中空结构的采用，确保了所有的居住空间都可享受自然照明和通风。

对该历史中心建筑的限制来自各个方面。在竖直方向上，天际线限制了建筑的高度增长。在水平方向上，受保护历史建筑阻碍了任何大规模的开发。

The limitations to building in the Historic Centre are in every direction. Vertically, the colonial skyline is the limit to the growth of buildings. Horizontally, protected buildings crowd out any possible development of significant size.

The Historic Centre is frozen in its status as a colonial city. More than 70% of its buildings are considered historical patrimony and cannot be altered or expanded. Its parks are equally protected. The small patches of land that are not protected by the INAH (National Institute of History and Anthropology) cannot rise above the height of their historical neighbors as the skyline is strictly controlled.

这一历史中心停滞在作为一个殖民城市的状态。这里 70% 的建筑都被视为历史遗产，不能改造或扩建。这里的公园也受到同等保护。一些不受国家历史和人类学研究所保护的小块土地，在上面新建的建筑不能比周围的历史建筑高，天际线受到了严格的控制。

Public Space
公共空间

Protected Historical Buildings
受保护的历史建筑

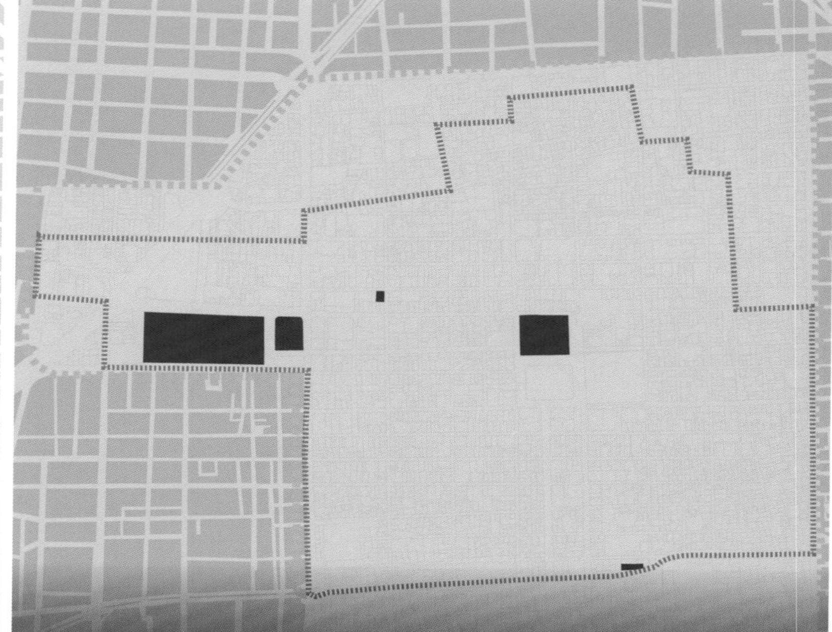

The Historic Centre is sinking. The silt of the lake bed which Mexico City was built on makes a poor foundation for an urban area and presents challenges when building high. The city sinks approximately 7cm a year and the Metropolitan Cathedral has sunk more than 9m since 1895 and now lean dangerously to one side. These geological problems compound the height restrictions already in place. The Historic Centre is essentially fixed at its current height.

历史中心正在下陷。墨西哥城建在湖泊淤泥堆积而成的河床上，对于一个城市区域来说，其基础非常不牢靠，而且建筑高层建筑也面临很大的挑战。该城市每年约下沉 7 厘米，自 1895 年以来，大都会教堂下降超过 9 厘米，目前向一侧倾斜，十分危险。这些地质问题限定了该地区的建筑高度。因此，这一历史中心基本上固定在了目前的高度。

Sinking Zones (7cms/year)
下沉区（7cm/ 年）

Zones of Greater Sinkage
大下沉区

Restrictions on growth create large programmatic issues because these buildings do not have the scale or flexibility of modern commercial or office spaces. Even though they are permitted, these types of functions have moved to other parts of the city .The Historic Centre was in decline from the 1970s onwards, culminating in the destruction of the 1985 earthquake which left it to be taken over by informal commerce. Recently this has begun to turn around due to policies designed to repopulate housing in the centre, clear informal business, improve security, and revitalize its nightlife.

这些限制条件形成了大的规划问题，因为这些规划的建筑在规模上和灵活性方面不能满足商业或办公空间的要求。尽管他们的开发经过了许可，但是这些功能空间还是转移到了城市的其他地方。从二十世纪七十年代开始，历史中心就一直在衰退，1985 年的地震让这一地区的衰败达到顶峰，也使非正式贸易盛行起来。近期，这一局面正在改观，政策设计在这一中心重新建立了住房建筑，清除了非正式商业，提升了地区的安全性，恢复了它的夜生活。

Formal Trade
正式贸易

Office Buildings
办公大楼

Informal Trade
非正式贸易

Regardless of height restrictions, there is very little free space in the Historic Centre to develop. Empty lots, dilapidated buildings and parking lots account for a small portion of the total area and are not of the dimensions required for significant expansion of commercial or office space.

先不管高度的限制，这一历史中心本身可开发的自由空间非常少。空地、危房和停车场只占据了总面积的一小部分，不够进行商业或是办公空间的大面积扩展。

Parking
停车场

Disused Buildings
废弃建筑

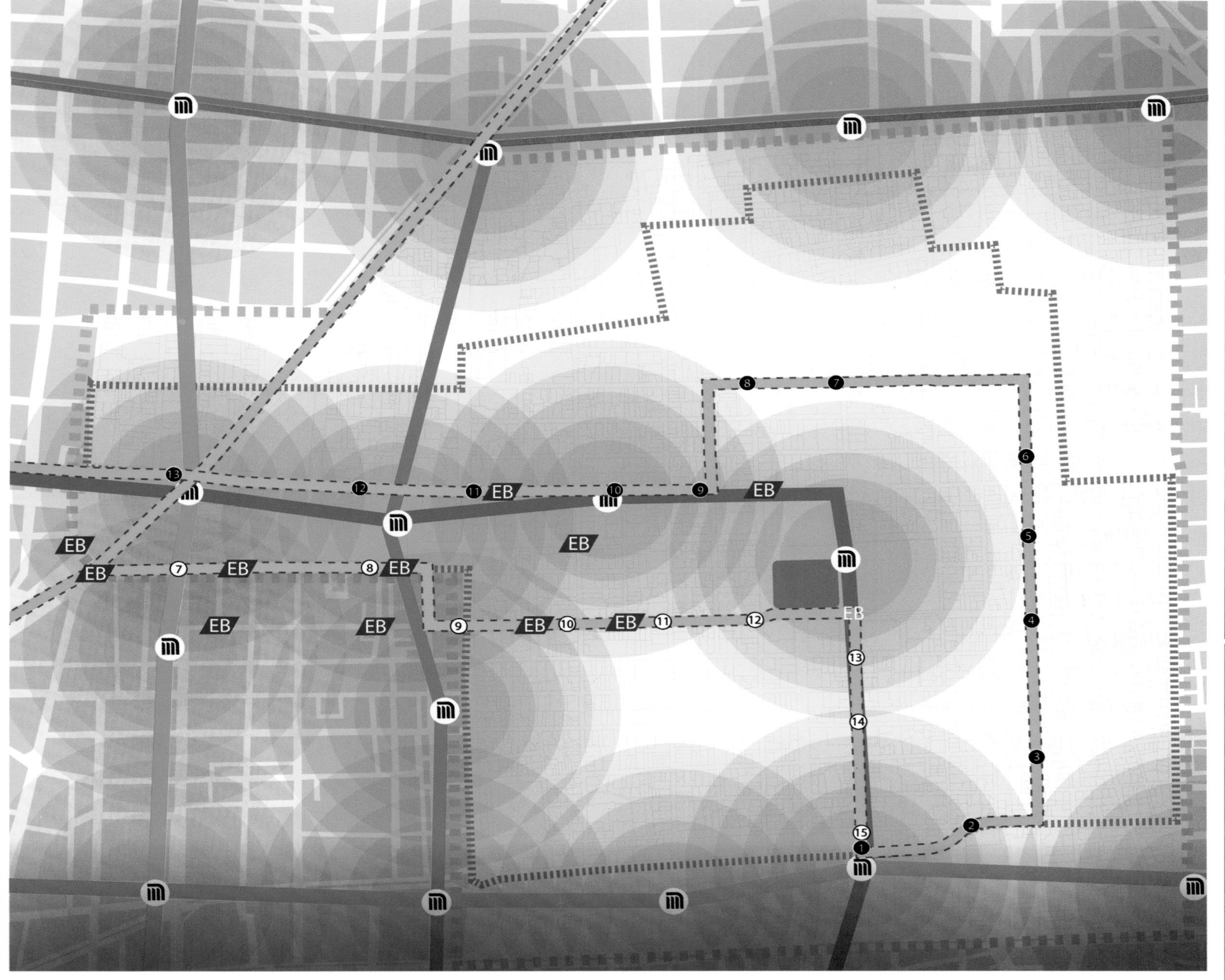

A concerted effort is being made to turn the Historic Centre into an efficient transport hub. A network of metro stations already works fairly well in this area and new programmes including bicycle stations and a project for a tram (currently on hold) are making it an increasingly attractive location not just for tourists but also for businesses. But where to grow?

项目致力于将这一历史中心打造成一个高效的交通枢纽。这一地区现有的地铁站网络已经运转得相当良好，新的规划部分，包括自行车站和电车轨道项目（目前搁置）将会使其成为越来越富有吸引力的场地，吸引更多的游客及商家。但问题是向哪里发展呢？

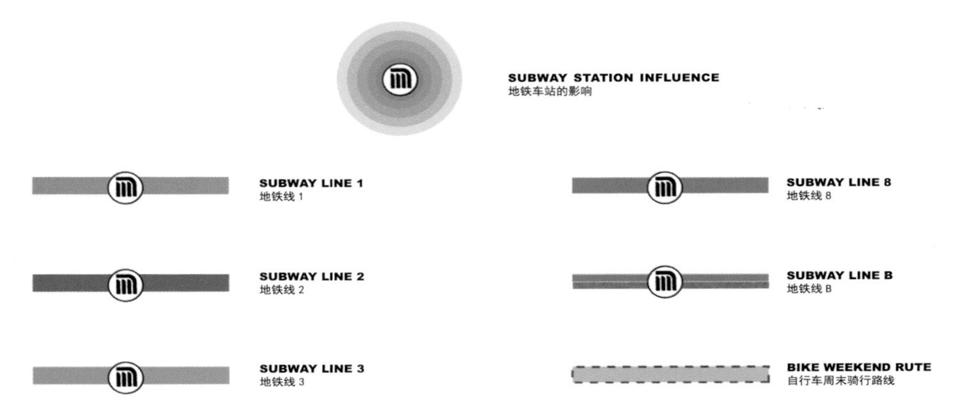

Profile

BNKR Arquitectura has designed a 300 meters underground inverted pyramid "The Earthscraper". This building will be built on the Constitution Square in the middle of Mexico. It provides solutions for population growth, land scarcity, historic building protection, building height restriction and etc.

Architectural Design

This inverted pyramid has 65 floors. The upper 10 floors are used for museum and visitor's center. Visitors can enjoy Aztec Culture here. The middle part is for shopping center and apartments. And the lower part accommodates 35 floors office space.

Since the building is constructed underground, it heads off the problem of building height restriction of Mexico City. The "Roof" of "The EarthScraper" coincides with street level. A glass plate of 240 meters × 240 meters on the top allows natural light penetrating deep inside. In the meantime, a central void allows all habitable spaces to enjoy natural lighting and ventilation.

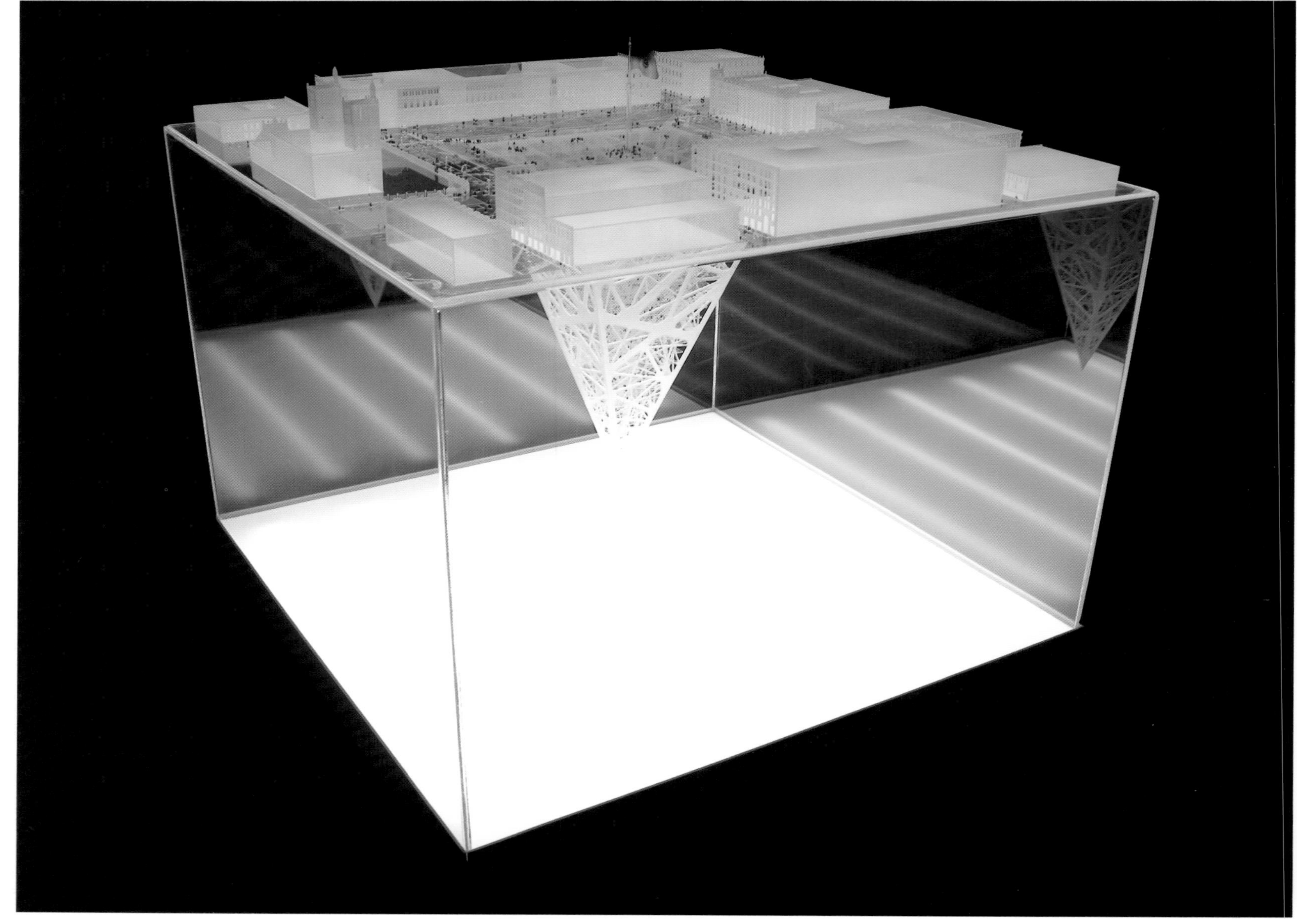

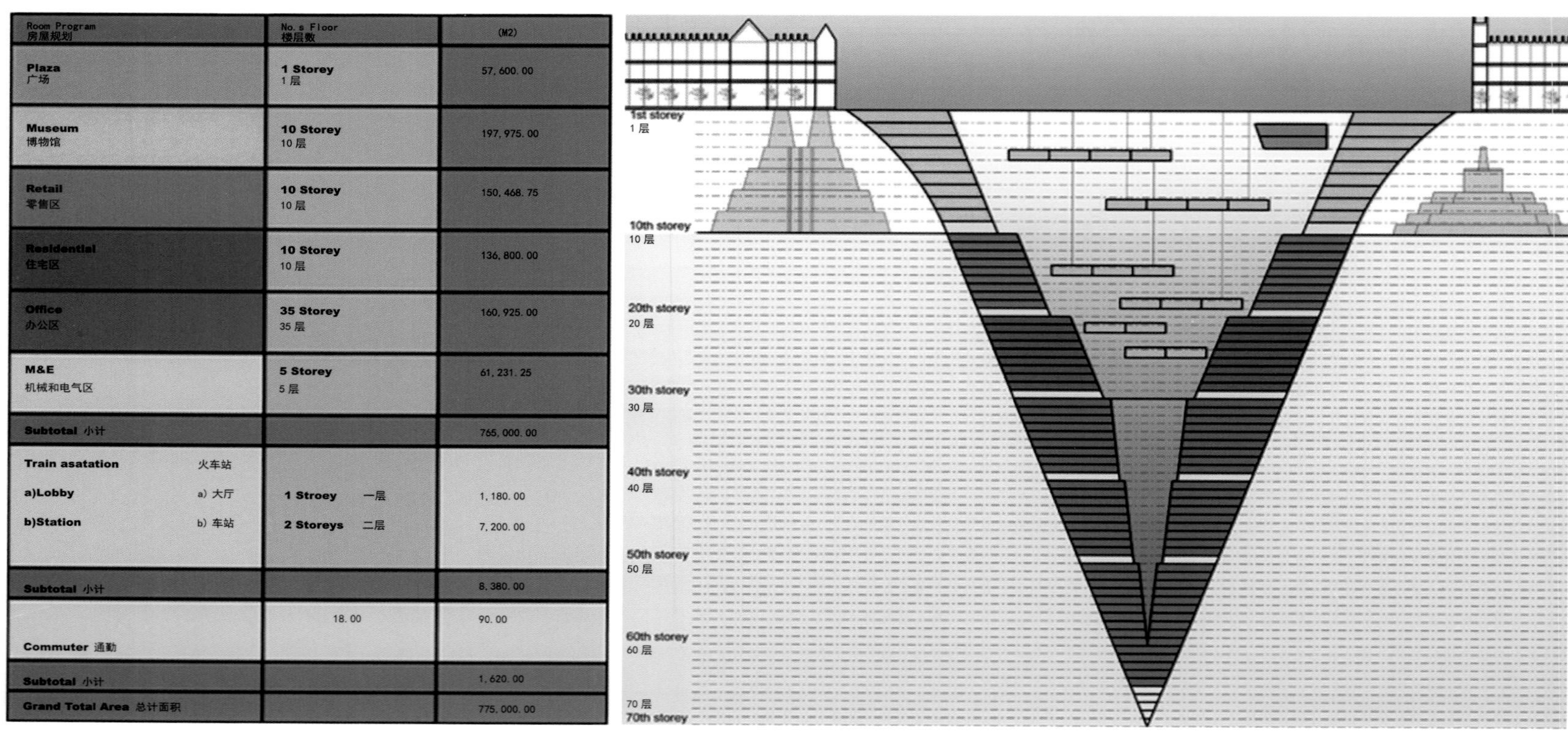

Room Program 房屋规划	No. s Floor 楼层数	(M2)
Plaza 广场	1 Storey 1 层	57, 600. 00
Museum 博物馆	10 Storey 10 层	197, 975. 00
Retail 零售区	10 Storey 10 层	150, 468. 75
Residential 住宅区	10 Storey 10 层	136, 800. 00
Office 办公区	35 Storey 35 层	160, 925. 00
M&E 机械和电气区	5 Storey 5 层	61, 231. 25
Subtotal 小计		765, 000. 00
Train asatation 火车站 a)Lobby a) 大厅 b)Station b) 车站	 1 Stroey 一层 2 Storeys 二层	 1, 180. 00 7, 200. 00
Subtotal 小计		8, 380. 00
Commuter 通勤	18. 00	90. 00
Subtotal 小计		1, 620. 00
Grand Total Area 总计面积		775, 000. 00

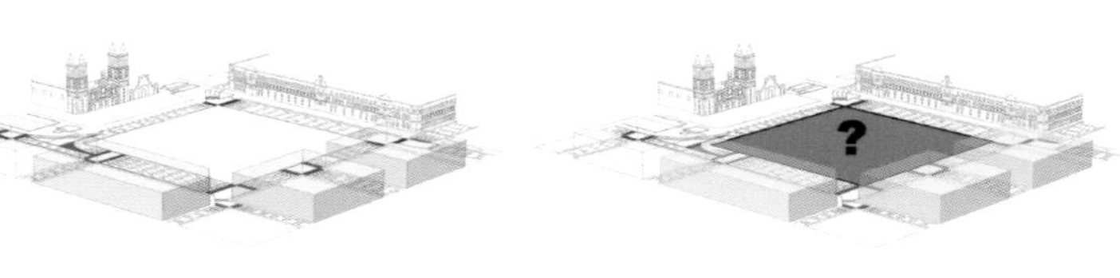
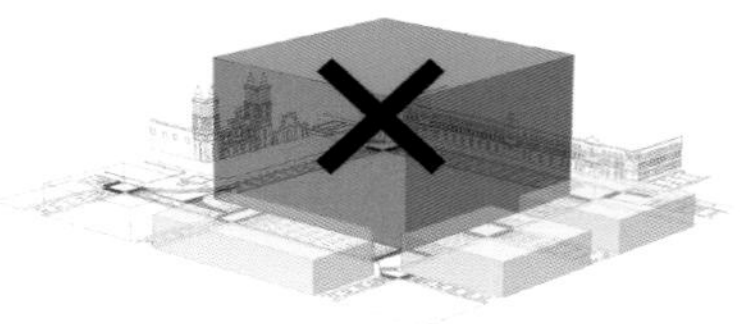
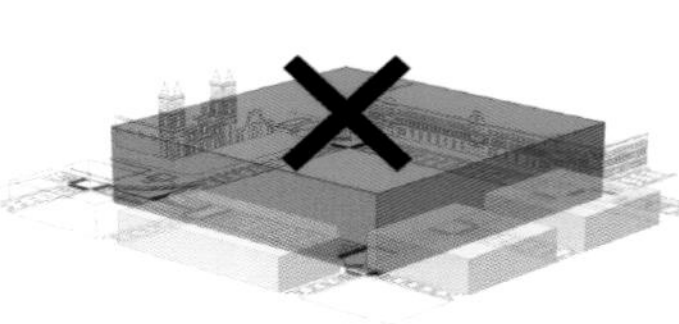
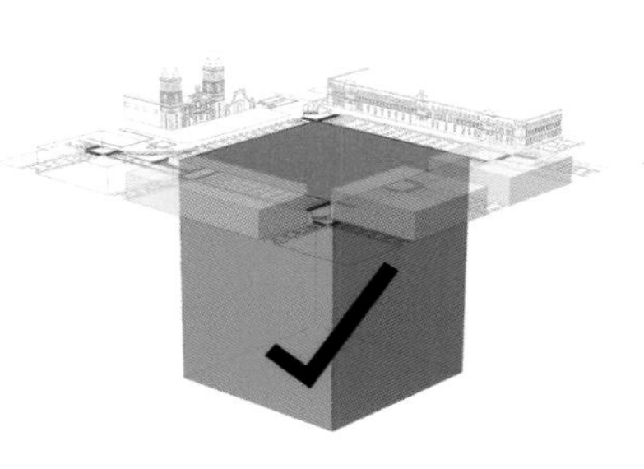
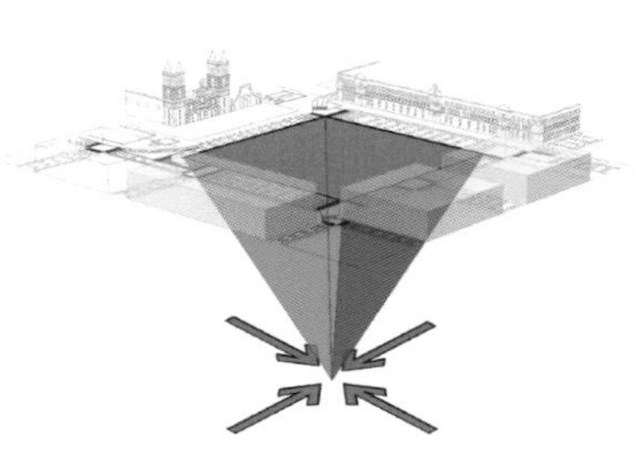
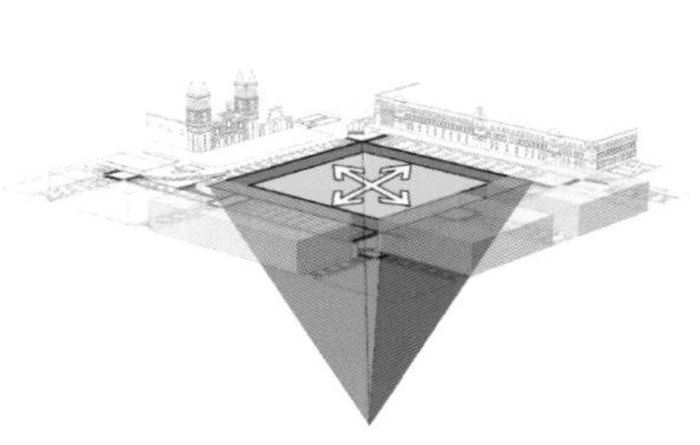
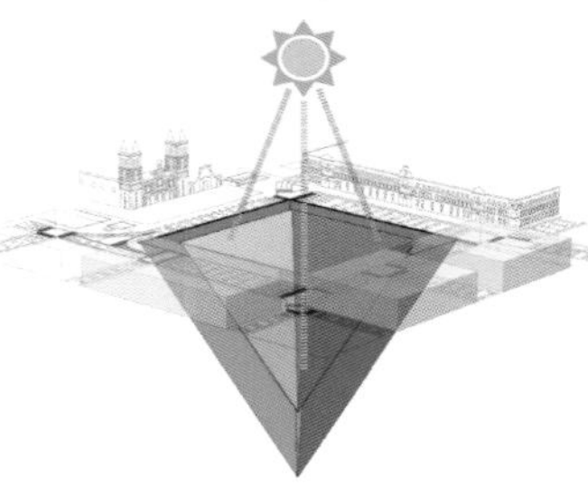
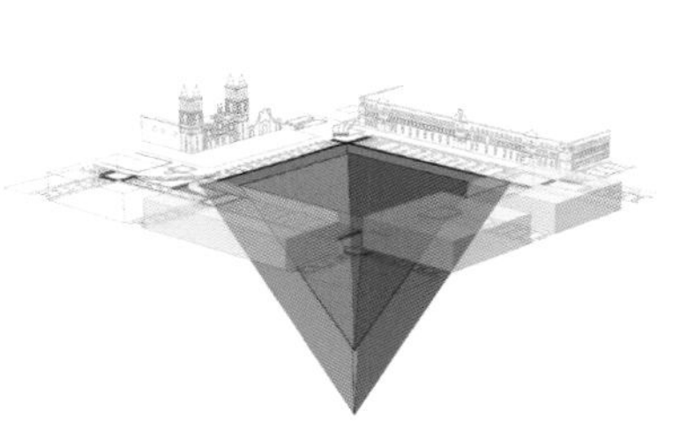
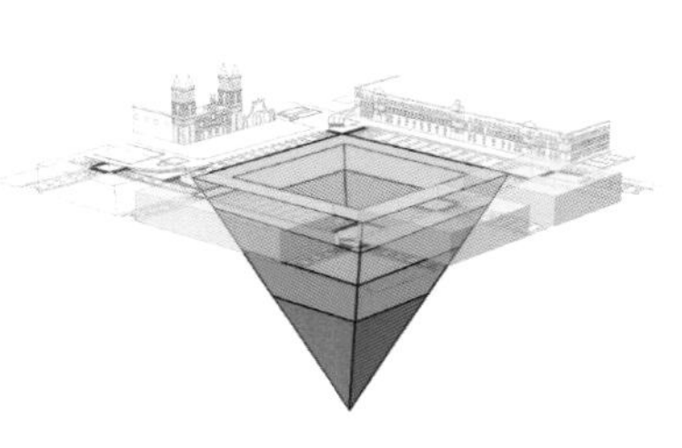
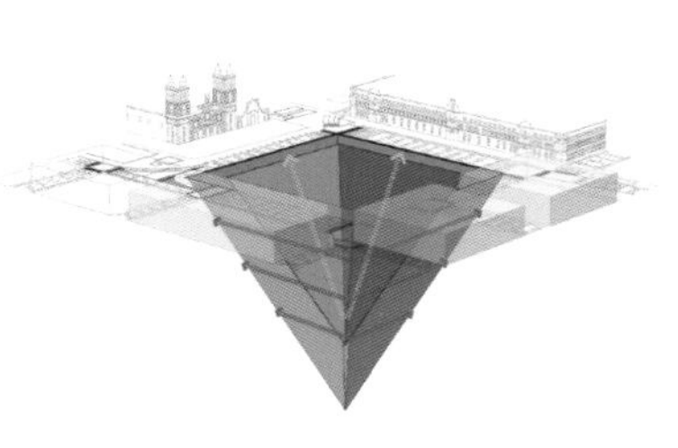

It is not our intention to displace the Zocalo's importance as a cultural and political hub. Furthermore, height restrictions and lack of developable land impede any significant development upwards. We build downwards. Taking advantage of the interest in unearthing the buried Aztec city that lies beneath this plaza we initiate an archaeological dig in the Zocalo. We dig at an angle that maintains the structural integrity of the earth and allows us to reach a depth of 300m, creating an empty inverted pyramid.

我们并没有打算取代 Zocalo 作为一个重要的文化和政治枢纽的地位。同时，高度限制和有限的场地面积也限制了项目向上长足发展。因此，我们选择向下建设。利用对挖掘被掩埋的阿芝特克市（位于这一广场的下方）的兴趣，我们在 Zocalo 发动了考古挖掘。我们以一定的角度开展挖掘活动，保持了地面结构的完整性，同时也使我们能到达 300 米深度的地下，营造出一个空的倒金字塔。

- **MUSEUM** 博物馆
- **RETAIL** 零售区
- **RESIDENTIAL** 住宅区
- **OFFICES** 办公区

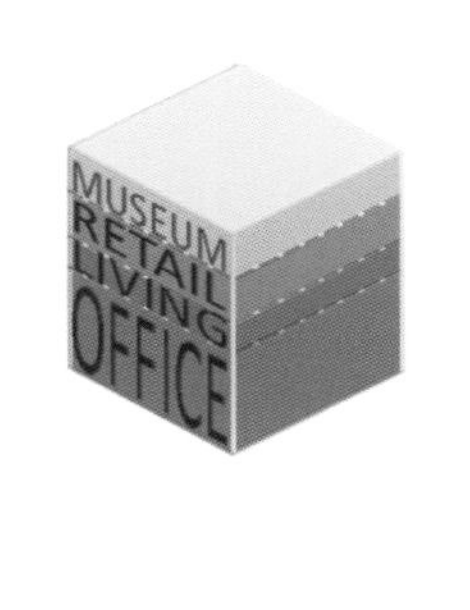

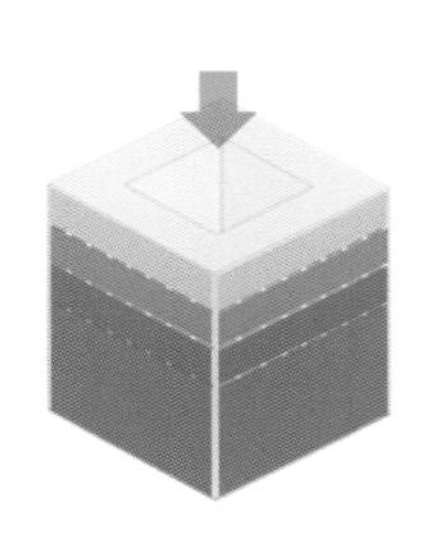
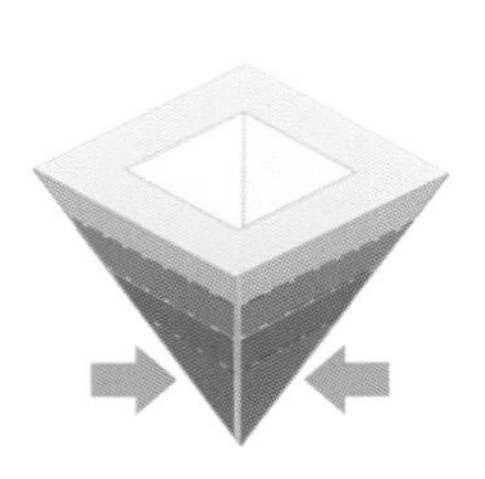
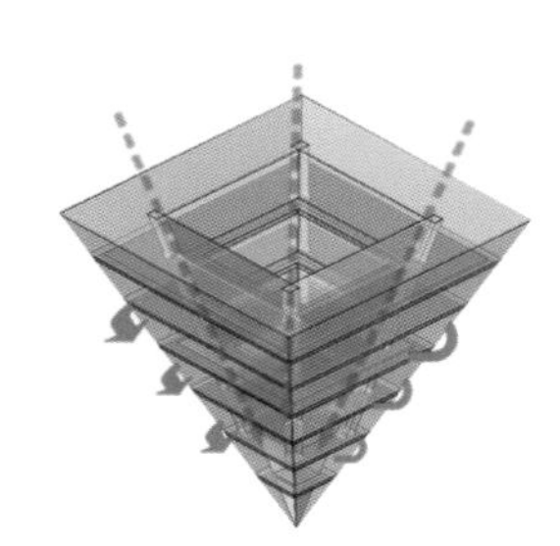
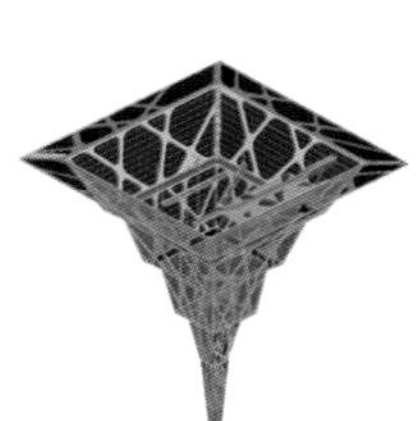
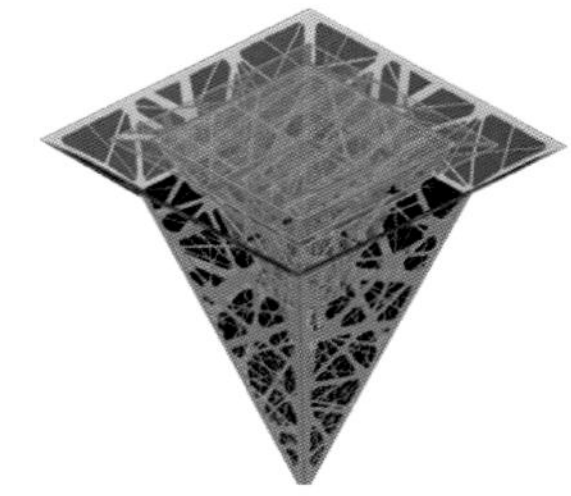
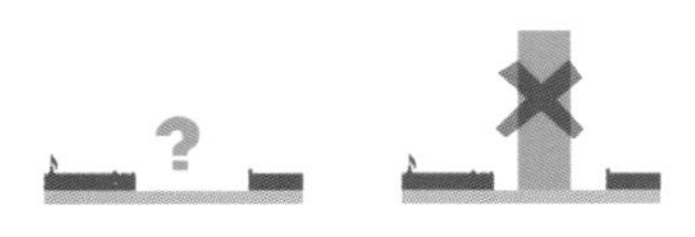

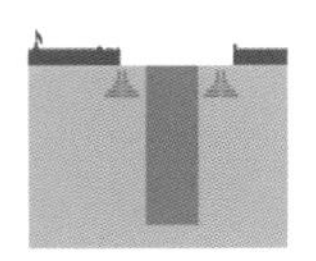
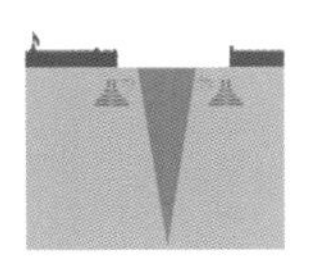
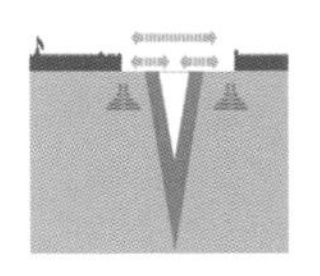
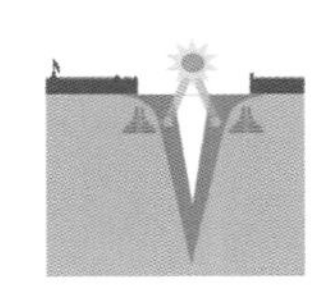
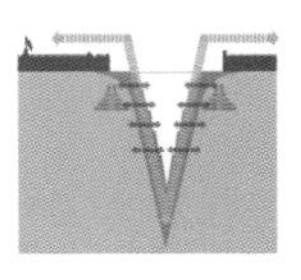

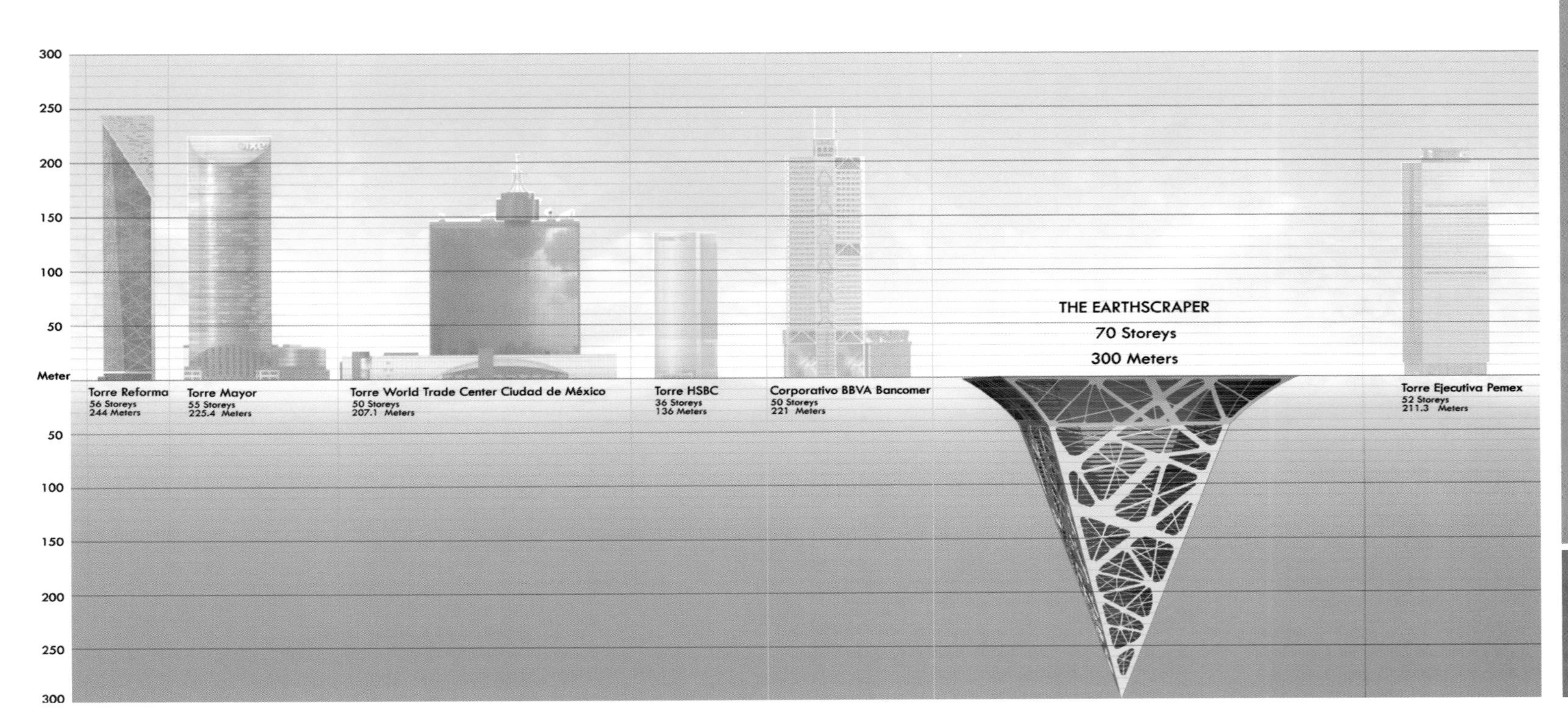

印度尼西亚雅加达 South Quarter

South Quarter

设计单位：阿特金斯 ATKINS
开发商：PT Intiland Development tbk
项目地址：印度尼西亚雅加达
用地面积：136 600 ㎡
建筑面积：472 120 ㎡

Designed by: Atkins
Client: PT Intiland Development tbk
Location: Jakarta, Indonesia
Site Area: 136,600 m^2
Built Up Area: 472,120 m^2

项目概况

项目位于印度尼西亚雅加达南部地区，其在节能、可持续性方面的设计使之成为雅加达最具可持续发展特色的建筑之一。

设计特色

设计的灵感来源于印尼的自然风光，设计采用几何图形勾勒建筑形态，并将建筑体量与建筑元素简化，从而在自然环境中实现建筑与景观的和谐。

可持续性设计是整个项目的核心。设计分析了光与影的平衡关系，采用了一种特定的太阳能定位遮蔽计算器，该测量仪器应用到每一个玻璃构件中，依照可变的垂悬部分和太阳能曝光的程度安装适量的遮阳百叶窗。该措施可确保吸收的太阳能维持在一定的限度内，使建筑能够采用低能耗环境控制系统。设计将悬臂结构和百叶窗依照建筑的定位进行了调整，并以栏杆覆盖层对其进行进一步补充。

设计的主要目的是最大限度地降低水资源的消耗。大楼配备有低流量的高品质配件，结合当地多雨的气候和高效的废水回收系统，最低可减少 25% 的淡水需求。

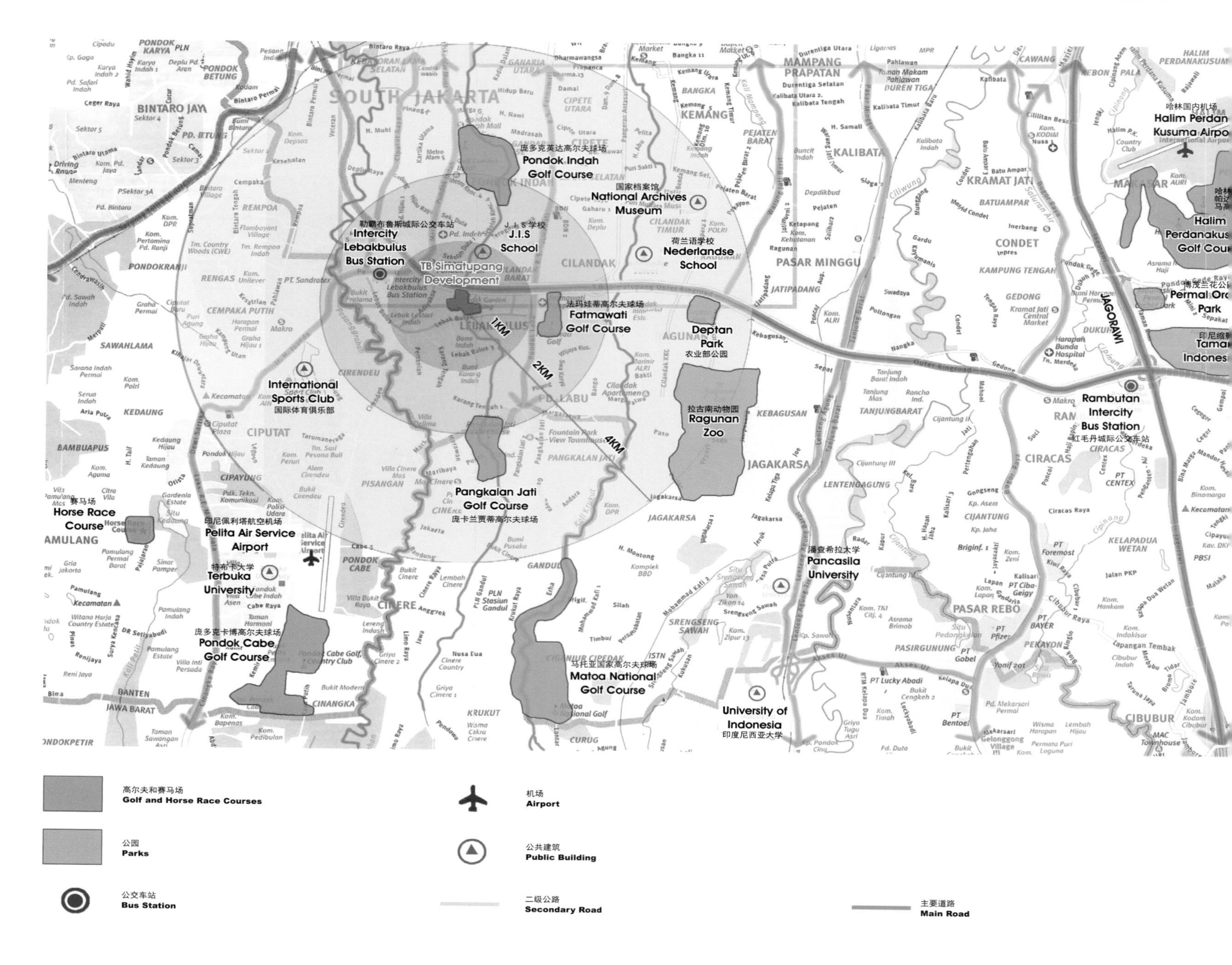

Profile

The project is located in the south region of Jakarta, Indonesia. The design on energy saving and sustainability makes the project one of the most distinctive sustainable buildings in Jakarta.

Design Feature

The concept design takes inspiration from natural Indonesia and elements of architecture are reduced to minimal shapes and geometry creating harmony between buildings and landscape in a natural environment.

Sustainability is central throughout the design approach. Analyzing the balance of light and shade, a solar orientation specific, shading calculator has been developed for the project and applied to every glazed element to establish the correct number of shading louvers with respect to the variable overhang and solar exposure. This ensures that the solar gains are kept within limits to enable the building to use low energy environmental control systems. The overhangs and louvers are tuned according to the orientation of the buildings and further complemented by an innovative balustrade overlay.

The primary aim is to reduce water consumption with high efficiency, low flow and high quality fittings. In combination with the high rainfall rates and a grey water recycling system, the fittings have reduced fresh water demand in excess of 25%.

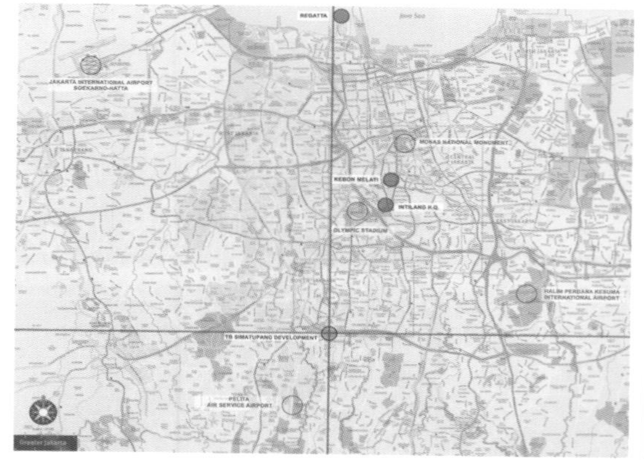

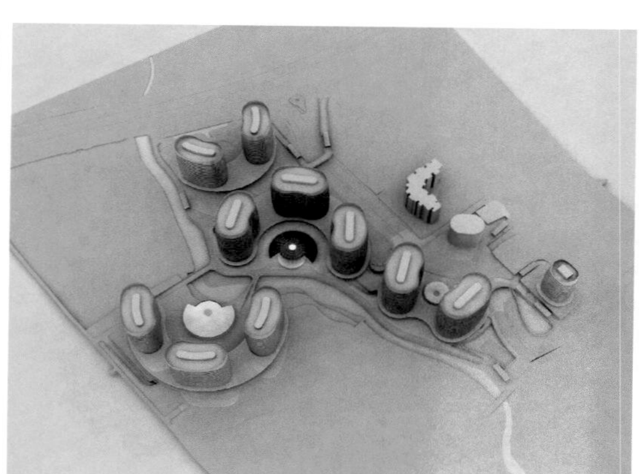

JAKARTA

上海五洲国际广场

Shanghai Wuzhou International Plaza

设计单位：Synthesis Design + Architecture
合作单位：Shenzhen General Architectural Design Institute
OneView CG
开发商：香港五洲国际集团有限公司
项目地址：中国上海市
项目面积：180 000 m²

Designed by: *Synthesis Design + Architecture*
Collaboration: *Shenzhen General Architectural Design Institute;*
OneView CG
Client: *Hong Kong Wuzhou International Group Co., Ltd.*
Location: *Shanghai, China*
Area: *180,000 m²*

项目概况

上海五洲国际广场位于上海大都会三环的华泰路，设计体现了上海这个独特的、充满了生机与活力的城市环境。

设计特色

受中国传统的阴阳概念的激发，这个“都市峡谷”被设计成两个嵌入式的岩石体量，这两个体量被分割开来，展示了一个流动的“峡谷”空间，这一建筑形态也与城市肌理形成呼应。

这个流动的“峡谷”空间将两个入口广场通过如“河流”状分布的独立的零售空间连接起来，这些零售空间可经由网状的天桥到达。一系列“岛屿状”的绿色空间分布在“河流”沿线，既为建筑提供了自然的遮蔽，同时也优化了城市环境。

在这个“峡谷”的出口处，都有一个景观化的入口广场，这些广场由各自所在位置的塔楼围合而成，广场的动态模式通过条纹的连接结构表现出来。这种将立面与屋顶、裙楼与塔楼整合的模式和这个由“河流”雕琢出来的“峡谷”的概念，体现了城市活动的节奏，激发了城市的活力。

Profile

Situated along Huatai Road in the third ring of the urban metropolis of Shanghai, the Shanghai Wuzhou International Plaza embodies the energy and vibrancy of the city's distinct urban environment.

Design Feature

Inspired by traditional Chinese concepts of Yin and Yang, the "Urban Canyon" is organized as two nested rock-like volumes which have been broken apart to reveal a flowing canyon condition which connects the project to the urban fabric of the city. The fluid canyon condition connects the two entry plazas of the site with a "river" of free-standing detached retail units with a network of connective sky bridges. A series of green space "islands" are distributed within the river

ENTRANCE PLAZA
入口广场
主力店
Anchor store
主力店
Anchor store
OFFICE办公
PLAZA广场
主力店
Anchor store
HOTEL
酒店
总部
HEADQUARTERS
办公
OFFICE
次级商店
Sub-anchor store
GREEN AREA
绿化区
TOWER 大楼
SHOPPING BLOCKS 购物街区
STORES 商店
SLABS 板材
TOWER 大楼
SHOPPING BLOCKS 购物街区
STORES 商店
SLABS 板材

ENTRANCE 入口
下乘区
DROP OFF
广场
PLAZA
OFFICE
办公
ENTRANCE PLAZA
入口广场
办公
OFFICE
购物街区
SHOPPING BLOCKS
PARKING
停车场
总部
HEADQUARTERS
办公
OFFICE
办公广场
OFFICE PLAZA
PARKING
停车场

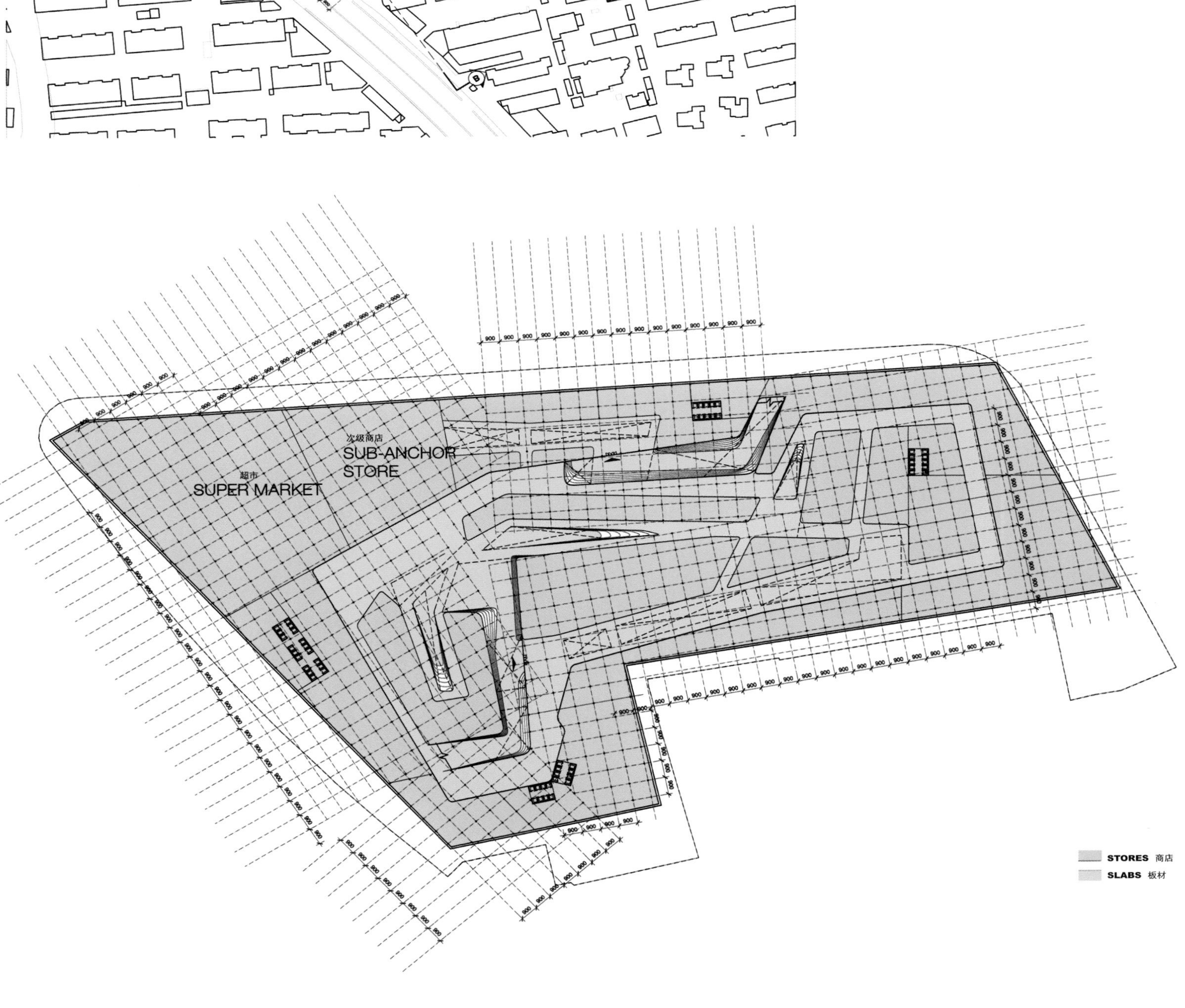

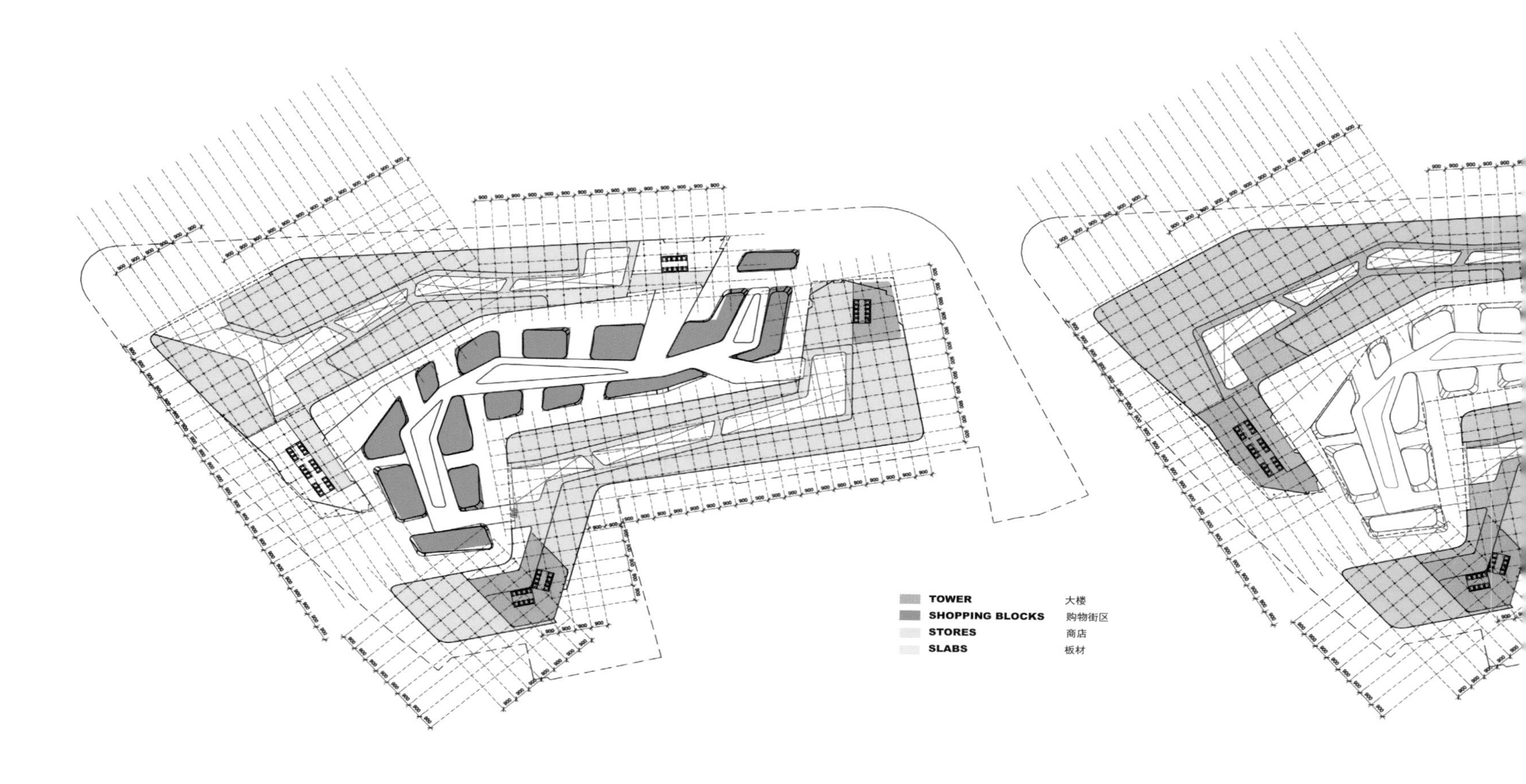

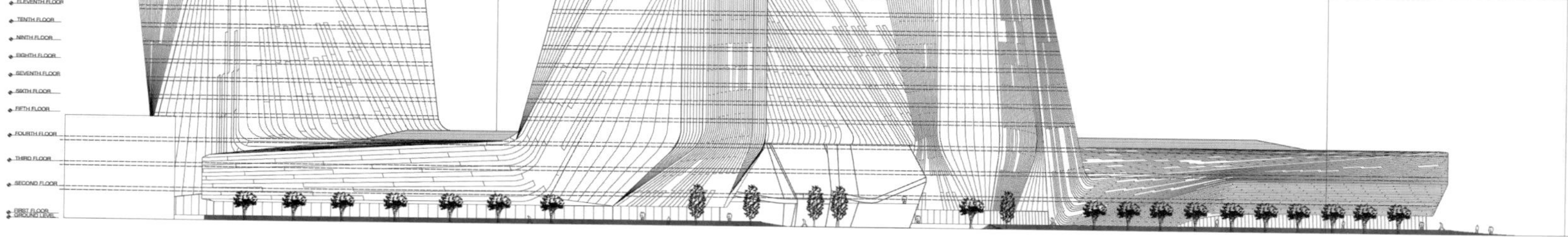
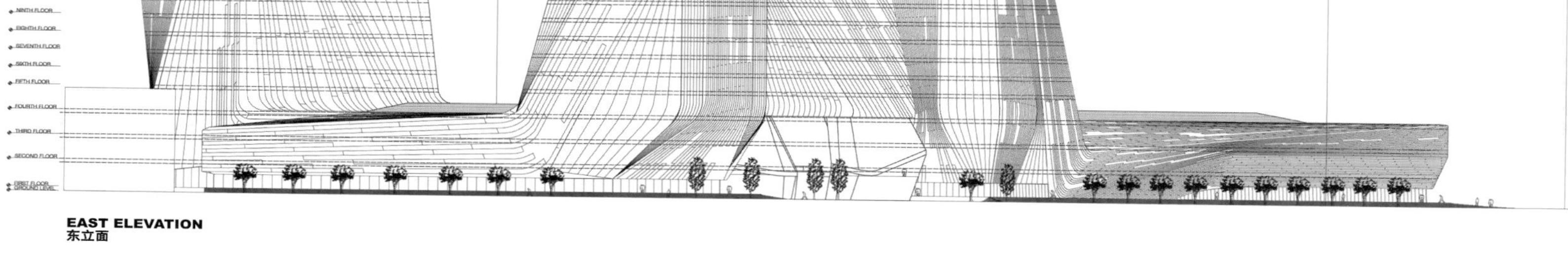

EAST ELEVATION
东立面

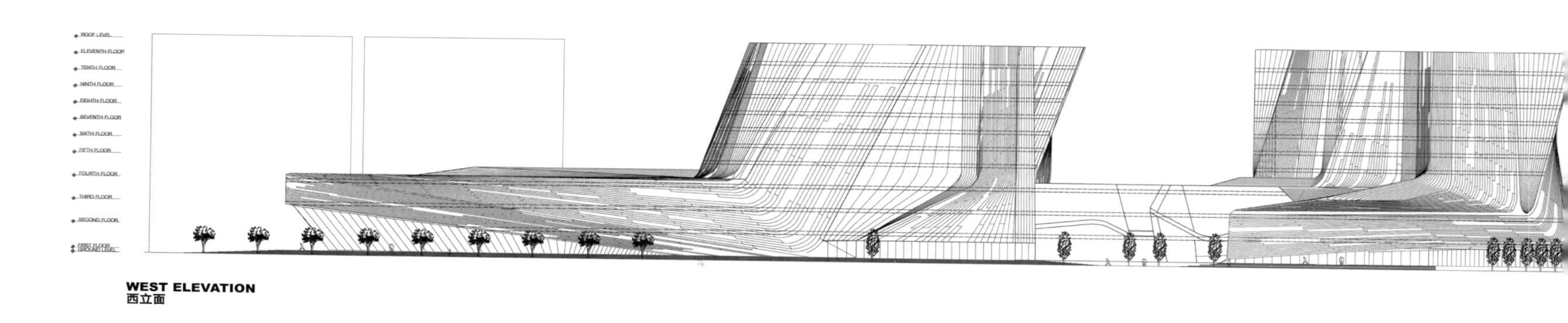

WEST ELEVATION
西立面

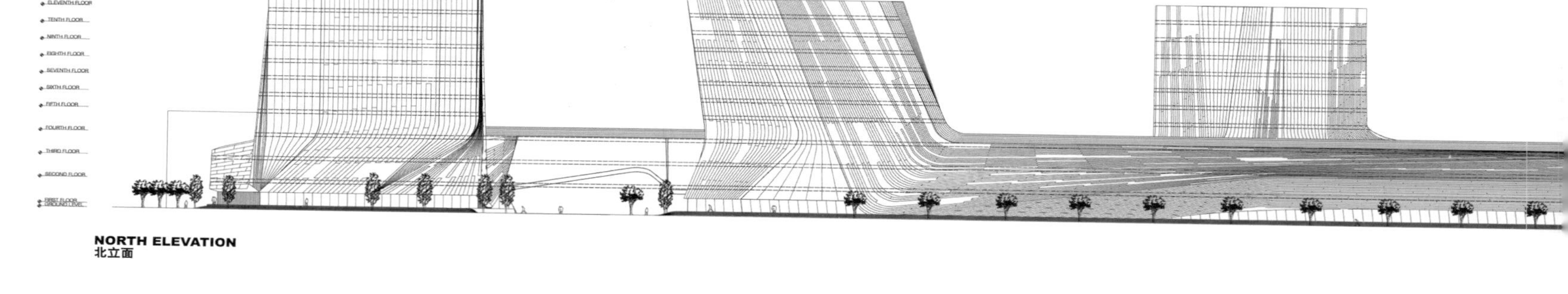

NORTH ELEVATION
北立面

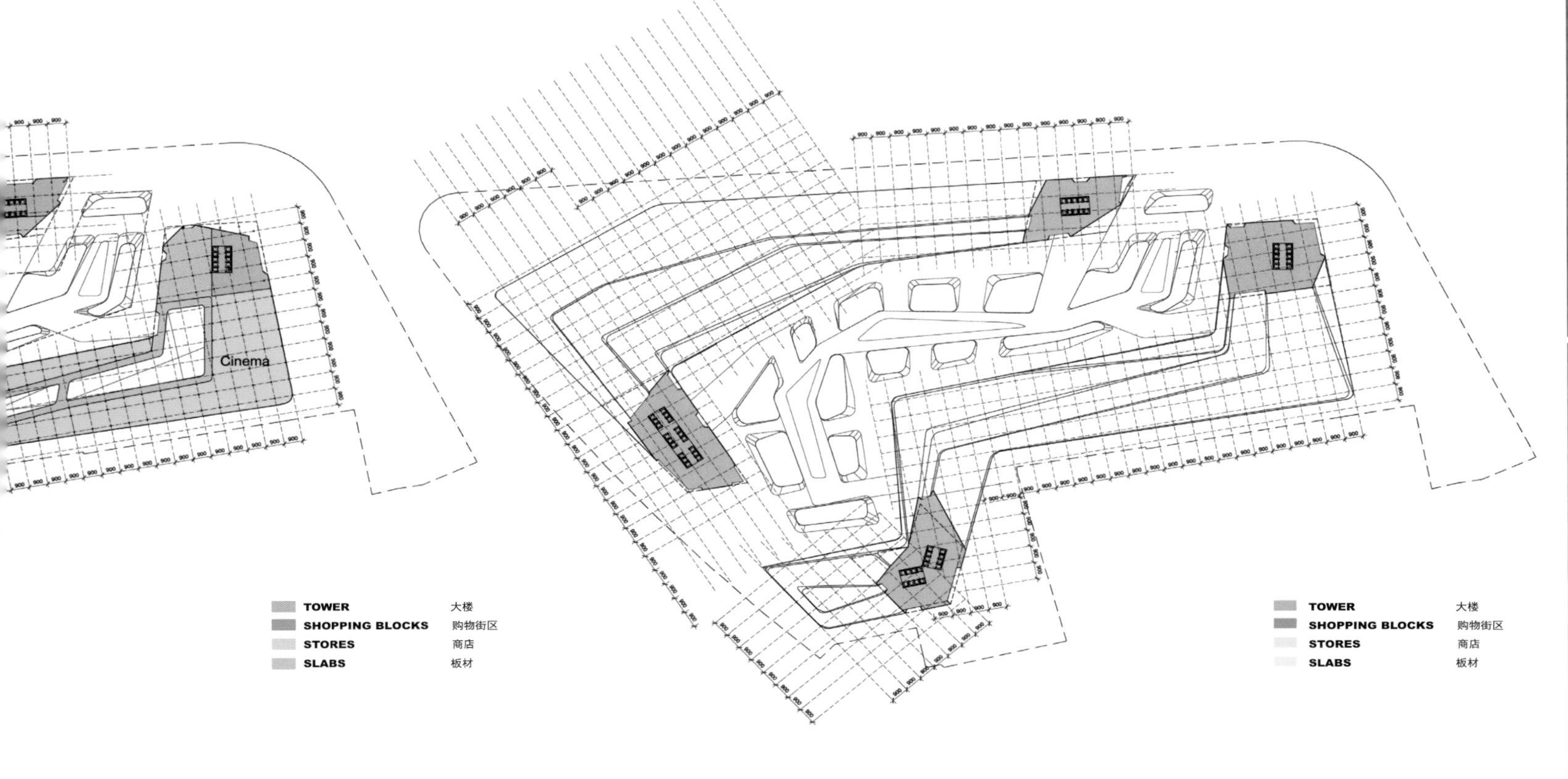

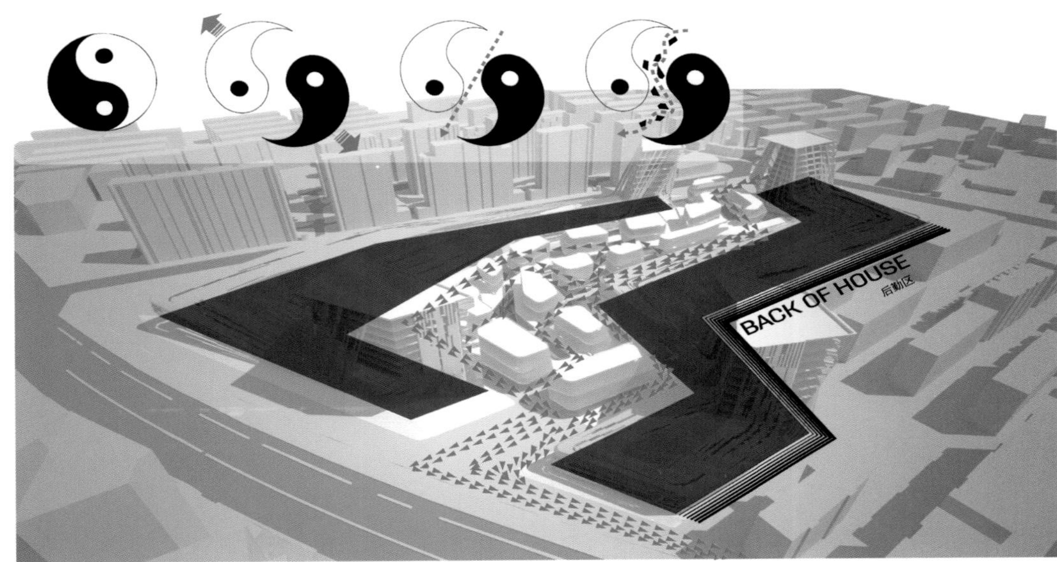

SITE FLOW
场地流动

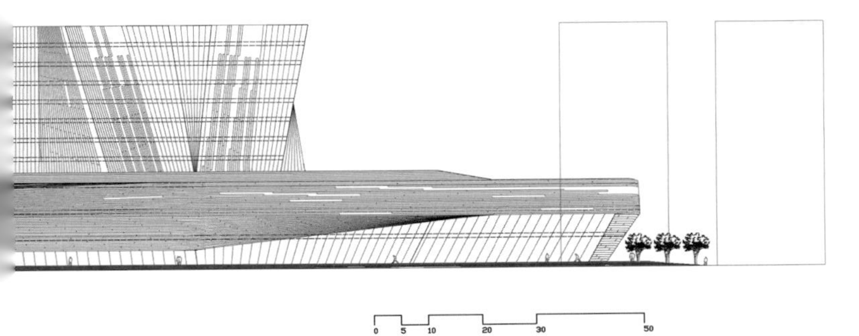

to provide natural shading and to soften the urban condition.

At the mouth of each canyon is a landscaped entry plaza framed by the portal created by its respective towers. The dynamic patterning of the plaza is further expressed in the striated articulations. This pattern embodies the pulses of activity and urban energy of the city to merge façade with roof and podium with tower, which is conceptualized as the river that has carved the canyon.

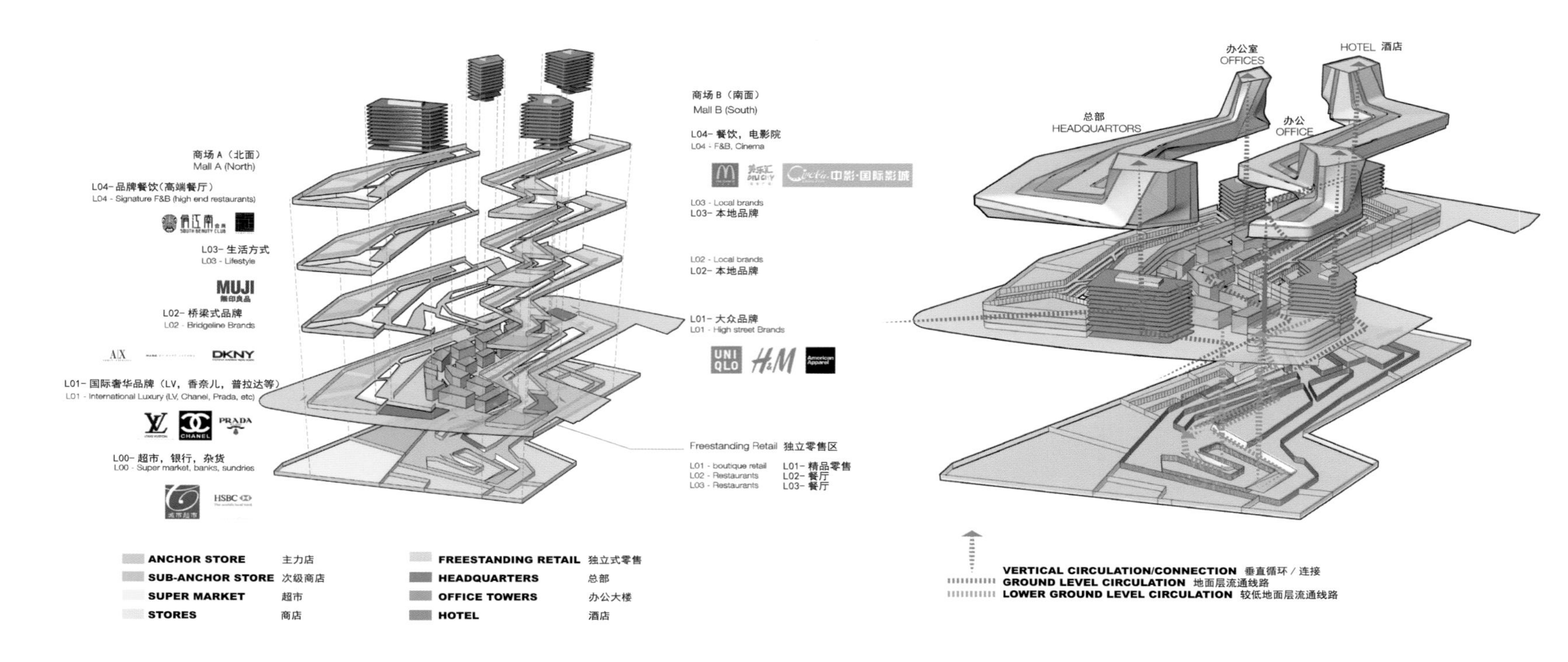

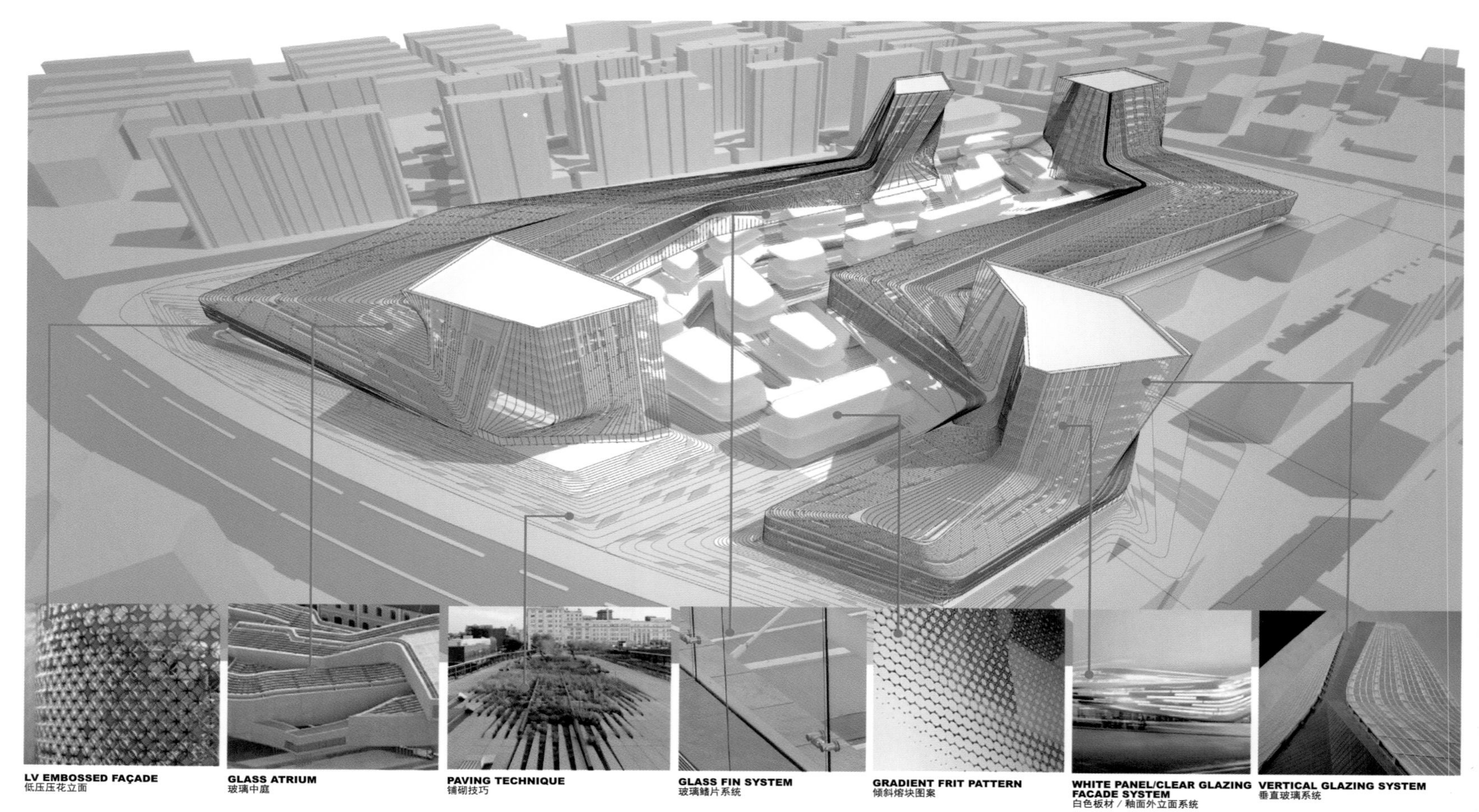

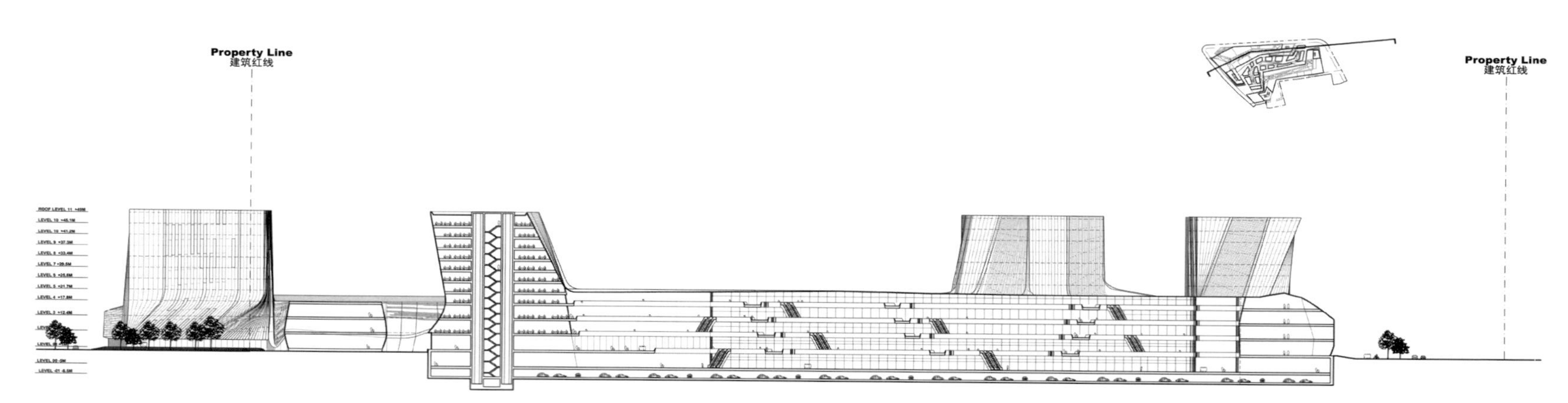

SECTION AA
剖面 AA

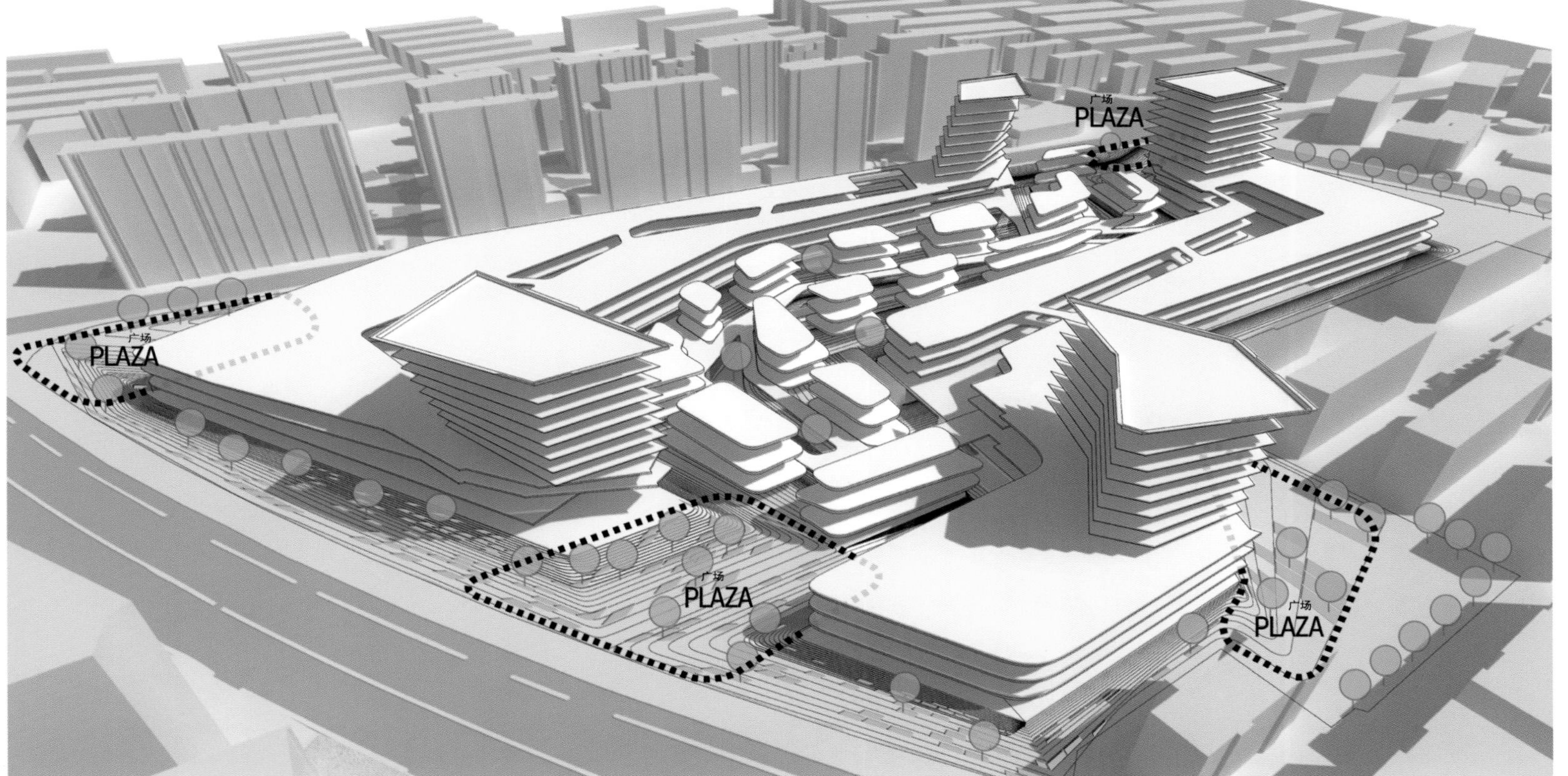

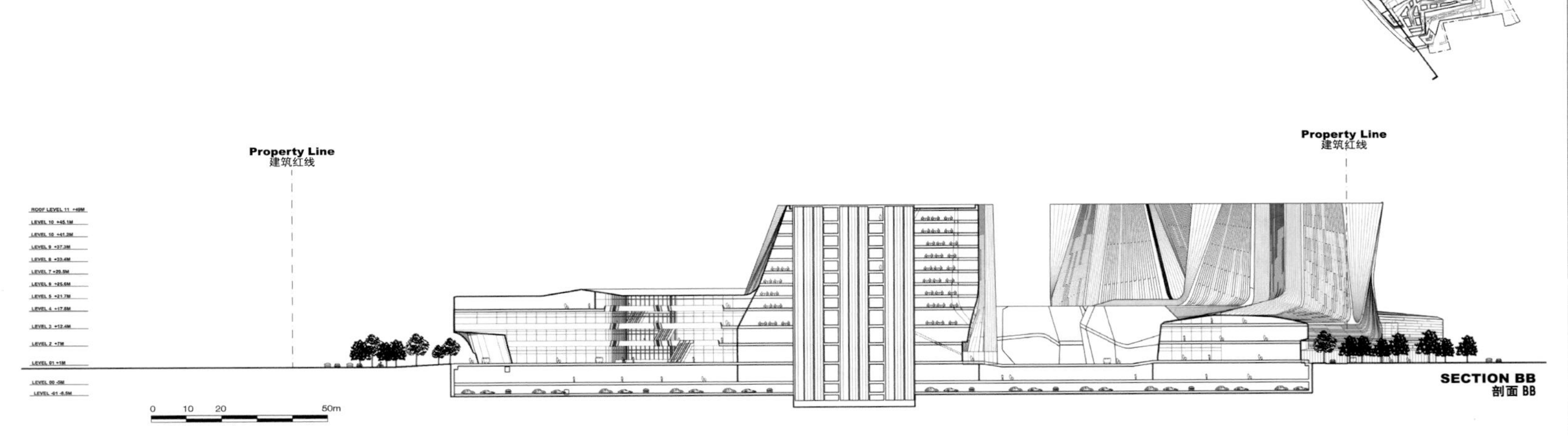

SECTION BB
剖面 BB

四川成都合力达商业综合体

Sichuan Chengdu Helida Commercial Complex

设计单位：OAD 欧安地建筑设计事务所
开发商：成都合力达房地产开发公司
项目地址：中国四川省成都市
总用地面积：30 891 m²
总建筑面积：338 941 m²
建筑密度：52%
容积率：9.0

Designed by: OAD Office for Architecture & Design
Developer: Chengdu Helida Real Estate Development Co., Ltd.
Location: Chengdu, Sichuan, China
Land Area: 30,891 m²
Gross Floor Area: 338,941 m²
Building Density: 52%
Plot Ratio: 9.0

项目概况

项目位于四川省成都市北二环，与西南交通大学隔路相望。项目定位为集办公、酒店、商业、娱乐于一体的超高层商业综合体，旨在在成都北二环老城区打造一个现代、时尚、生态的高端商务中心区。

建筑设计

立面造型形成自下而上的有机变化，疏密有致的柱子形成了树的意向，赋予建筑自然生长的张力与生命力。在平面及立面形态设计中，设计参照了“叶”的形态，其流畅动感的建筑线条，赋予了建筑自然而又优美的外观。

建筑主要涵盖了办公、酒店、商业、行政酒廊等功能区间。项目主要分为两个部分：场地东面的基座是一栋扇形的裙房，主要以商业空间为主，裙房弧线两端，矗立着两栋相对而立、形态相似的高层建筑；场地西面是两栋较低的建筑，主要以办公空间为主。

Profile

The project is located in the North 2nd Ring Road of Chengdu, Sichuan Province, facing Southwest Jiaotong University oppositely. The project is oriented as a super high-rise commercial complex integrating office, hotel, commerce and recreation. It aims to build a modern, fashionable, ecological high-end CBD in the Old Town of the North 2nd Ring Road of Chengdu.

Architectural Design

The building façade changes organically from bottom to top. Pillars set in different densities create an image of “Tree”, which endow the building with the tension and vitality of natural growth. In plane and elevation design, the project refers to the form of “Leaf”, using smooth dynamic lines to give the building natural gorgeous shape.

The building has mainly included office, hotel, business, executive lounge and other functional areas. The project can be divided into two parts, the east part and the west part. The podium in the east is mainly occupied by commercial spaces. At both ends of the podium arc erect two similar high rises. The two relatively low rise buildings in the west are primarily for offices.

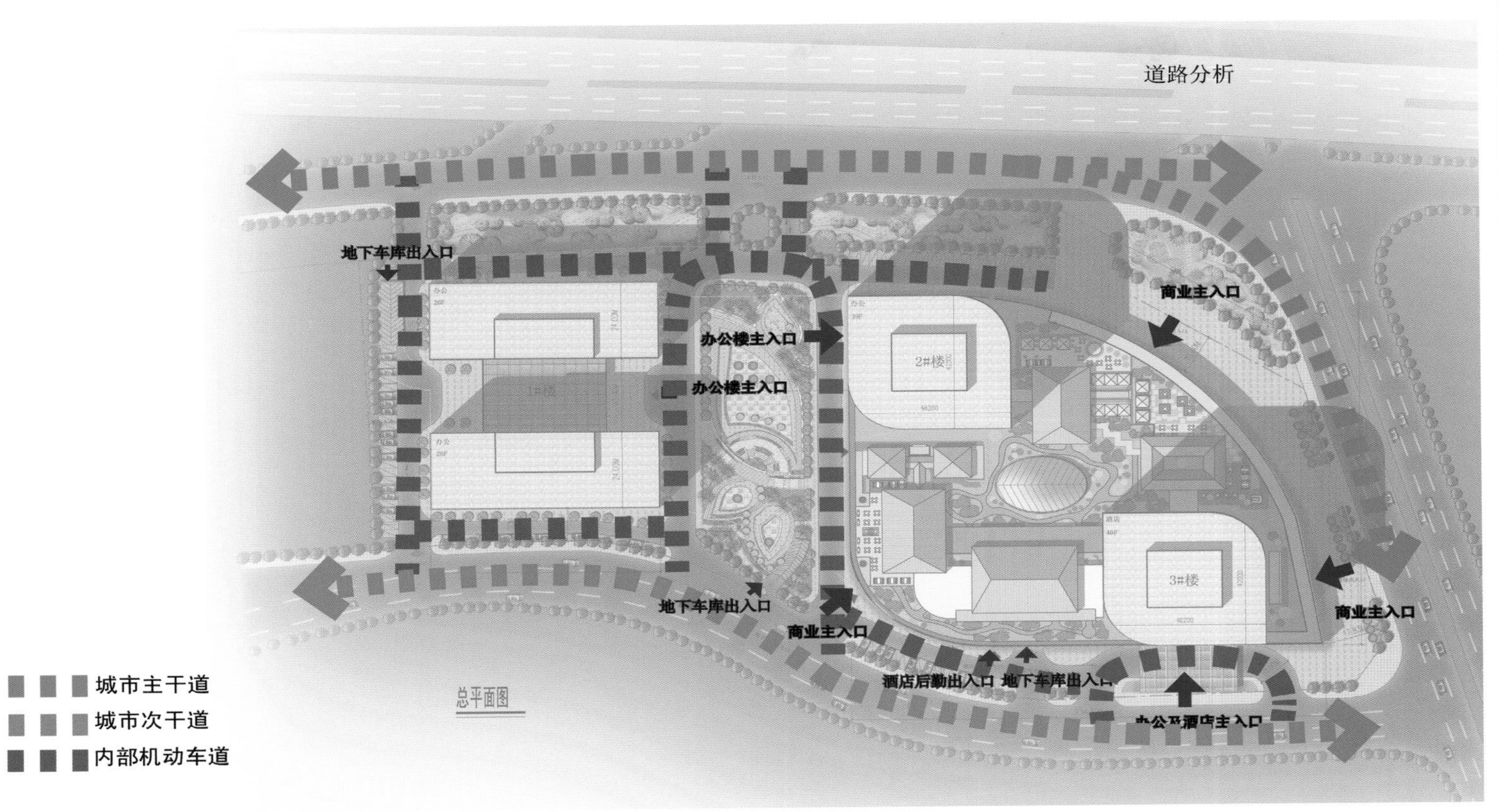
道路分析
地下车库出入口
办公楼主入口
办公楼主入口
商业主入口
1#楼
2#楼
3#楼
地下车库出入口
商业主入口
商业主入口
酒店后勤出入口
地下车库出入口
总平面图
城市主干道
城市次干道
内部机动车道

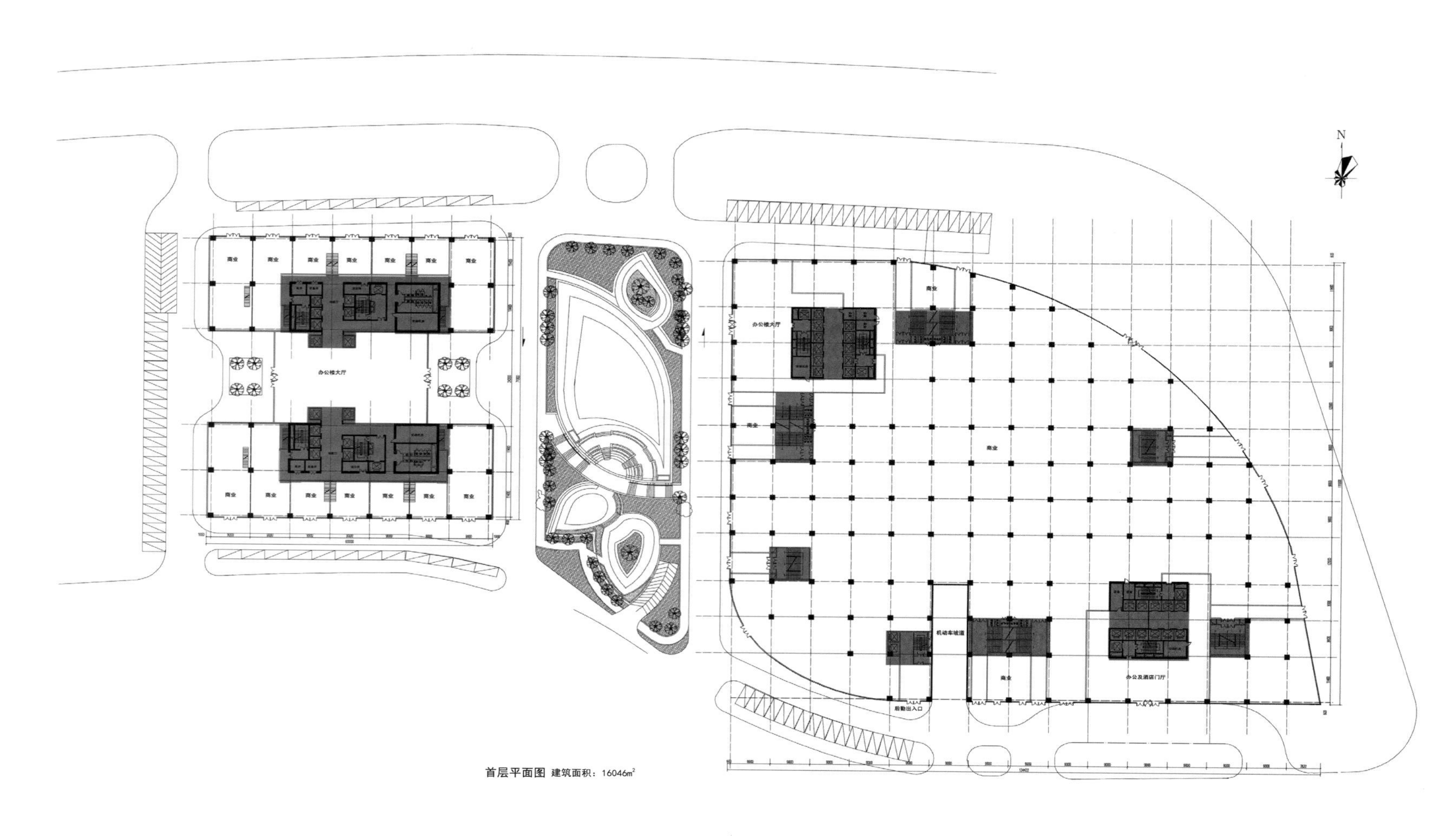

首层平面图 建筑面积：16046m²

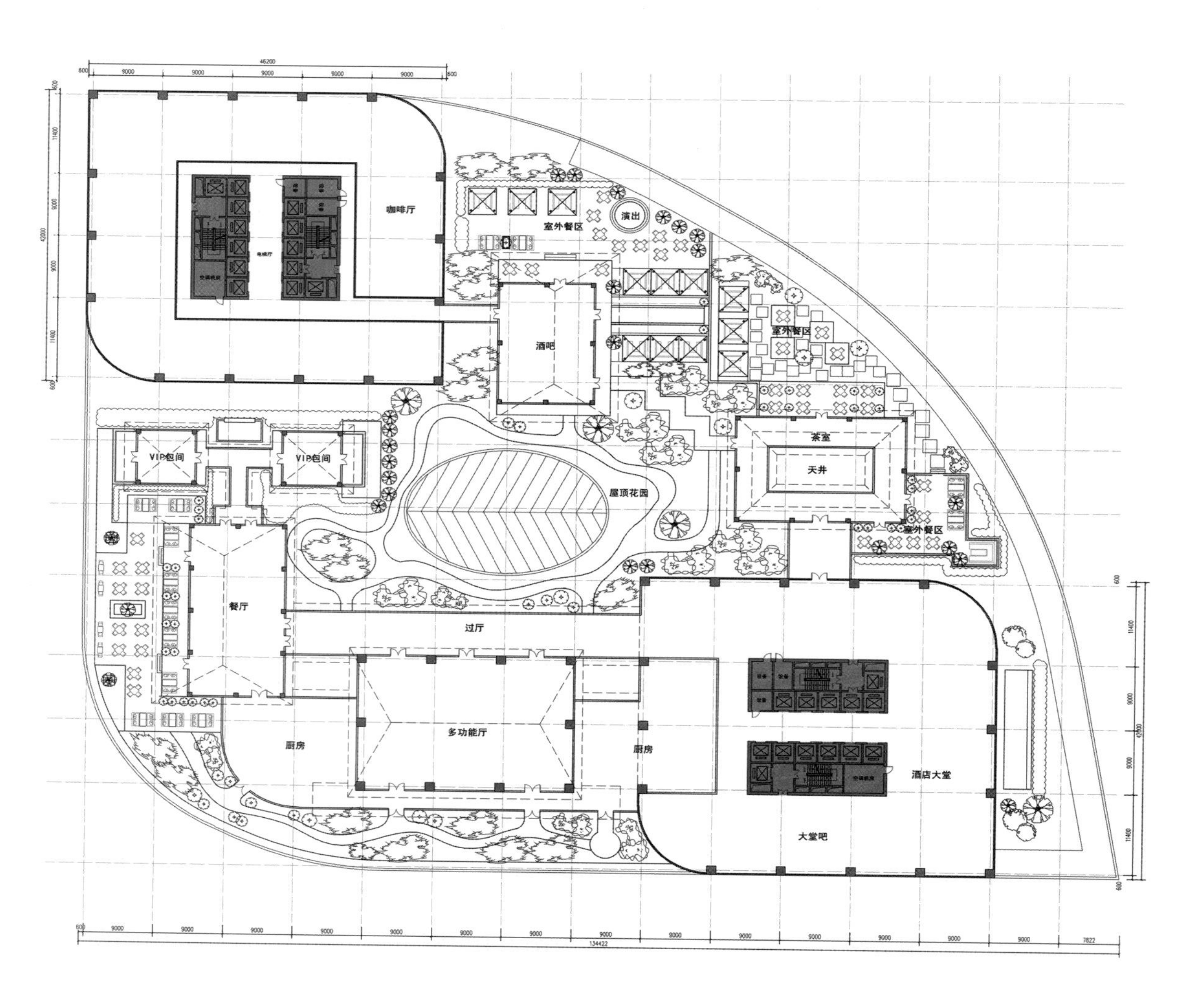

2# 楼、3# 楼 7 层平面图

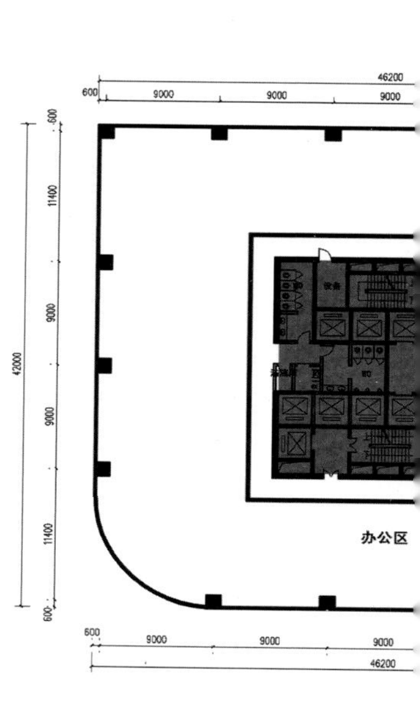

3# 楼办公区标准层平

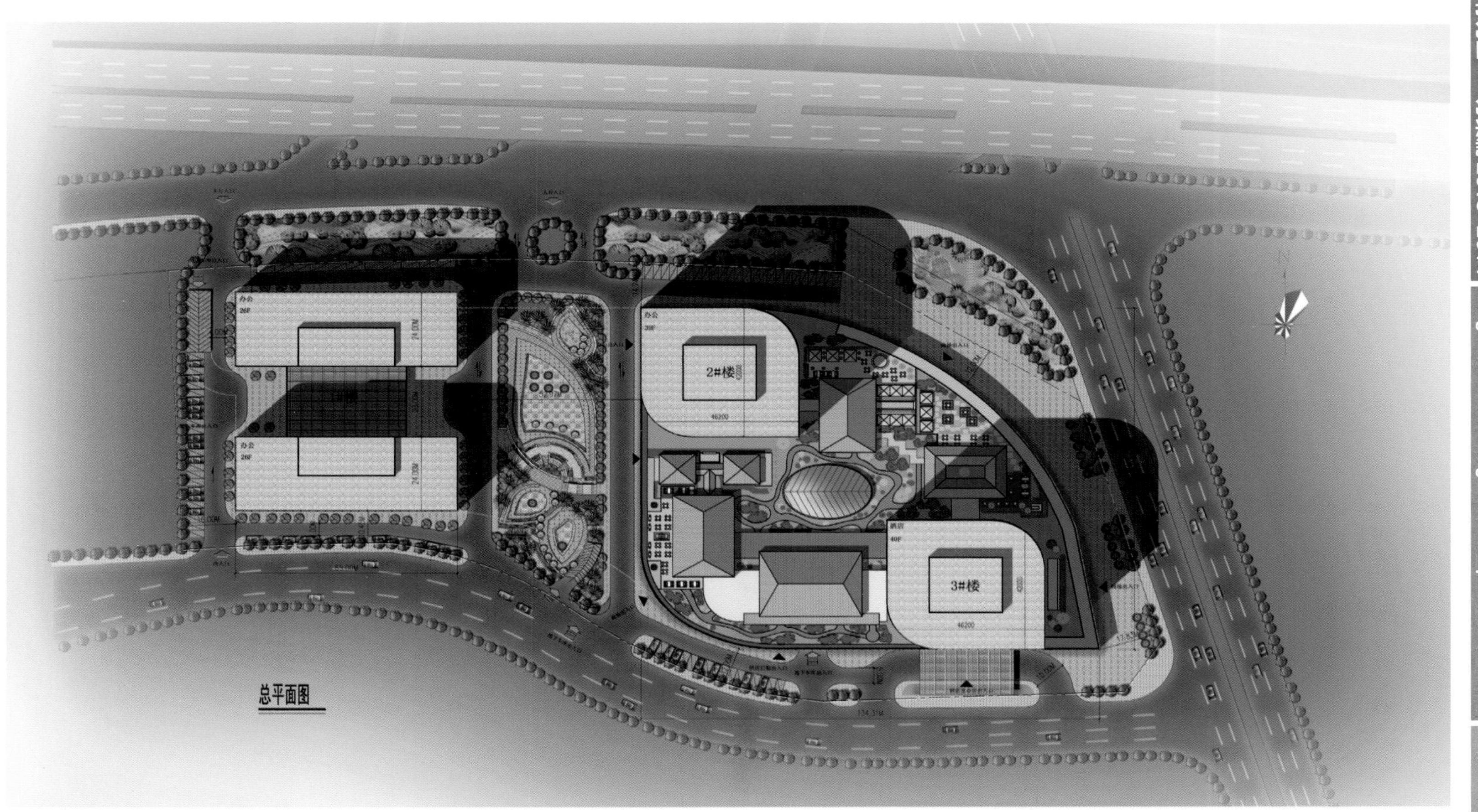

总平面图

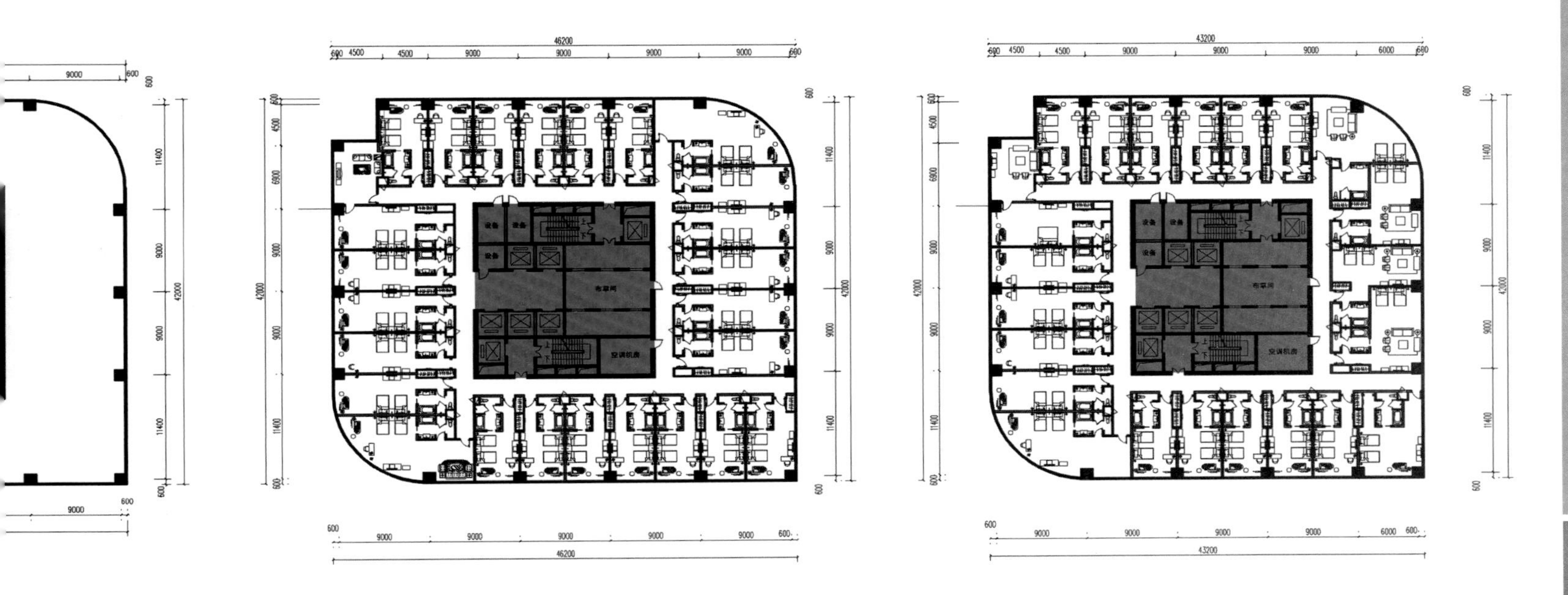

3# 楼客房区标准层平面图 1

3# 楼客房区标准层平面图 2

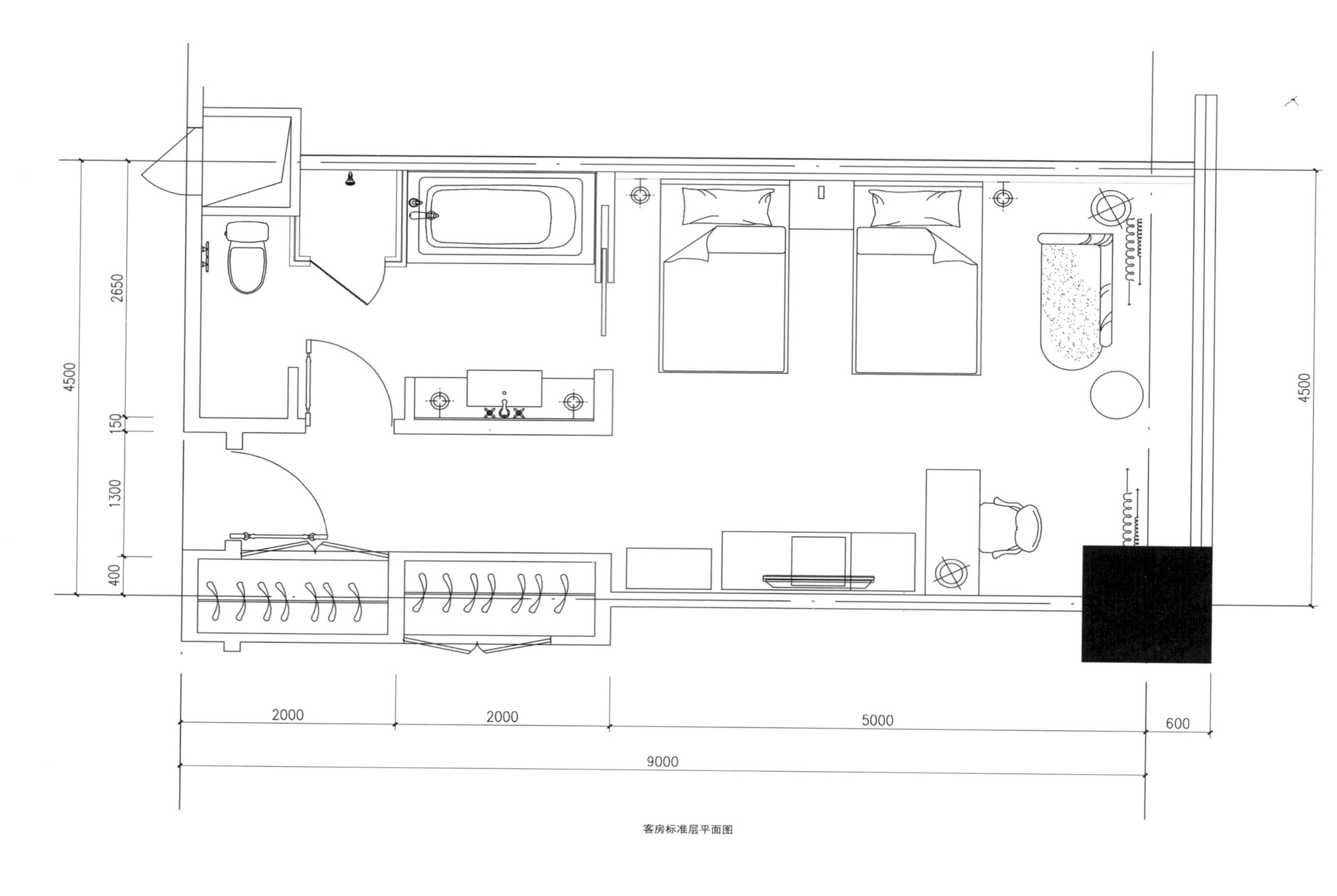

客房标准层平面图

主要经济技术指标

一、总规划用地面积:	30891 ㎡
二、规划总建筑面积:	338941 ㎡
地上建筑面积:	278540 ㎡
地下建筑面积(3层):	60401 ㎡
三、各部分建筑面积:	
1、办公建筑面积:	169426 ㎡
1#楼建筑面积(26F):	81920 ㎡
2#楼建筑面积(含1层大堂):	63464 ㎡
3#楼建筑面积(含1层大堂):	34042 ㎡
2、酒店建筑面积	40000 ㎡
3、商业建筑面积:	85285 ㎡
地上	66516 ㎡
地下一层	18769 ㎡
四、容积率	9.0
五、建筑覆盖面积:	16046 ㎡
六、建筑密度:	52%
七、机动车位:	约850个
1、地上机动车位:	约90个
2、地下机动车位:	约760个
八、非机动车位:	约2560个
1、地上非机动车位:	约200个
2、地下非机动车位:	约2360个

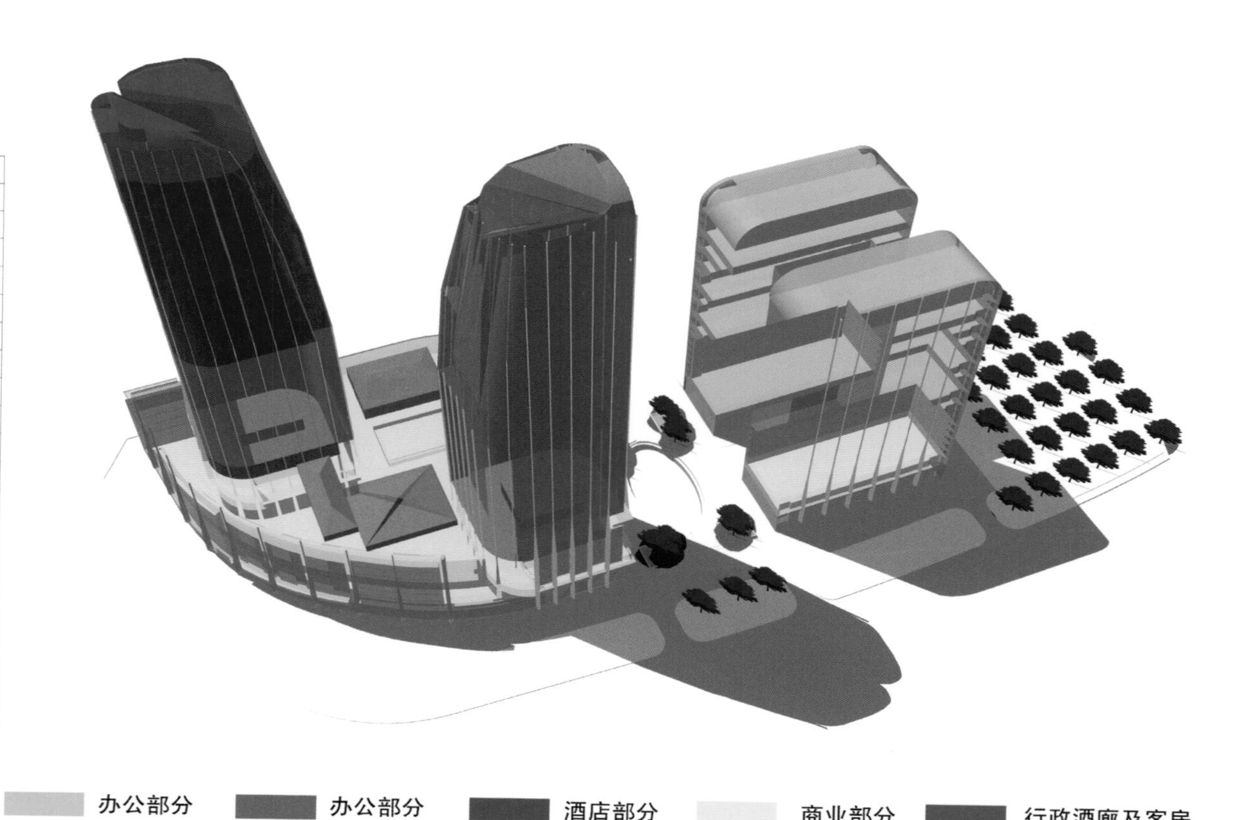

办公区
商业
中庭
餐饮/小型商业
机动车库/设备用房

1# 办公楼剖面

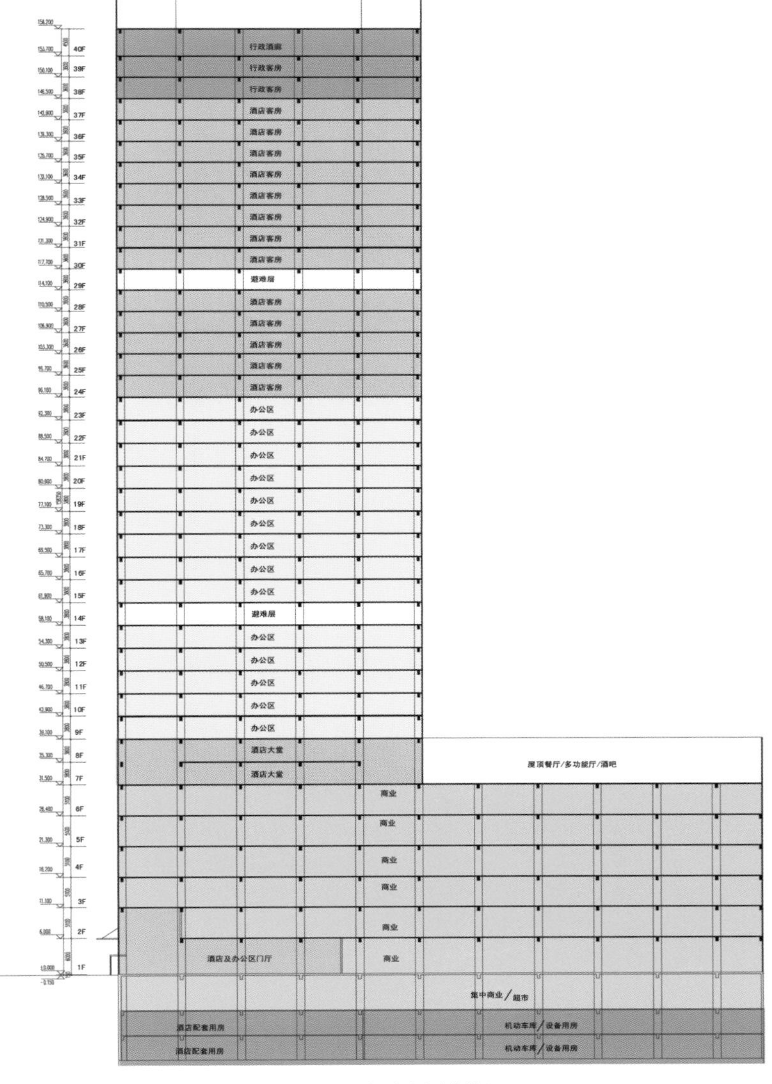

3# 酒店、办公综合楼剖面

四川成都中信昊园

Sichuan Chengdu CITIC-HY

设计单位：北京世纪安泰建筑工程设计有限公司
开发商：中信昊园成都置业有限公司
项目地址：中国四川省成都市
总用地面积：277 658 ㎡
总建筑面积：698 668 ㎡
建筑密度：40%
绿化率：37.2%
容积率：2.52

Designed by: Beijing SJAT Architecture & Engineering Design Company
Developer: CITIC-HY Chengdu Real Estate Co., Ltd.
Location: Chengdu, Sichuan, China
Site Area: 277,658 m^2
Gross Floor Area: 698,668 m^2
Building Density: 40%
Greening Ratio: 37.2%
Plot Ratio: 2.52

项目概况

项目位于成都高新技术区内，北至百业路，东至新业路，西至新创路，南至百川路，是一个以办公产品为主，商业为辅的大型复合项目，也是中国西南地区一个高端地产的标志性项目。

建筑设计

项目包括了商业、办公、高档定制公寓、独栋别墅等功能区。商业空间和办公空间主要沿街布置，住宅区和别墅区则位于地块内部。建筑整体布局高低错落，极富层次感。建筑造型简约，立面色彩淡雅，细节处理精细，显得现代而又时尚。

基地内分布着不同宽度的车行道，设计充分考虑了交通的便捷性，既保证了交通流线的直达性，同时也使道路总长最小。地下停车场结合天井、天窗采光等综合手段，既改善了地下车库的采光条件，又提高了地下停车场的安全性和舒适度。

Profile

The project is located in Chengdu Hi-Tech Industrial Development Zone adjoining Baiye Road to the north, Xinye Road to the east, Xinchuang Road to the west and Baichuan Road to the south. It is a large complex dominated by offices and supported by business spaces. The project is a symbolic project of high-end real estate in Southwest China.

Architectural Design

The project has included commercial areas, offices, high-end customized apartments, villas and etc. The commercial spaces and office spaces are set along the street. Residential areas and villa districts are arranged inside of the site. The buildings' overall layout is staggered and rich of levels. Simple shape, simple elegant colored façade plus finely treated details contribute to modern and fashion buildings.

Vehicle lanes of different widths distributed in the site have fully considered the convenience of transportation. It ensures directly accessible traffic lines and minimum road length. Patio and skylight of underground parking have improved lighting condition of underground garage and enhanced security and comfort of the underground parking.

昊园 CITIC PARK

昊园 CITIC PARK

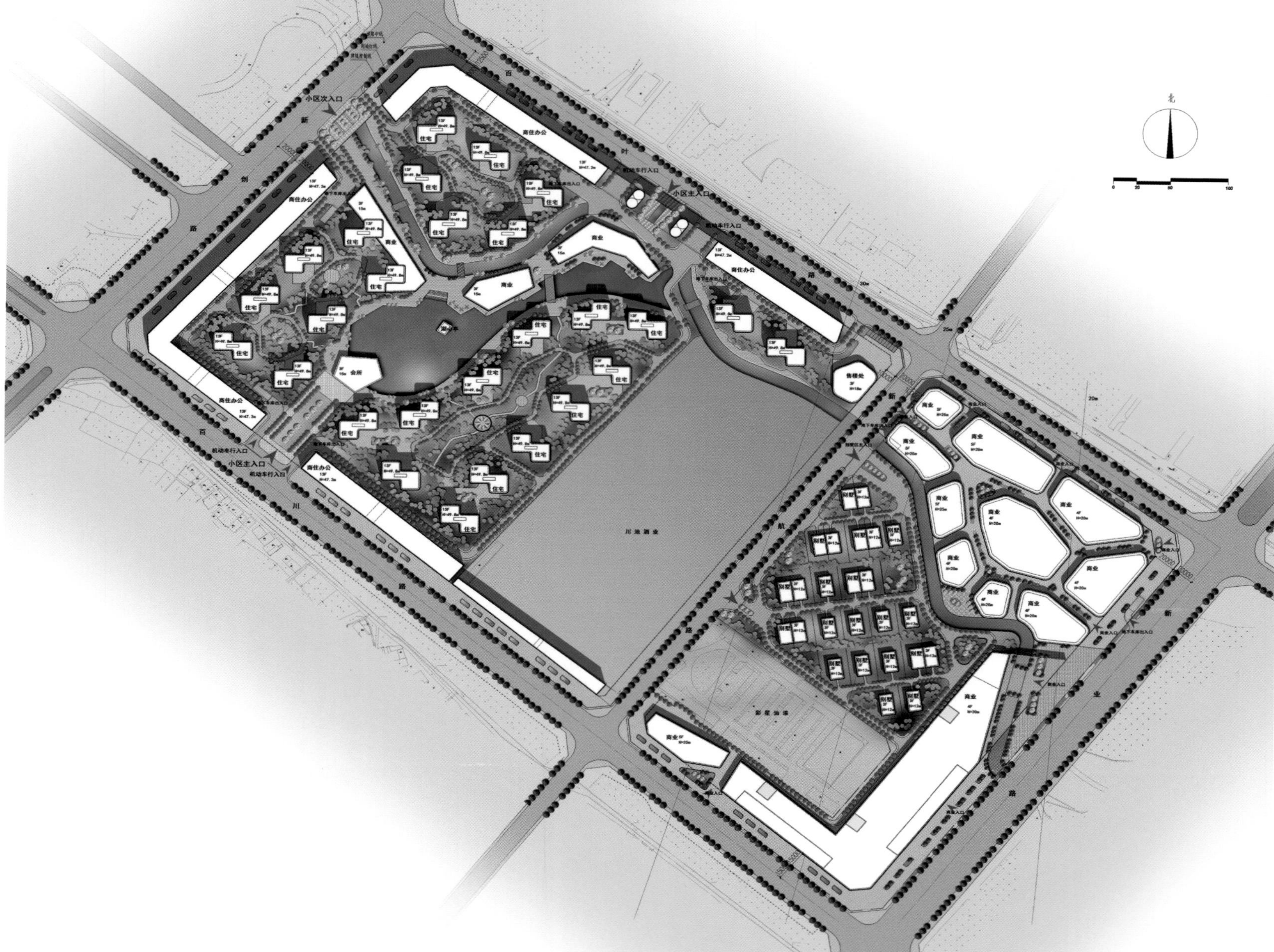

小区次入口
新
叶
路
剑
商住办公
小区主入口
机动车行入口
西
川
会所
商业
住宅
川池酒业
售楼处
北

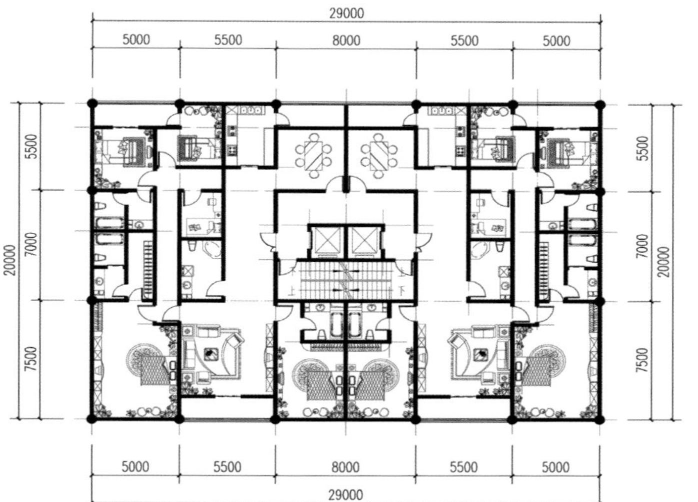

商住办公楼标准层平面图（住宅用途） 1：200

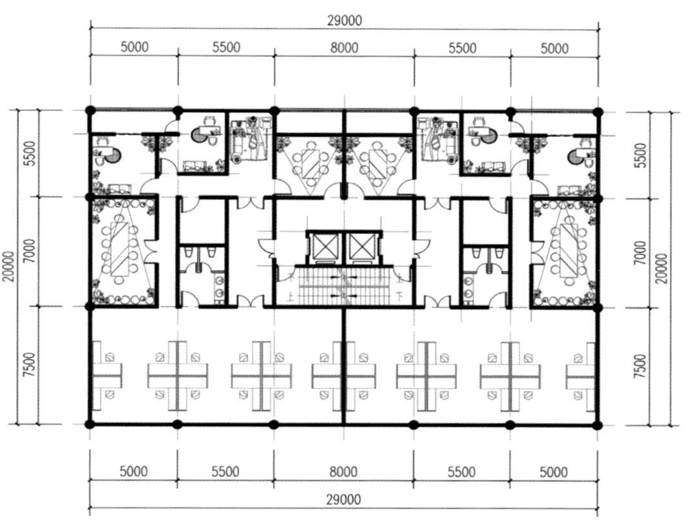

商住办公楼标准层平面图（办公用途） 1：200

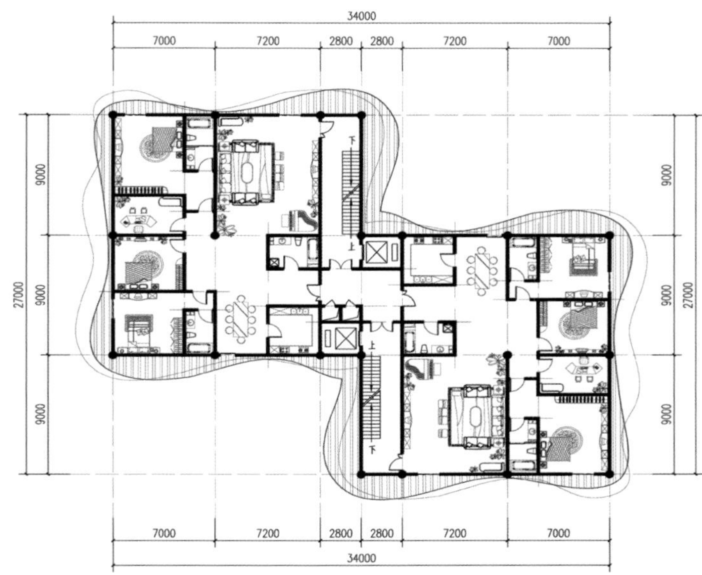

小高层住宅标准层平面图 1：200

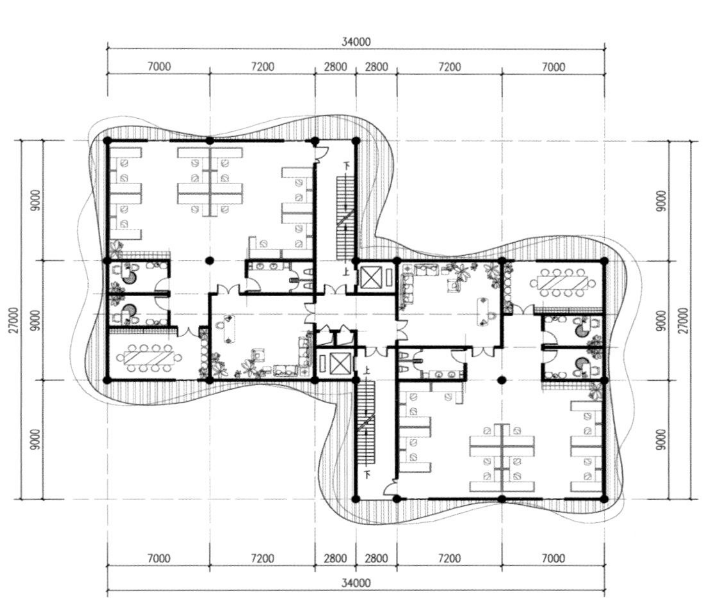

小高层办公楼标准层平面图 1：200

Dior

上海新江湾城办公园区

New Jiangwan Office Park

设计单位：10 DESIGN（拾稼设计）
项目地址：中国上海市
用地面积：160 000 ㎡
建筑面积：400 000 ㎡
设计团队：Barry Shapiro　Dai Yu Cheng　Tania Conforti
Chen Yue Yue　Dawen Qin　Zhu Qi
Scott Findley　Garry Phillips　Nigel Height
Adrian Boot　Philip Gray　Robert Rodriguez（建筑）
Ewa Koter　Fabio Pang（景观）　Shane Dale（多媒体）
Alex Wong　Kylie Chow　Nick Chow
Winky Siu（室内）

Designed by:10 DESIGN
Location: Shanghai, China
Site Area: 160,000 m^2
Floor Area: 400,000 m^2
Design Team: Barry Shapiro, Dai Yu Cheng, Tania Conforti, Chen Yue Yue, Dawen Qin, Zhu Qi, Scott Findley, Garry Phillips, Nigel Height, Adrian Boot, Philip Gray, Robert Rodriguez (Architecture Design); Ewa Koter, Fabio Pang (Landscape Design); Shane Dale (CGI); Alex Wong, Kylie Chow, Nick Chow, Winky Siu (Interior Design)

项目概况

新江湾城办公园区位于上海市杨浦区新江湾城，是一个约 16 公顷的商务园区发展项目，该设计为客户提供了一个符合未来市场发展需要和期望的灵活方案。

建筑设计

项目规划以一系列大型庭院和一座中央社区公园为核心，旨在建造一座大型的办公园区，为在此上班的人们提供一个优良的工作环境。

建筑立面的设计符合成本效益，灵活且具有很强的建筑风格。立面深度和窗户的方向将考虑到利用太阳光的照射等问题，以降低运营成本，创造一个更好的室内工作环境。

该办公园区以干净、现代和精致为标准，营造具有强大吸引力的工作环境，将成为上海新江湾城区一个重要组成部分。

N
0 5 15 25 50M
C7
2
9
C8
13
B7
13
C6
13
C5
13
C4
18
C1
7
5
C2
7
5
C3
7
5
B4
5
7
B3
5
7
国
权

+77.60
59.60
59.60
RF
+55.40
13F
+51.20
12F
+47.00
11F
+42.80
10F
+38.60
9F
+34.40
8F
+30.20
7F
+26.00
6F
+21.80
5F
+17.60
4F
+13.40
3F
+9.20
2F
+5.00
1F
-0.150
±0.000
59900
+34.40
+34.40
+30.20 RF
+26.00 7F
+21.80 6F
+17.60 5F
+13.40 4F
+9.20 3F
+5.00 2F
±0.000 1F
34700

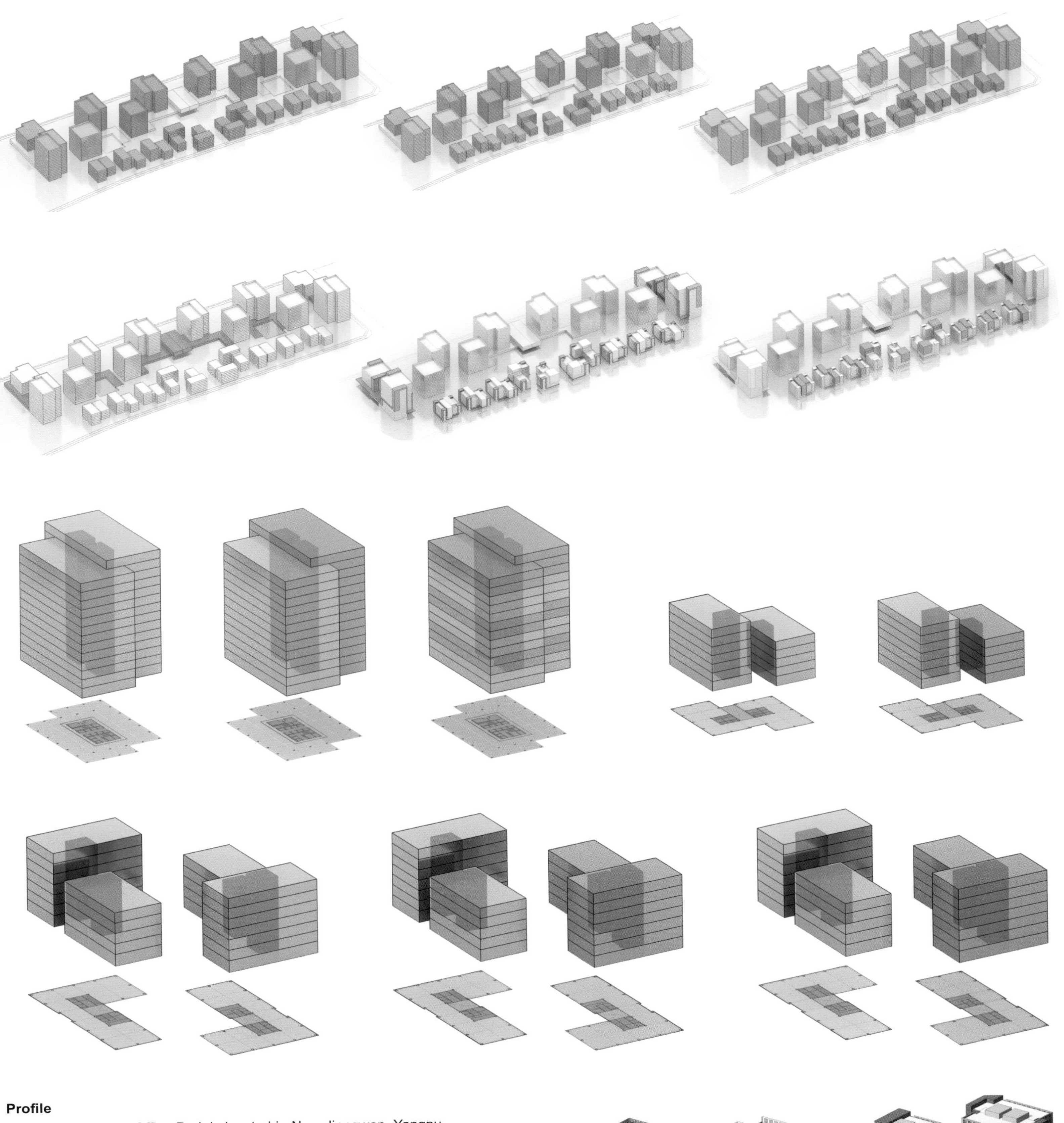

Profile

The New Jiangwan Office Park is located in New Jiangwan, Yangpu District of Shanghai. It is a 16 hectare business park development offering a flexible design that meets the client's future market expectations.

Architectural Design

The project is organized around a series of large courtyards and a central community garden. It aims to build a large office park providing a great environment for people to work in.

The building facades are designed to be cost-effective, flexible and have a strong architectural character. The depth of the facades and the orientation of windows take solar radiation into consideration to help reduce operating costs and create a better working environment within the buildings.

The overall image of the Office Park is meant to be clean, modern and sophisticated to attract future tenants and create a strong working environment. It will be an important addition to the New Jiangwan District of Shanghai.

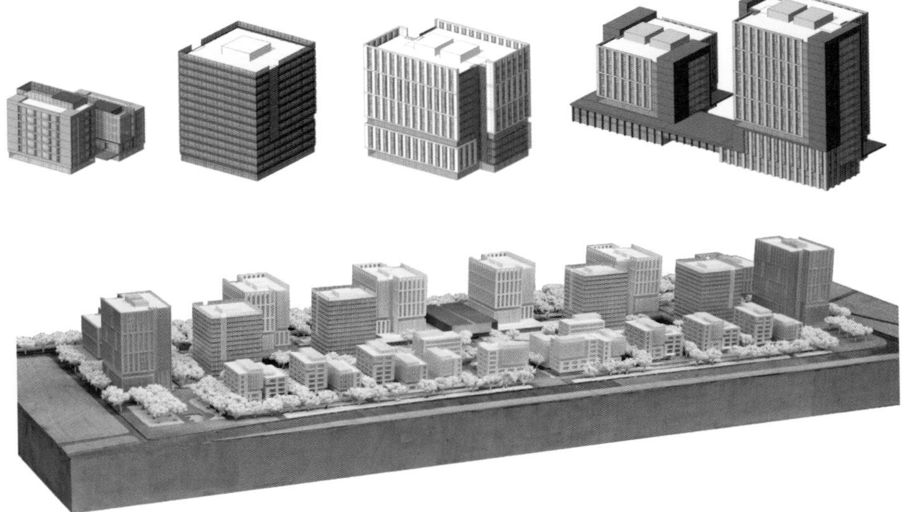

阿联酋迪拜 Pearl

Dubai Pearl

设计单位：Oppenheim Architecture+Design
项目地址：阿联酋迪拜
项目面积：204 386 ㎡

Designed by: Oppenheim Architecture + Design
Location: Dubai, UAE
Area: 204,386 m^2

项目概况

这个近 204 386 平方米的总体规划和建筑设计项目包括了豪华住宅区、酒店、办公楼、零售商场、文化设施和公共空间等功能区，是一个高性能的可持续性发展项目。

建筑设计

项目设计新颖独特，堪称世界图标，其展现的视觉效果将使之成为迪拜一个独特的、具有标志性意义的象征。项目包括裙楼的重建以及位于高层建筑之间的中央商场的设计。设计将裙楼设想为一个有机的城市，而不是一个纯粹的人造环境。

项目包括米高美大酒店、住宅、空中阁楼、大型的零售商场和餐厅、配备有多功能宴会厅的会议中心、有 10 个放映厅的电影院、媒体创作和后期制作工作室、Pearl 健康中心和音乐 / 舞蹈 / 电影创作学院、幼儿园，以及 Pearl 歌剧院和音乐厅——一个有三个艺术表演中心、能容纳 1 800 个席位的艺术空间。

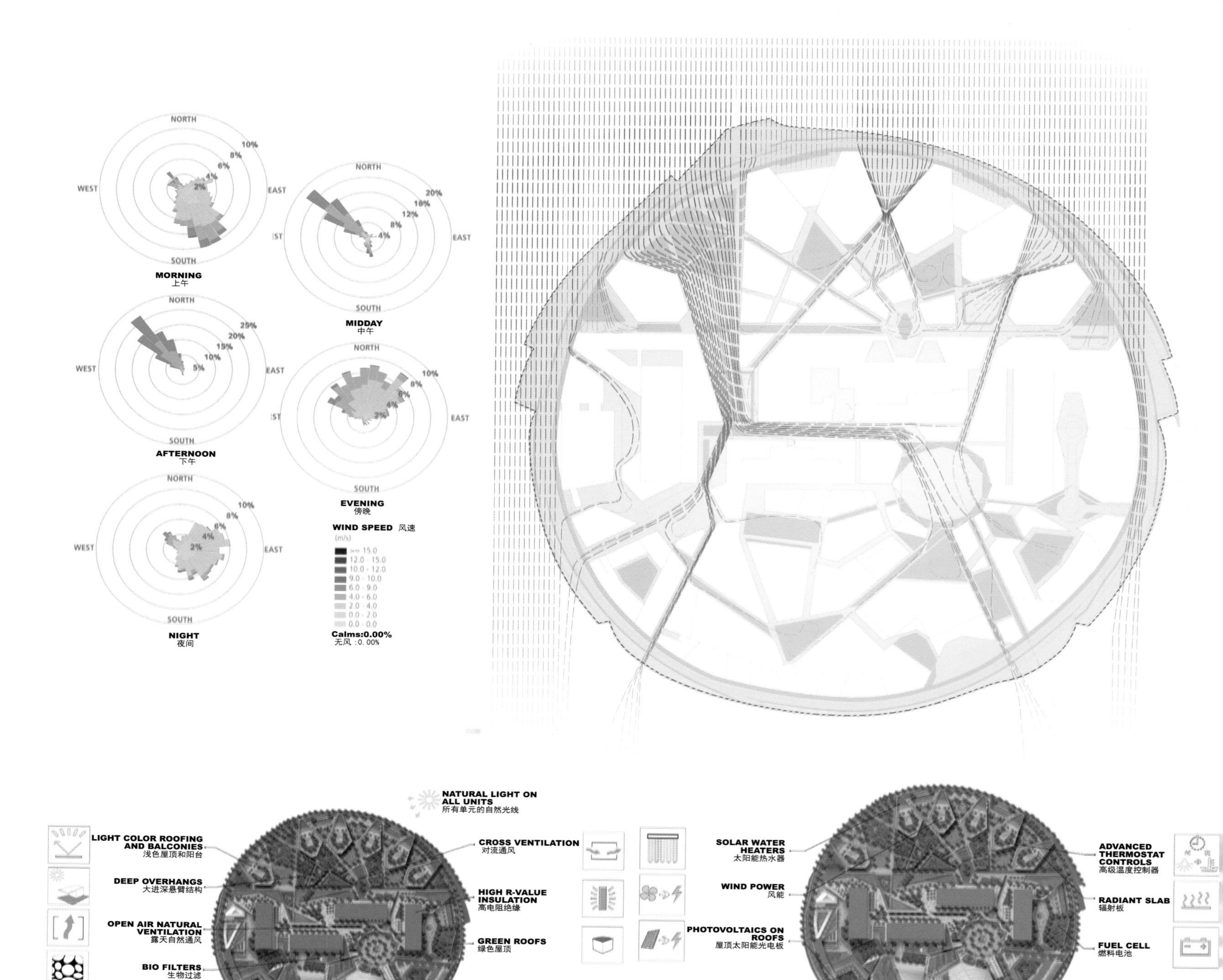

Profile

Master planning and architecture for a 204,386 square meters mixed-use development comprised of luxury residential area, hotels, office, retail, cultural facilities and public spaces is a high-performance sustainable project.

Architectural Design

With a creative and innovative design befitting a worldwide icon and bestowing a visual effect, the project becomes a unique, landmark symbol for Dubai. The project involved the re-design of the Podium Buildings and the Central Mall area between the high-rise towers. The podium development was designed as an organic city, rather than a contrived environment.

The project includes an MGM Grand Hotel and residences, Sky Lofts, large collection of retail and restaurants, conference center with multipurpose ballrooms, 10-theater cinema, media creation and post production studio, the Pearl Health Clinic & Creative Society Academy for Music/Dance/Film, child nursery and kindergarten, and the Pearl Opera & Music House – a 3-venue performing arts center totaling 1,800 seats.

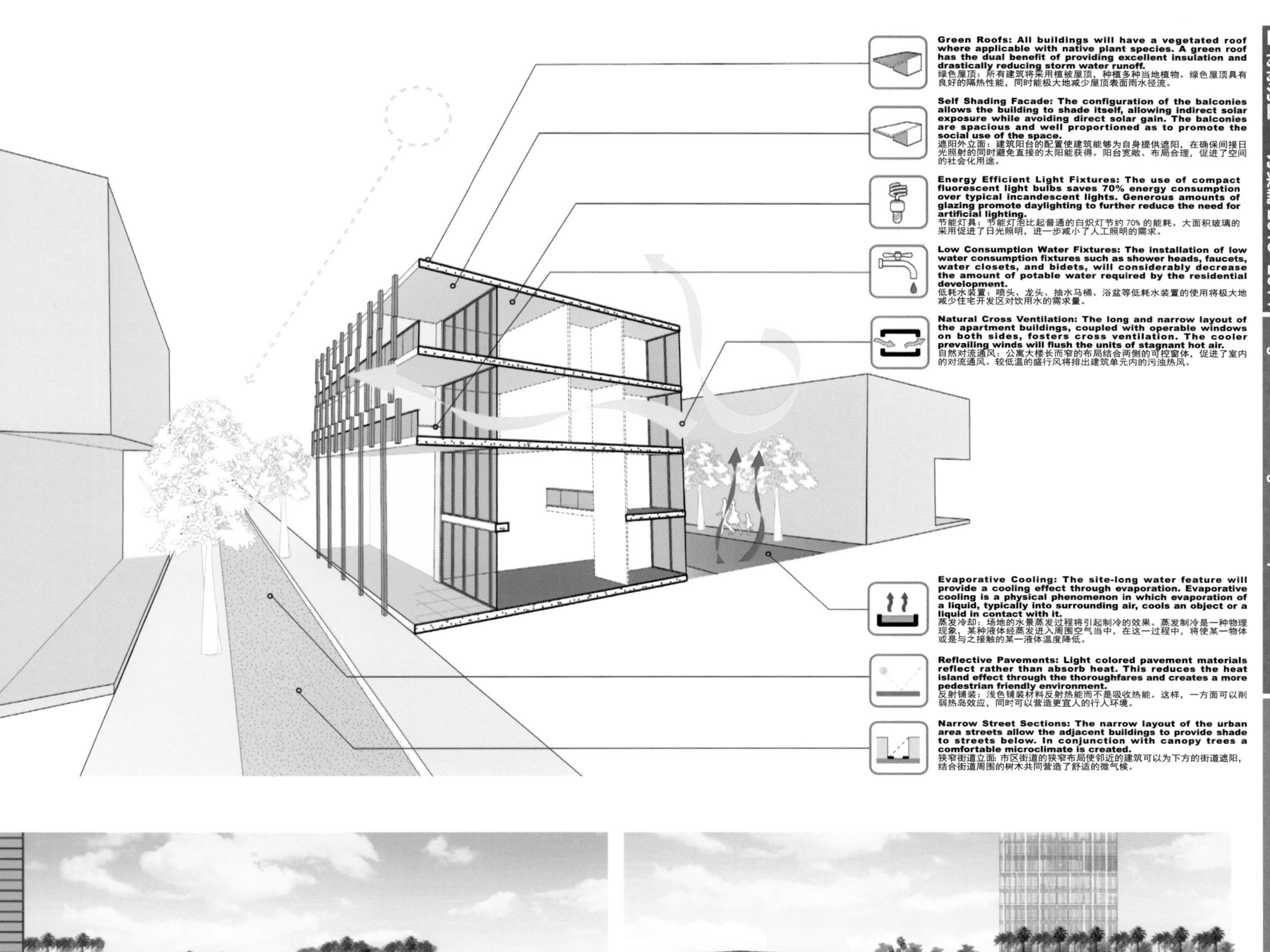
Green Roofs: All buildings will have a vegetated roof where applicable with native plant species. A green roof has the dual benefit of providing excellent insulation and drastically reducing storm water runoff.
绿色屋顶：所有建筑将采用植被屋顶，种植多种当地植物。绿色屋顶具有良好的隔热性能，同时能极大地减少屋顶表面雨水径流。
Self Shading Facade: The configuration of the balconies allows the building to shade itself, allowing indirect solar exposure while avoiding direct solar gain. The balconies are spacious and well proportioned as to promote the social use of the space.
遮阳外立面：建筑阳台的配置使建筑能够为自身提供遮阳，在确保间接日光照射的同时避免直接的太阳能获得。阳台宽敞、布局合理，促进了空间的社会化用途。
Energy Efficient Light Fixtures: The use of compact fluorescent light bulbs saves 70% energy consumption over typical incandescent lights. Generous amounts of glazing promote daylighting to further reduce the need for artificial lighting.
节能灯具：节能灯泡比起普通的白炽灯节约 70% 的能耗。大面积玻璃的采用促进了日光照明，进一步减小了人工照明的需求。
Low Consumption Water Fixtures: The installation of low water consumption fixtures such as shower heads, faucets, water closets, and bidets, will considerably decrease the amount of potable water required by the residential development.
低耗水装置：喷头、龙头、抽水马桶、浴盆等低耗水装置的使用将极大地减少住宅开发区对饮用水的需求量。
Natural Cross Ventilation: The long and narrow layout of the apartment buildings, coupled with operable windows on both sides, fosters cross ventilation. The cooler prevailing winds will flush the units of stagnant hot air.
自然对流通风：公寓大楼长而窄的布局结合两侧的可控窗体，促进了室内的对流通风。较低温的盛行风将排出建筑单元内的污浊热风。
Evaporative Cooling: The site-long water feature will provide a cooling effect through evaporation. Evaporative cooling is a physical phenomenon in which evaporation of a liquid, typically into surrounding air, cools an object or a liquid in contact with it.
蒸发冷却：场地的水景蒸发过程将引起制冷的效果。蒸发制冷是一种物理现象，某种液体经蒸发进入周围空气当中，在这一过程中，将使某一物体或是与之接触的某一液体温度降低。
Reflective Pavements: Light colored pavement materials reflect rather than absorb heat. This reduces the heat island effect through the thoroughfares and creates a more pedestrian friendly environment.
反射铺装：浅色铺装材料反射热能而不是吸收热能。这样，一方面可以削弱热岛效应，同时可以营造更宜人的行人环境。
Narrow Street Sections: The narrow layout of the urban area streets allow the adjacent buildings to provide shade to streets below. In conjunction with canopy trees a comfortable microclimate is created.
狭窄街道立面: 市区街道的狭窄布局使邻近的建筑可以为下方的街道遮阳，结合街道周围的树木共同营造了舒适的微气候。

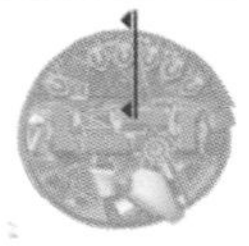

俄罗斯莫斯科 Kuntsevo 中心

Kuntsevo Plaza

设计单位：捷得国际建筑师事务所
开发商：ENKA TC
项目地址：俄罗斯莫斯科
建筑面积：235 000 ㎡

Designed by: The Jerde Partnership
Client: ENKA TC
Location: Moscow, Russia
Gross Floor Area: 235,000 m²

项目概况

Kuntsevo 中心是捷得国际建筑师事务所在莫斯科设计的一个新的居住、生活、工作和购物休闲空间，一个根植于艺术、天然和城市连接性的现代化社区。

建筑设计

Kuntsevo 中心包括 7 万平方米的零售和娱乐空间、3 座 A 级办公楼和 2 座高层住宅楼。这个以步行为主导的项目将设置一系列休闲、购物和商业住宅等设施，并将历史性区域连接起来，为城市提供一座新的地标和一个城市居住空间。

在该项目中，主要的设计元素来自于自然，这给人们提供了独特的参观体验。天然的味道和随着四季而不断变化的景观吸引游客前往参观，区域内自然流通的步行街景、广场和室内外的绿色露台，给予人们探索和发现的乐趣。

项目也是一个高标准的可持续性建筑项目。设计十分重视当地材料的使用以及对自然光的利用。楼层顶部植被葱郁的平台，也为未来可持续性的开发起到了催化剂的作用。

Profile

Kuntsevo Plaza is a new living, work, and retail leisure space in Moscow designed by Jerde. It is a modern community gathering destination rooted in art, nature and urban connectivity.

Architectural Design

Kuntsevo Plaza consists of 70,000 square meters of retail and entertainment spaces, plus three Classes-A office buildings, and two high-rise residential towers. The pedestrian-oriented center will establish a vibrant leisure, shopping, business, and residential complex reconnecting the urban fabric of the historic Kuntsevo District, while creating a new landmark and residential space for the city.

In this project, main design elements come from the nature. It serves as the fundamental element to host unique visitor experiences. The notion of nature and the celebration of the four seasons draw visitors into the complex. Once inside, the natural flow of the pedestrian streetscapes, plazas, courtyards and the spectacular indoor/outdoor green parks will create a sense of discovery and public exploration.

The project sets out to achieve the highest standard of sustainability. The design has attached great importance to the use of local materials and natural light. Green rooftops will contribute to promote future sustainable development.

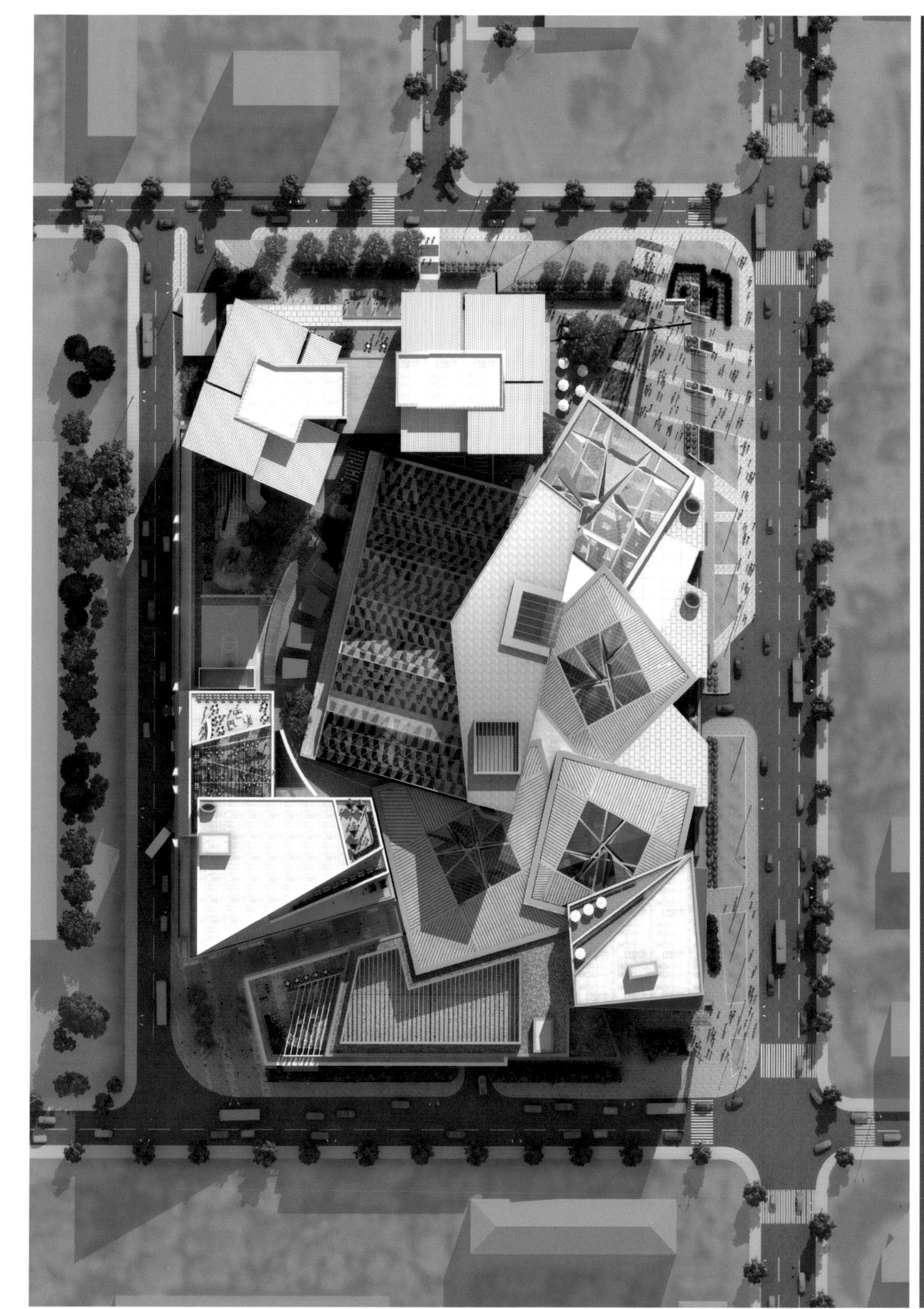

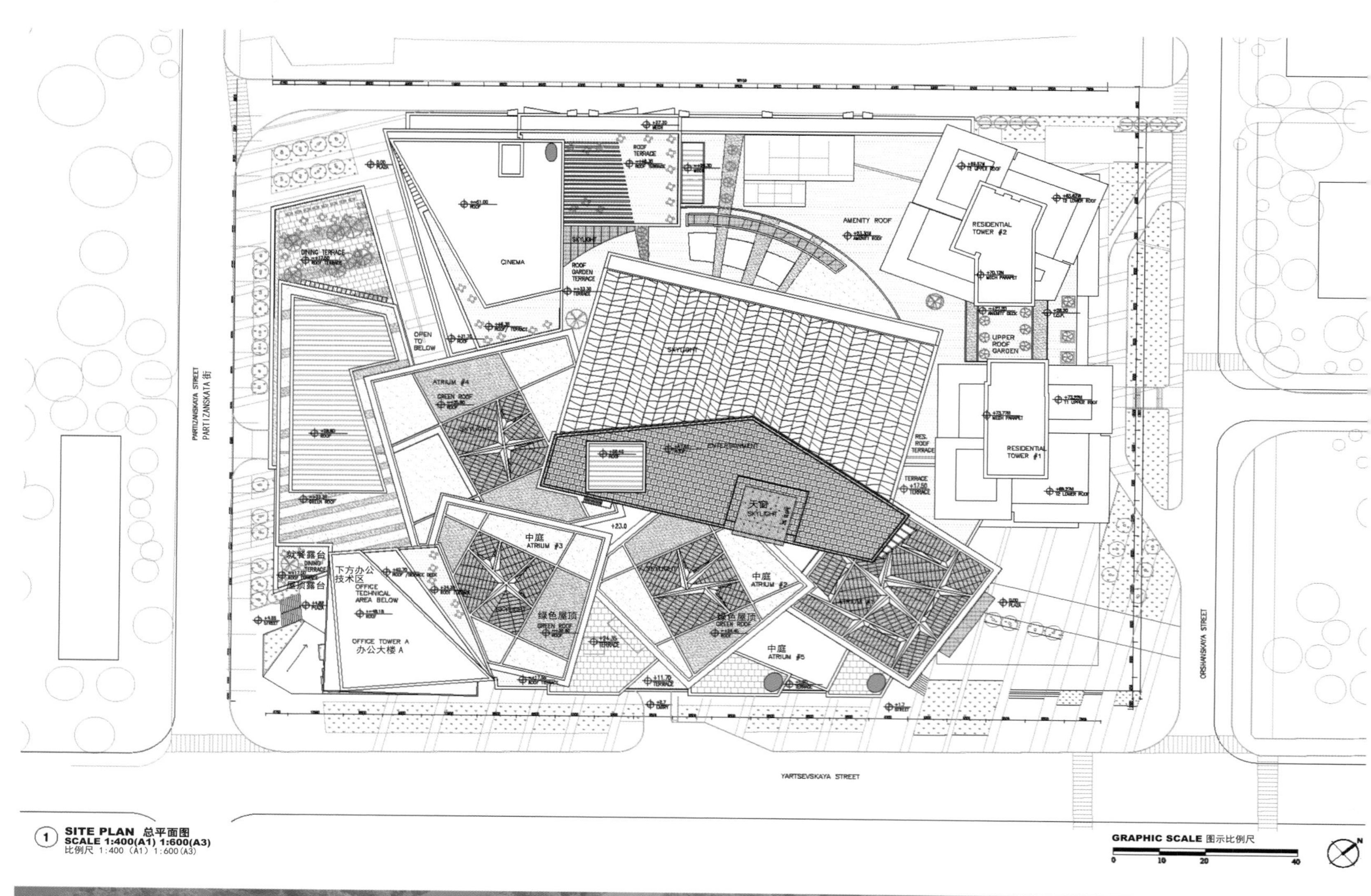

① **SITE PLAN 总平面图**
SCALE 1:400(A1) 1:600(A3)
比例尺 1:400（A1）1:600（A3）

COSMETICS
Ашан

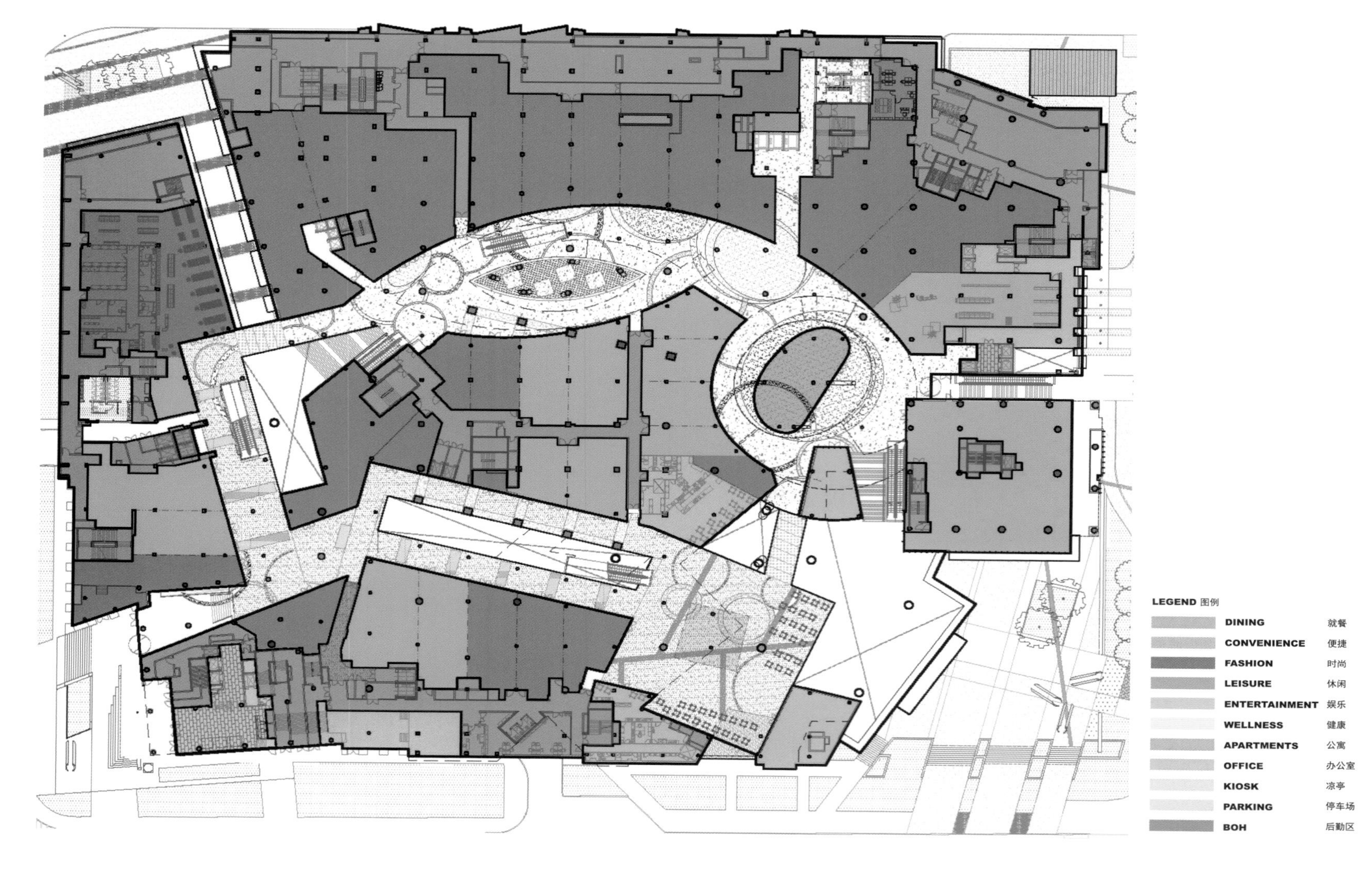
LEGEND 图例
DINING 就餐
CONVENIENCE 便捷
FASHION 时尚
LEISURE 休闲
ENTERTAINMENT 娱乐
WELLNESS 健康
APARTMENTS 公寓
OFFICE 办公室
KIOSK 凉亭
PARKING 停车场
BOH 后勤区

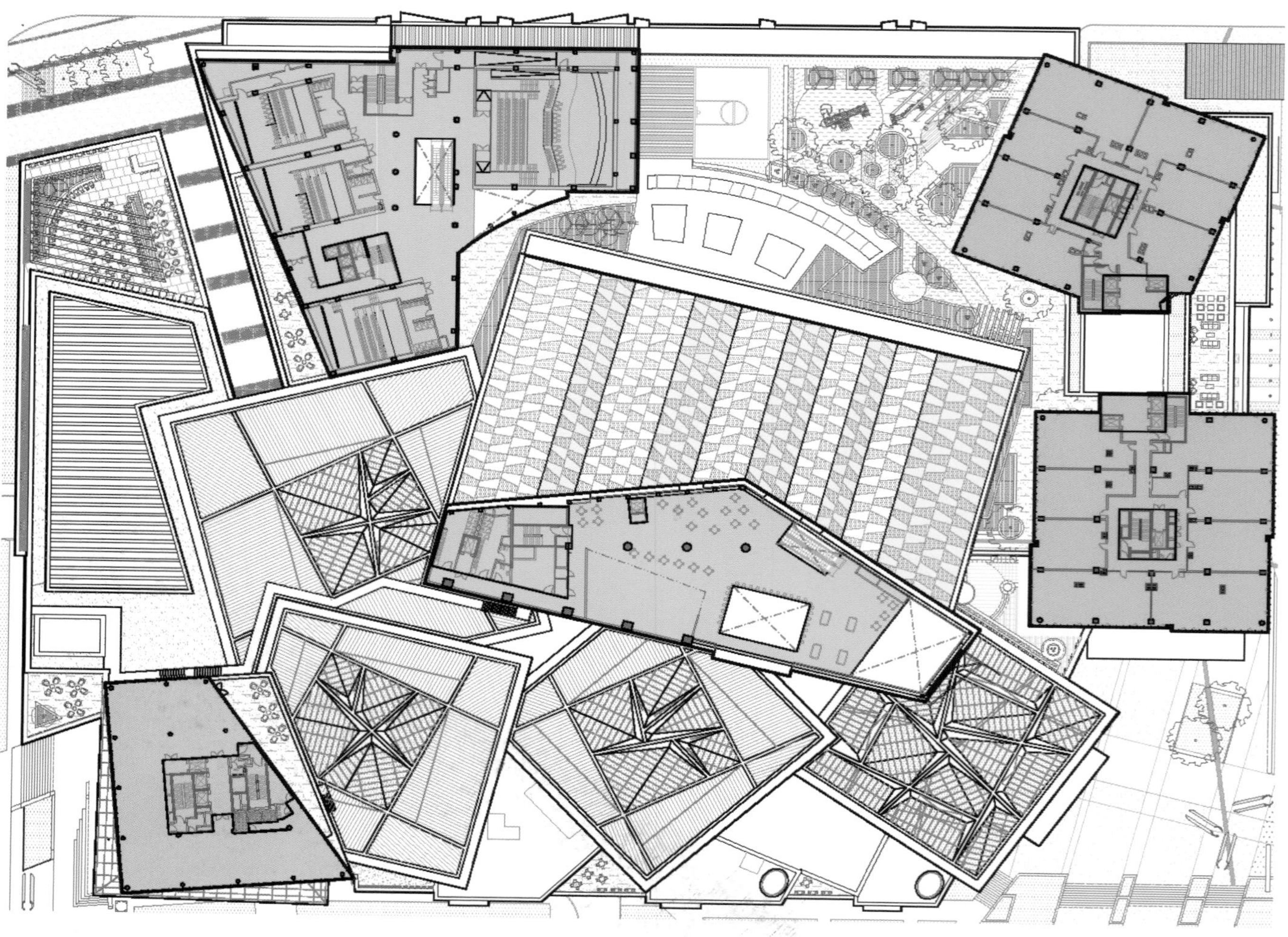

0 10 20 40

SCALE:800
比例：800
0 10 20 40

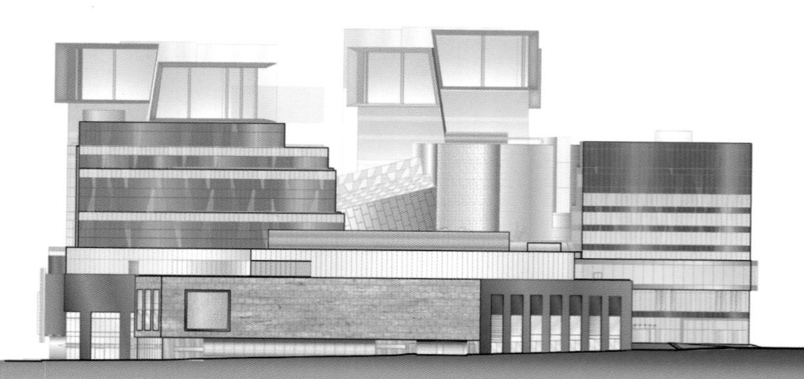

SCALE:800
比例：800
0 10 20 40

SCALE:800
比例：800
0 10 20 40

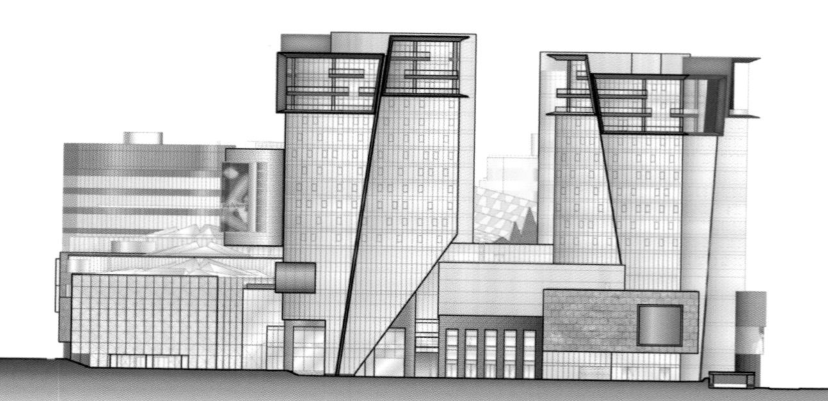

SCALE:800
比例：800
0 10 20 40

BA

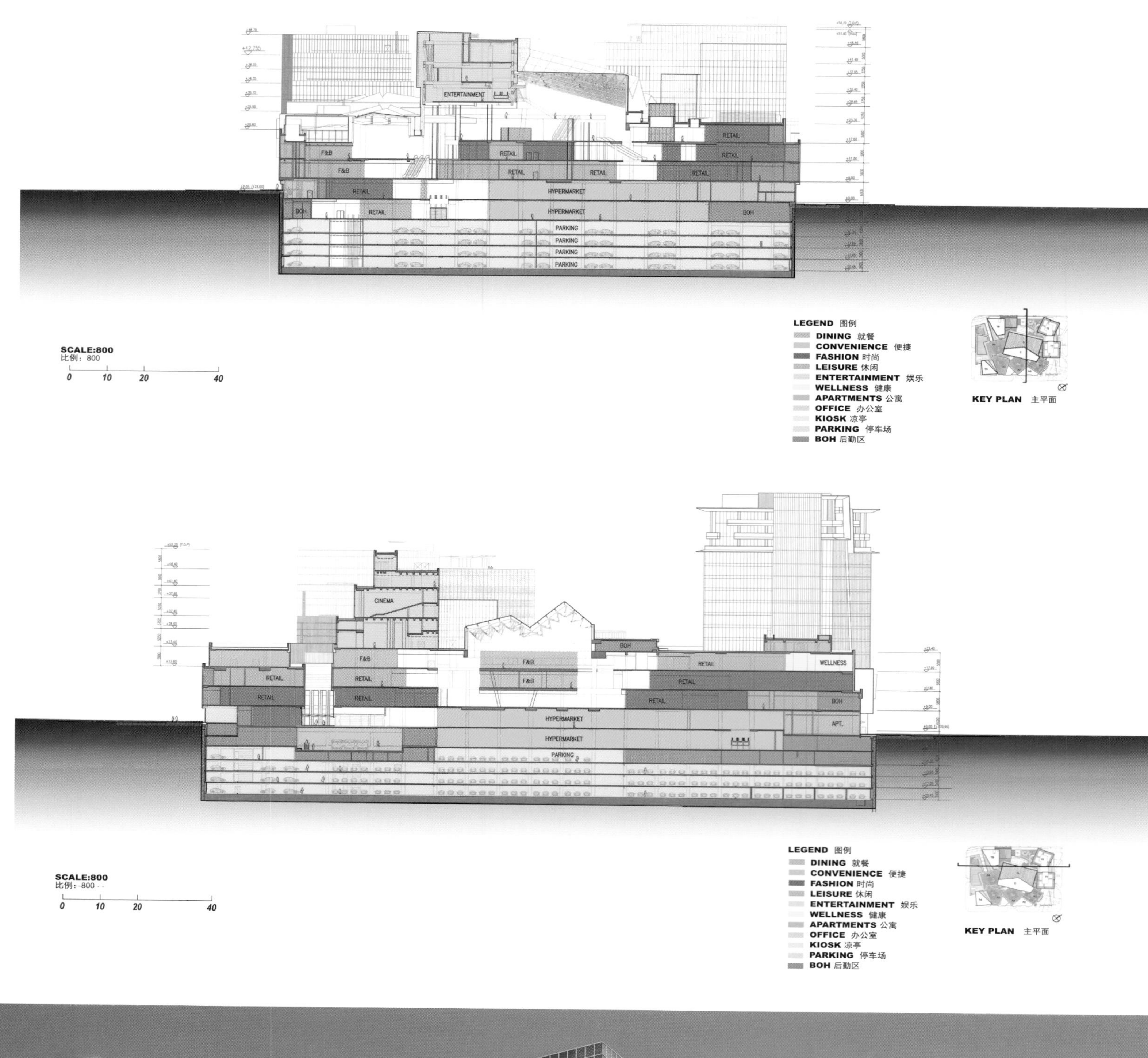
ENTERTAINMENT
F&B
RETAIL
HYPERMARKET
BOH
PARKING
SCALE:800
比例：800
0
10
20
40
LEGEND 图例
DINING 就餐
CONVENIENCE 便捷
FASHION 时尚
LEISURE 休闲
ENTERTAINMENT 娱乐
WELLNESS 健康
APARTMENTS 公寓
OFFICE 办公室
KIOSK 凉亭
PARKING 停车场
BOH 后勤区
KEY PLAN 主平面
CINEMA
F&B
RETAIL
WELLNESS
BOH
APT.
HYPERMARKET
PARKING
SCALE:800
比例：800
0
10
20
40
LEGEND 图例
DINING 就餐
CONVENIENCE 便捷
FASHION 时尚
LEISURE 休闲
ENTERTAINMENT 娱乐
WELLNESS 健康
APARTMENTS 公寓
OFFICE 办公室
KIOSK 凉亭
PARKING 停车场
BOH 后勤区
KEY PLAN 主平面

TEDDY BEAR

山东青岛南京路多功能混合使用项目

Shandong Qingdao Nanjing Road Mixed Use Area

设计单位：RTA-Office 建筑事务所
合作单位：Tengyuan Design Institute Co.
开发商： 中联建业股份有限公司
项目地址：中国山东省青岛市
场地面积：26 810 ㎡
总建筑面积：210 700 ㎡

Designed by: RTA-Office
Collaboration: Tengyuan Design Institute Co.
Client: Zhonglian Jianye Co., Ltd.
Location: Qingdao, Shandong, China
Site Area: 26,810 m^2
Total Construction Area: 210,700 m^2

项目概况

这个混合用途的项目位于青岛市一个紧凑的地块上，包括办公楼、住宅以及涵盖了酒店、购物中心、餐厅和体育中心在内的娱乐休闲设施，设计旨在构建一个新的现代的城市空间。

设计理念

青岛因青岛啤酒而闻名，这一品牌由德国人在这个城市建立，他们生产的啤酒闻名中国并远销世界。设计受这一品牌的启发，其形态和颜色均源自“青岛啤酒”这一极具地方特色的品牌，以构建一个如“青岛啤酒”般极富盛名的地标性建筑。

建筑设计

设计对当地环境和城市特征进行了深入了解和分析，以构建一个具有城市场所感和独特身份象征的项目。设计将建筑体量以分散式的形态布局，不仅获得了一个公共开放空间，同时也营造了一个亲切、宜人的环境。

设计十分注重建筑体量的形态和结构，设计师将建筑精心地布置在场地上，使之与周边环境和谐共存。建筑外立面颜色的选择，则赋予了建筑优雅而又温和的气质。

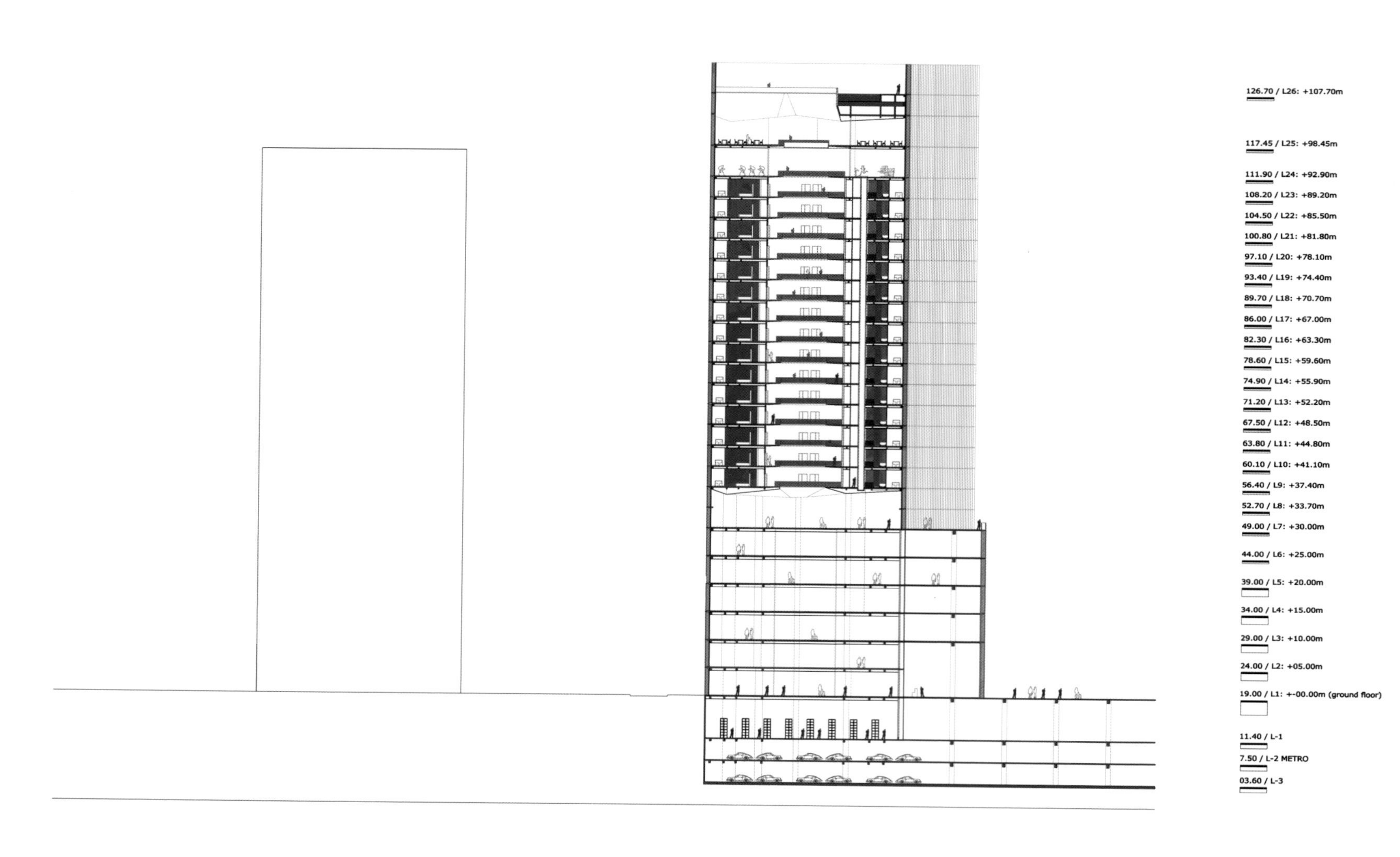

L1
GROUND FLOOR
底层

L2
SECOND FLOOR
二层

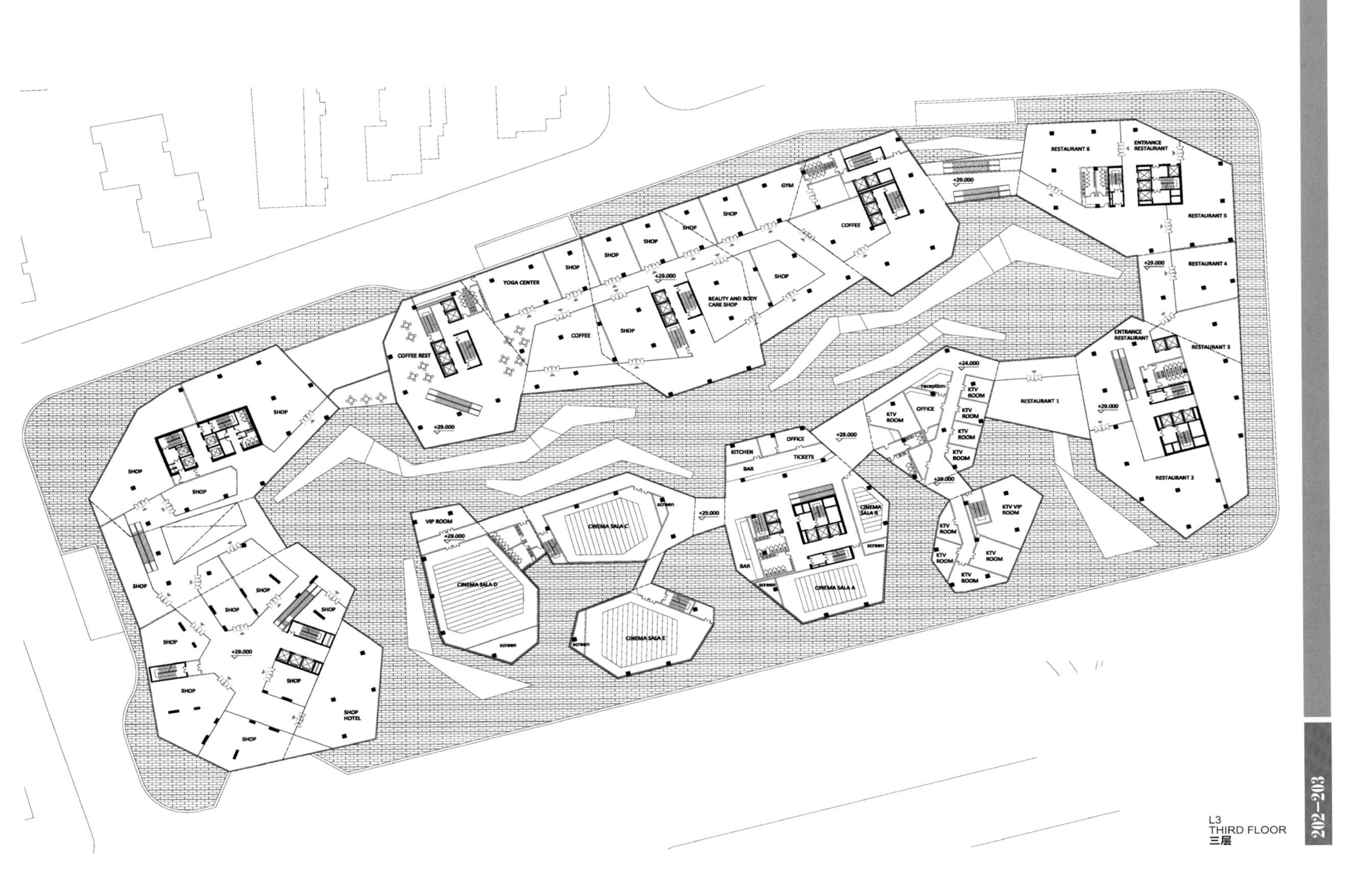

L3
THIRD FLOOR
三层

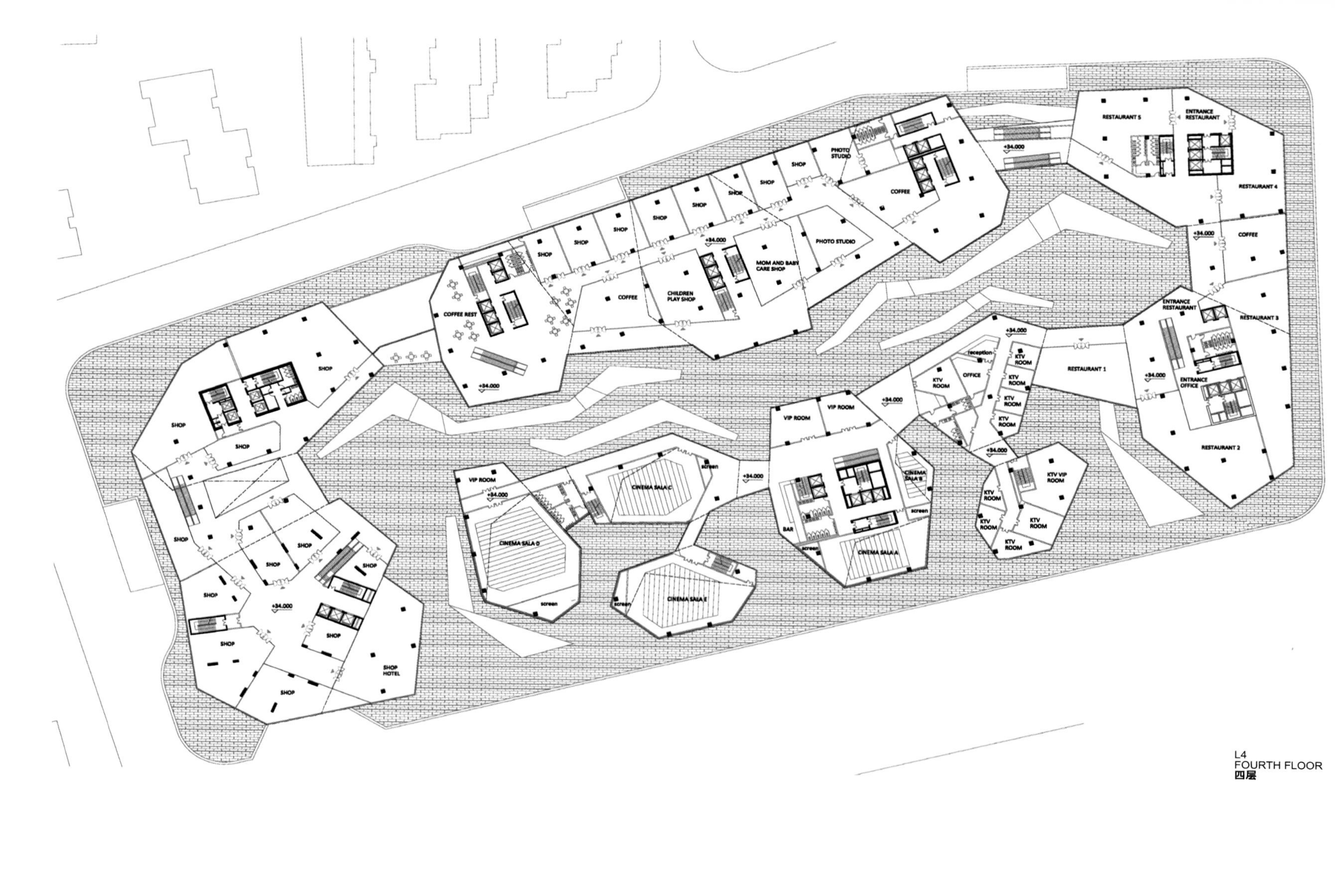

L4
FOURTH FLOOR
四层

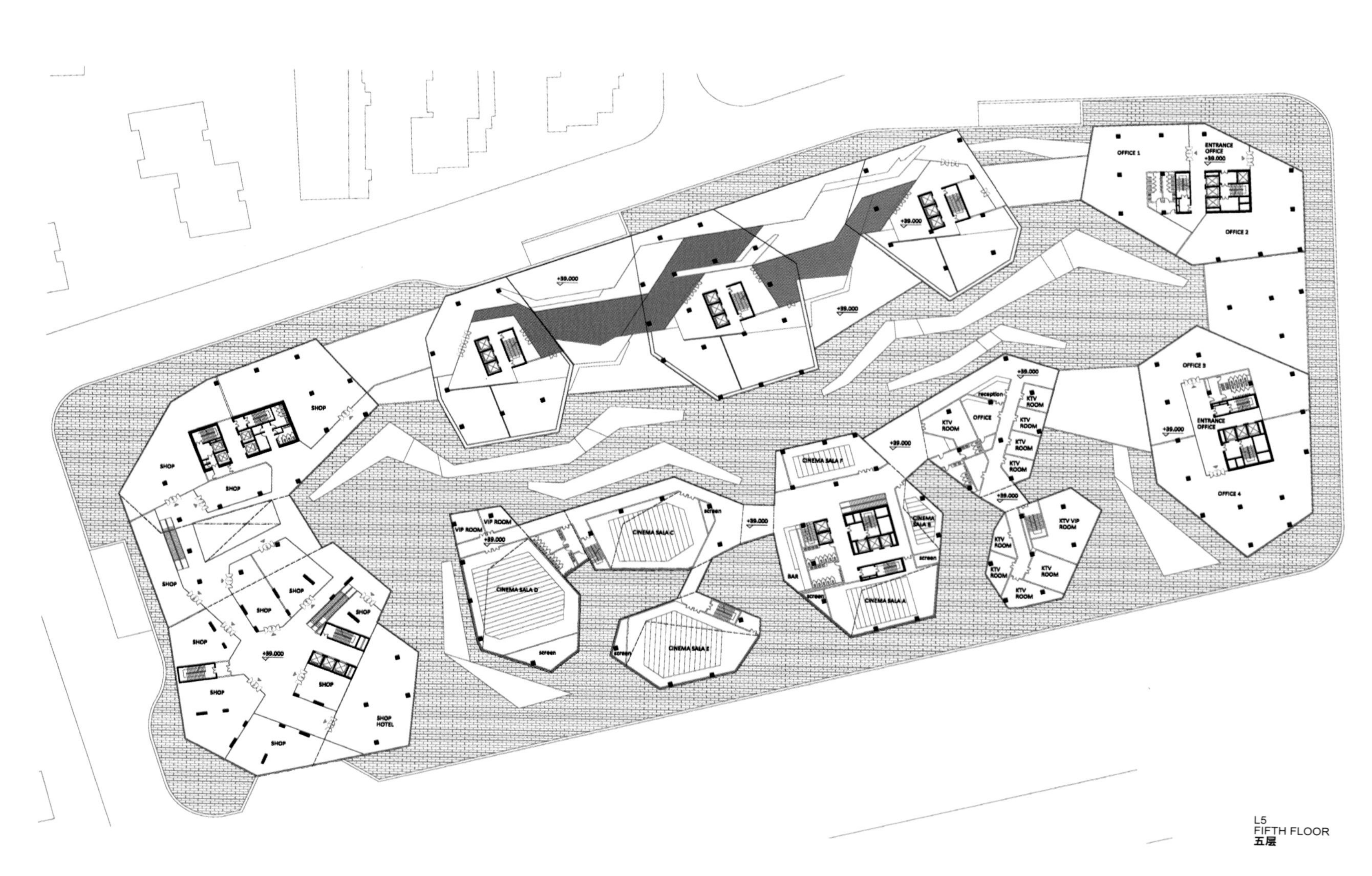

L5
FIFTH FLOOR
五层

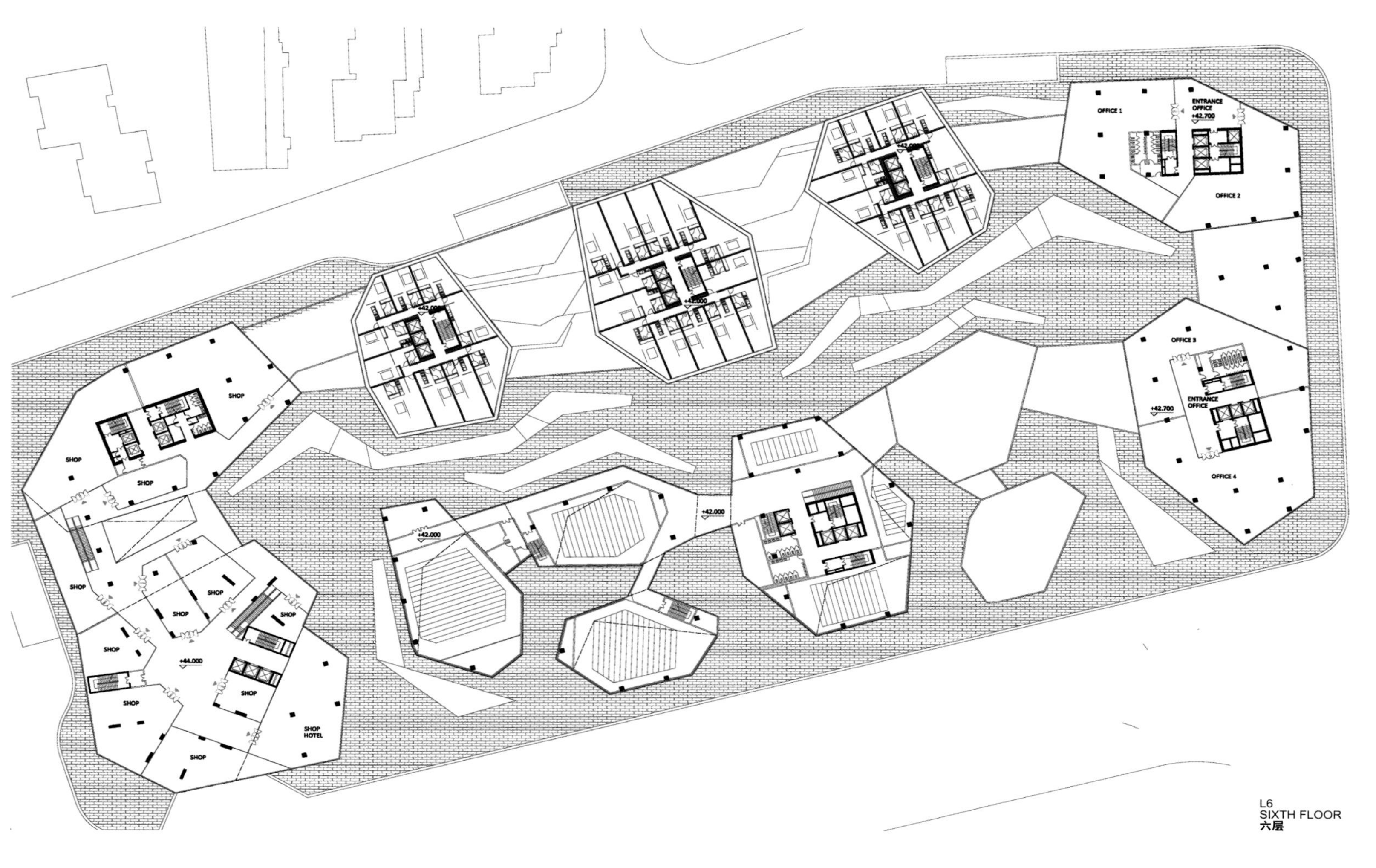

L6
SIXTH FLOOR
六层

L8
EIGHTH FLOOR
八层

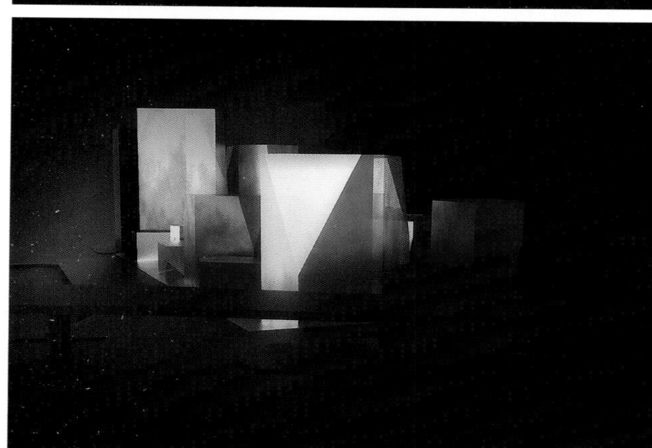

120.40 / L27: +101.40m
116.70 / L26: +97.70m
113.00 / L25: +94.00m
109.30 / L24: +90.30m
105.60 / L23: +86.60m
101.90 / L22: +82.90m
98.20 / L21: +79.20m
94.50 / L20: +75.50m
90.80 / L19: +71.80m
87.10 / L18: +68.10m
83.40 / L17: +64.40m
79.70 / L16: +60.70m
76.00 / L15: +57.00m
72.30 / L14: +53.30m
68.60 / L13: +49.60m
64.90 / L12: +45.90m
61.20 / L11: +42.20m
57.50 / L10: +38.50m
54.40 / L9: +34.80m
50.10 / L8: +31.10m
46.40 / L7: +27.40m
42.70 / L6: +23.70m
39.00 / L5: +20.00m
34.00 / L4: +15.00m
29.00 / L3: +10.00m
24.00 / L2: +05.00m
19.00 / L1: +-00.00m (ground floor)
11.40 / L-1
7.50 / L-2 METRO
03.80 / L-3

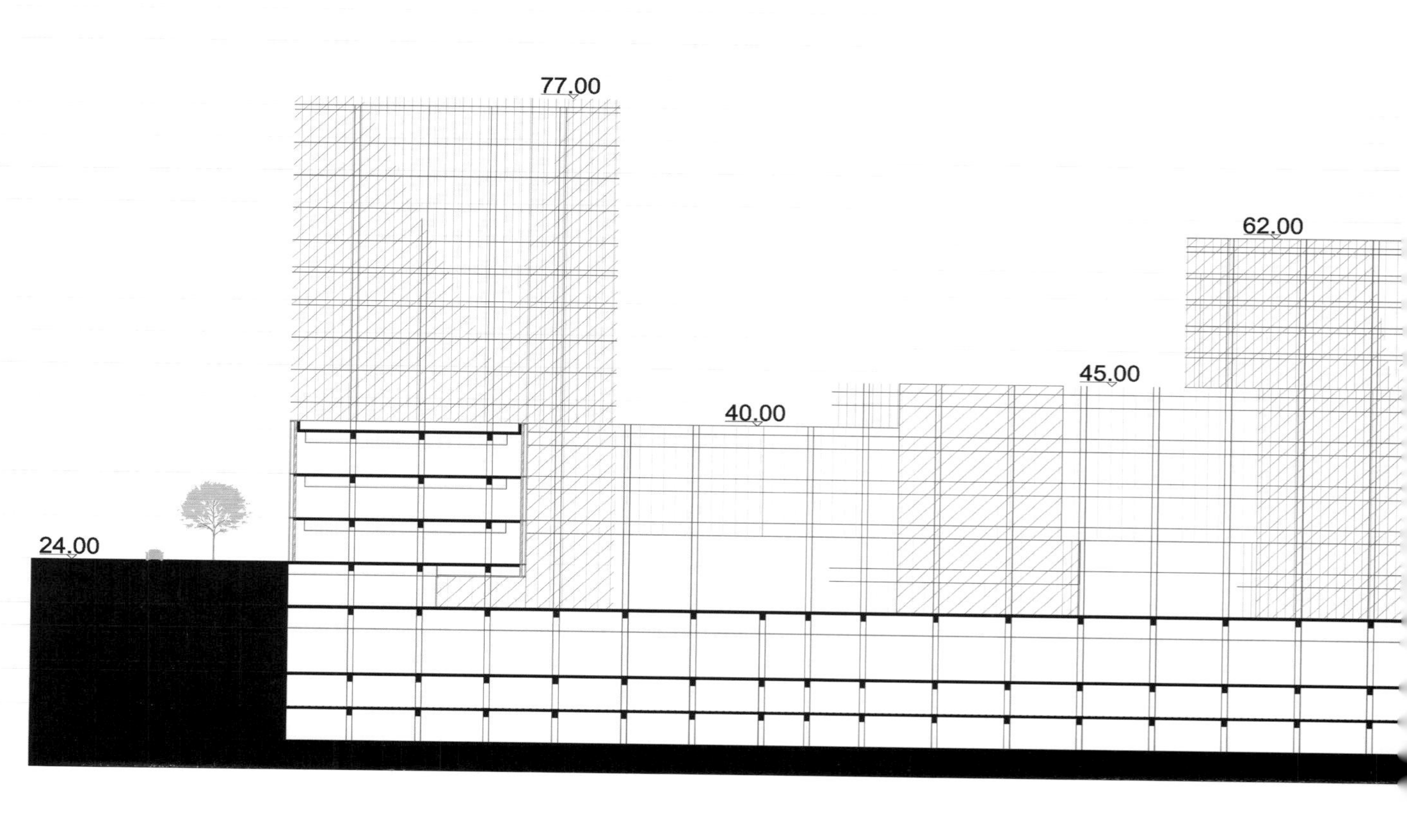
77.00
62.00
45.00
40.00
24.00

Profile

This mixed use project is located on a compact site of Qingdao City, comprising offices, residences, a hotel, a shopping centre, restaurants and a sports centre. It tries to construct a new modern urban space.

Design Concept

The city of Qingdao is also home to the Tsingtao brewery, one of China's largest lager manufacturers and the producer of the popular Tsingtao beer. RTA-Office chose to make reference to this by incorporating the shape and color of a glass of beer within each structure.

Architectural Design

Designers have thoroughly studied the environment and character of the city to propose a project with sense of place and a unique character. The scattered layout of building volumes creates an open public space as well as pleasant delightful atmosphere.

The design pays special attention to the form and configuration of the volumes. Designers set them very carefully and in a harmonious way. On the selection of the color of building facades, the result gives a touch of elegance and warmth to the complex.

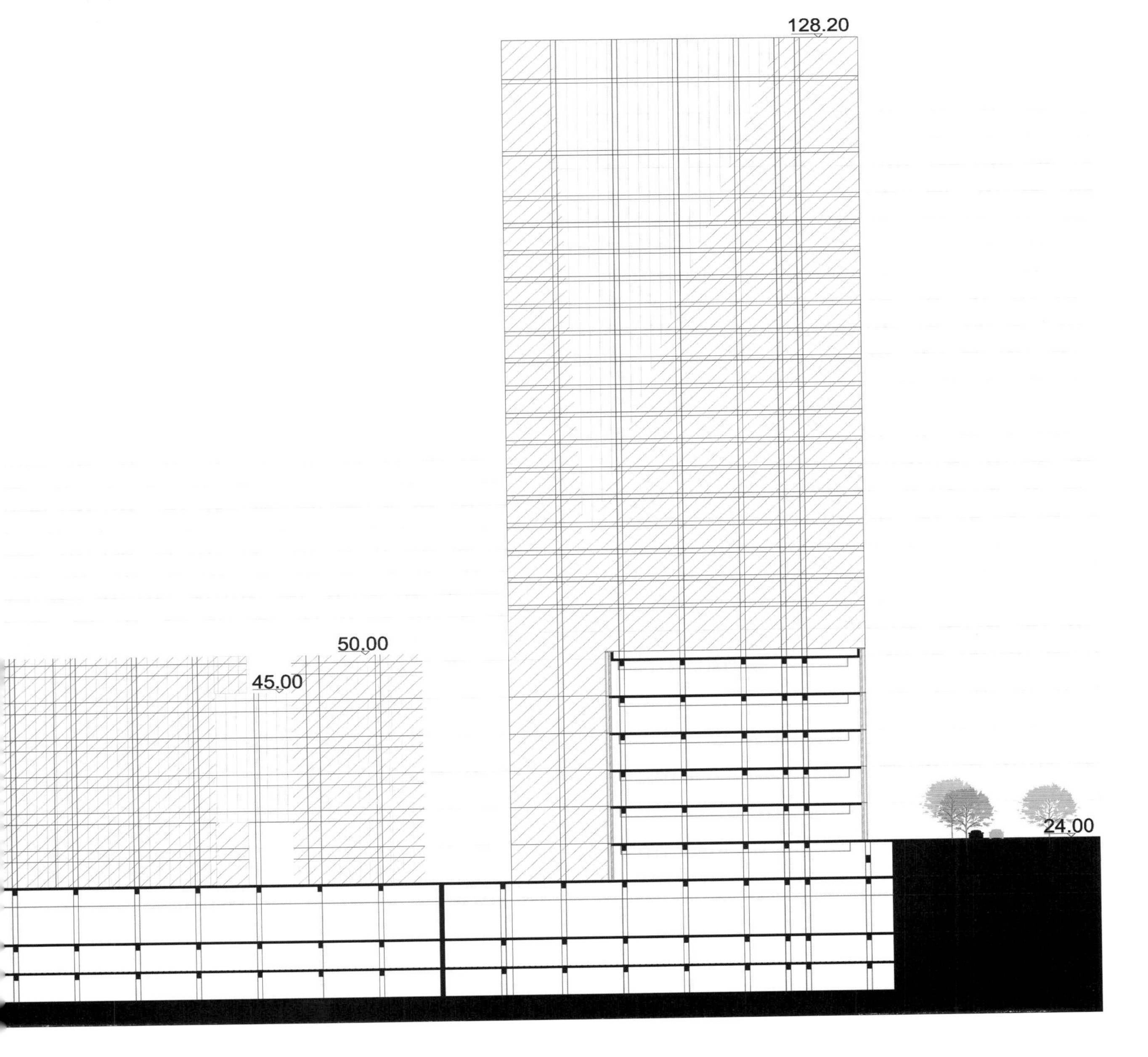

123.00 / L27: +104.00m
119.30 / L26: +100.30m
115.60 / L25: +96.60m
111.90 / L24: +92.90m
108.20 / L23: +89.20m
104.50 / L22: +85.50m
100.80 / L21: +81.80m
97.10 / L20: +78.10m
93.40 / L19: +74.40m
89.70 / L18: +70.70m
86.00 / L17: +67.00m
82.30 / L16: +63.30m
78.60 / L15: +59.60m
74.90 / L14: +55.90m
71.20 / L13: +52.20m
67.50 / L12: +48.50m
63.80 / L11: +44.80m
60.10 / L10: +41.10m
56.40 / L9: +37.40m
52.70 / L8: +33.70m
49.00 / L7: +30.00m
44.00 / L6: +25.00m
39.00 / L5: +20.00m
34.00 / L4: +15.00m
29.00 / L3: +10.00m
24.00 / L2: +05.00m
19.00 / L1: +-00.00m (ground floor)
11.40 / L-1
7.50 / L-2 METRO
03.60 / L-3

湖南长沙北辰新河三角洲

Northstar Xinhe Delta

设计单位：捷得国际建筑师事务所

项目地址：中国湖南省长沙市

总建筑面积：520 000 ㎡

Designed by: The Jerde Partnership

Location: Changsha, Hunan, China

Gross Floor Area: 520,000 m^2

项目概况

北辰新河三角洲临近湘江，毗邻当地的文化区域，设计以人文活动和社区互动为特色，构建了一个极具吸引力的现代滨水生活空间。

设计理念

北辰新河三角洲以水、花园和城市为主题，其零售娱乐和多用途发展的创新性理念打破了传统住宅和零售空间的束缚，为参观、购物生活、工作、娱乐提供了一个动态的场所。

建筑设计

设计以行人导向为基础，旨在激发人们探索的精神和加强人与人之间的社会联系。不管是沿着木质人行道漫步于林立的、可一览湘江风光的、独特的商店和餐馆之间，还是走在令人眼花缭乱的娱乐休闲区，整个滨水区将是一个充满着活力的空间。

错列排布的住宅楼、酒店、办公楼在湘江河畔拔地而起，建筑形态和自然之间的互动加强了场地的有机性和流畅性。住宅塔楼好比屹立在水平面上的帆船，现代而又有序的建筑设计既可将朝向西面河流和岳麓山的视野最大化，同时也可保证充足的太阳光照。

0 25 50 100
LRF

Profile

Rising out of the Xiang River and adjoining local cultural district, the project will thrive with human activity and social interaction, establishing an inviting modern waterfront living space.

Design Concept

Themed and inspired by the elements of water, gardens, and city, Northstar Xinhe Delta's new and innovative concept of retail entertainment and mixed-use development is transformed away from the typical housing and retail environment to deliver a dynamic destination to visit, shop, live, work, and play unlike any other.

Architectural Design

The project is designed to entice exploration and generate social interaction within its pedestrian-oriented foundation. Whether it is a stroll along the "Boardwalk" lined with unique specialty shops and restaurants that take full advantage of the sweeping river views, or the interjected dazzling and entertaining array of activities, the waterfront will be enlivened with energy.

As sweeping towers of residences, hotel and offices soar above a rooftop park along the Xiang River, the clear interaction between built form and nature enhances the site's organic and fluid quality. Symbolic of sails on the water's horizon, the sleek and contemporary design of the residential towers maximizes the westerly views to the river and Yuelu Mountain, while still capturing the ever precious southern sunlight.

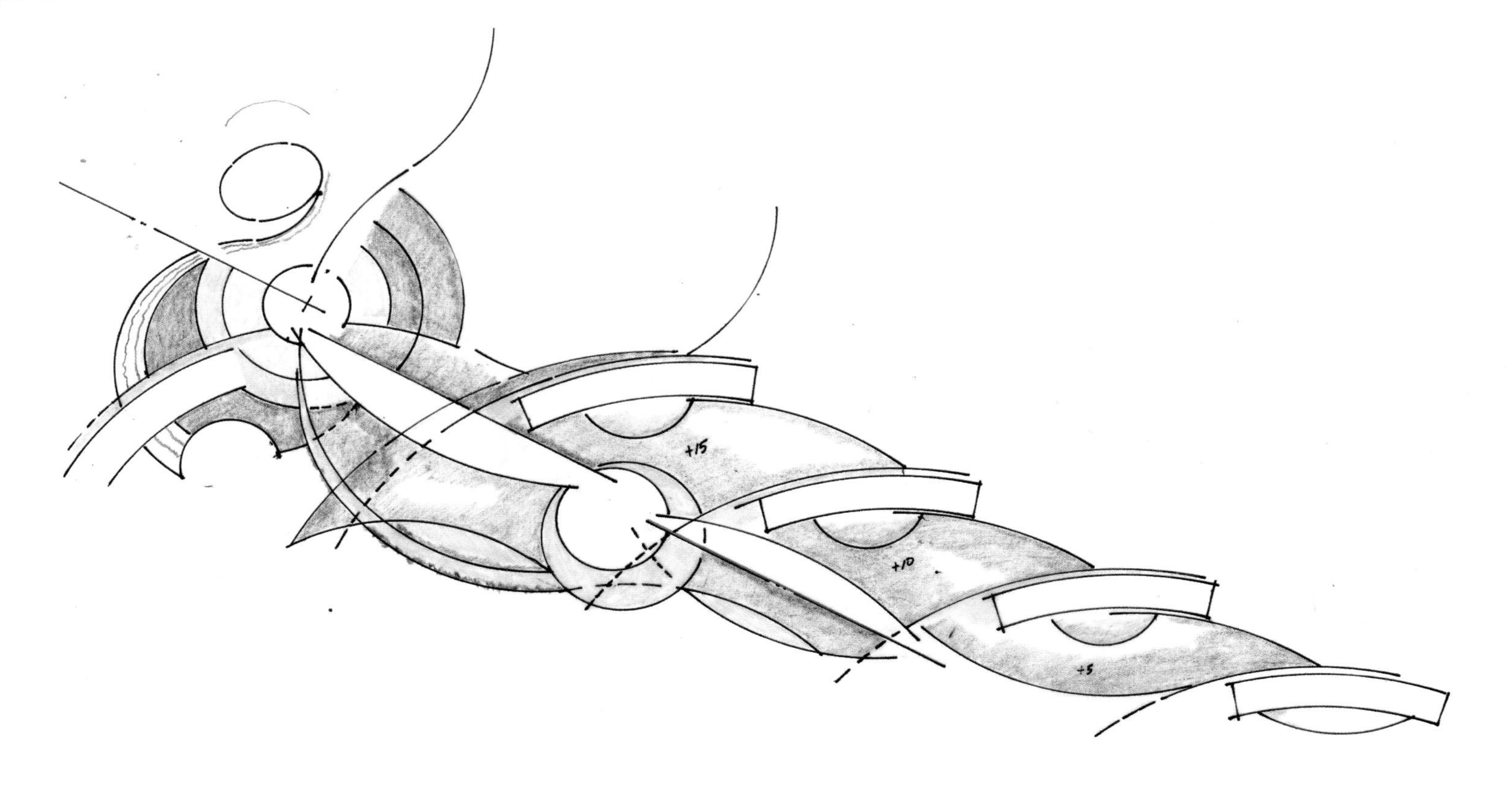

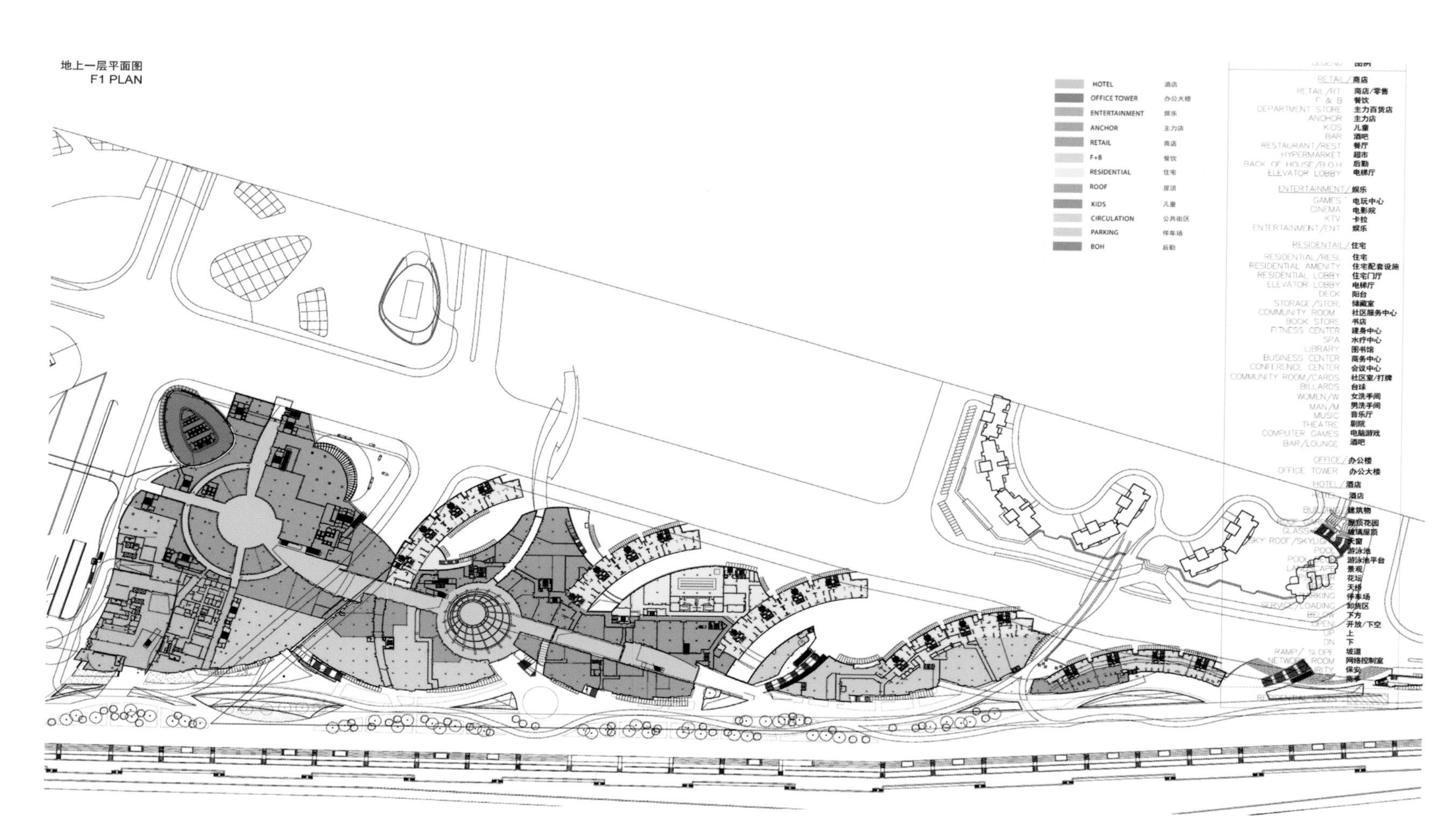

地上一层平面图
F1 PLAN
HOTEL 酒店
OFFICE TOWER 办公大楼
ENTERTAINMENT 娱乐
ANCHOR 主力店
RETAIL 商店
F+B 餐饮
RESIDENTIAL 住宅
ROOF 屋顶
KIDS 儿童
CIRCULATION 公共街区
PARKING 停车场
BOH 后勤
RETAIL/商店
RETAIL/RT 商店/零售
F & B 餐饮
DEPARTMENT STORE 主力百货店
ANCHOR 主力店
KIDS 儿童
BAR 酒吧
RESTAURANT/REST 餐厅
HYPERMARKET 超市
BACK OF HOUSE/B.O.H 后勤
ELEVATOR LOBBY 电梯厅
ENTERTAINMENT/娱乐
GAMES 电玩中心
CINEMA 电影院
KTV 卡拉
ENTERTAINMENT/ENT 娱乐
RESIDENTAIL/住宅
RESIDENTIAL/RESI. 住宅
RESIDENTIAL AMENITY 住宅配套设施
RESIDENTIAL LOBBY 住宅门厅
ELEVATOR LOBBY 电梯厅
DECK 阳台
STORAGE/STOR. 储藏室
COMMUNITY ROOM 社区服务中心
BOOK STORE 书店
FITNESS CENTER 建身中心
SPA 水疗中心
LIBRARY 图书馆
BUSINESS CENTER 商务中心
CONFERENCE CENTER 会议中心
COMMUNITY ROOM/CARDS 社区室/打牌
BILLARDS 台球
WOMEN/W 女洗手间
MAN/M 男洗手间
MUSIC 音乐厅
THEATRE 剧院
COMPUTER GAMES 电脑游戏
BAR/LOUNGE 酒吧
OFFICE/办公楼
OFFICE TOWER 办公大楼
HOTEL/酒店

地上五层平面图
F5 PLAN

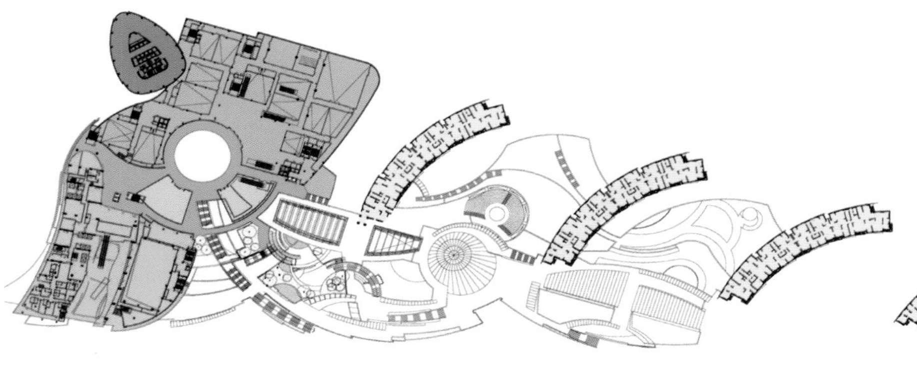

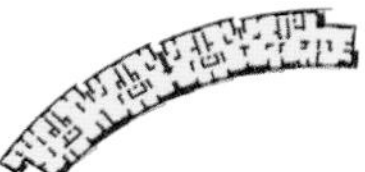

WEST ELEVATION
西立面

EAST ELEVATION
东立面

GLASS ATRIUM 玻璃中庭
WATER FEATURE WALL 景观水墙
WATER FEATURE WALL 景观水墙
GLASS GUARDRAIL 玻璃栏杆
STONE STAIR 石材楼梯
GREEN WALL 绿化墙
THEATER 电影院
GLASS SKYLIGHT 玻璃天窗
STONE WALL 石材墙面
GLASS STORE FRONT 玻璃店面
GLASS GUARDRAIL 玻璃栏杆
L05 EL. +59.5 M
L04 EL. +54.5 M
L03 EL. +49.5 M
L02 EL. +44.5 M
L01 EL. +39.5 M
L0M EL. +36.0 M
L00 EL. +33.5 M
B01 EL. +29.0 M
SCALE 1:400
0 5 10 20

1号住宅楼
RESIDENTAIL TOWER #1
STONE WALL
石材墙面
STAIR TO L02 楼梯
WATER FEATURE 水景
CONCRETE COLUMN (FACING T.B.D.)
混凝土柱
GLASS CURTAIN WALL 玻璃幕墙
STONE WALL 石材墙面
GLASS STORE FRONT 玻璃店面
RAMP TO L01 坡道
BRIDGES ON L01 AND L02
步行天桥
CINEMA
I-MAX
CROUCHING TIGER
HIDDEN DRAGON
L05
EL. +59.5 M
L04
EL. +54.5 M
L03
EL. +49.5 M
L02
EL. +44.5 M
L01
EL. +39.5 M
L0M
EL. +36.0 M
L00
EL. +33.5 M
B01
EL. +29.0 M
SCALE 1:400
0 5 10 20

浙江杭州西溪壹号

Zhejiang Hangzhou Xixi No.1

设计单位：日本 M.A.O. 一级建筑士事务所
开发商：杭州永茂商务咨询有限公司
项目地址：中国浙江省杭州市
占地面积：47 249 m²
总建筑面积：93 598.41 m²
绿化率：35.7%
容积率：1.2
设计团队：叶建为 谢 非
陈霞敏 甘丽凤

Designed by: M.A.O.
Developer: Hangzhou Yongmao Business Consulting Co., Ltd.
Location: Hangzhou, Zhejiang, China
Land Area: 47,249 m²
Gross Floor Area: 93,598.41 m²
Greening Ratio: 35.7%
Plot Ratio: 1.2
Design Team: Ye Jianwei, Xie Fei, Chen Xiamin, Gan Lifeng

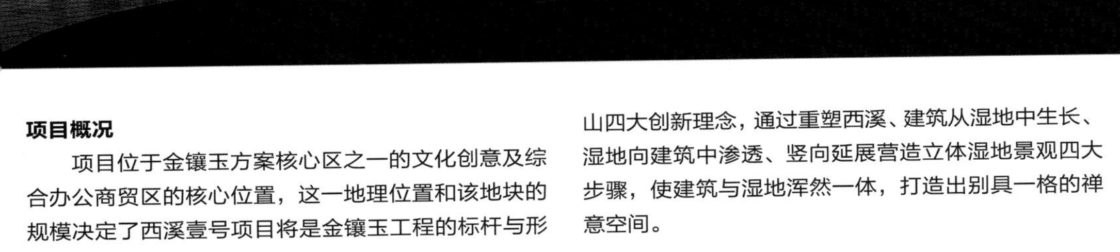

项目概况

项目位于金镶玉方案核心区之一的文化创意及综合办公商贸区的核心位置，这一地理位置和该地块的规模决定了西溪壹号项目将是金镶玉工程的标杆与形象工程之一。

设计理念

当前，城市快速路的存在和西溪湿地周边地块不合理的开发模式，导致城市与西溪湿地之间形成了生硬的界面，阻碍了湿地效应的延展。本方案站在城市设计的角度，从西溪湿地出发，通过物理手法将西溪湿地景观延续和移植到该地块中，使之得到进一步的提升。同时，以人的行为体验为支撑，模糊西溪湿地与城市之间的界面，使西溪湿地的效益最大化。

设计特色

设计遵循西溪湿地的自然肌理与原生态景观特色，通过对湿地景观进行抽象化处理和提升，重塑西溪，使之在“冷、野、淡、雅”的意境上萌生温润、优雅、丰富多彩的生活格调。设计以飞地、静卧、矗立、堆山四大创新理念，通过重塑西溪、建筑从湿地中生长、湿地向建筑中渗透、竖向延展营造立体湿地景观四大步骤，使建筑与湿地浑然一体，打造出别具一格的禅意空间。

Profile

The project is located at a prominent position of Jinxiangyu Scheme which is dominated by cultural innovation and comprehensive office & business area. Its location and scale contribute to this landmark and iconic project of Jinxiangyu Engineering.

Design Concept

Currently, city expressways and unreasonable development of the surrounding areas of Xixi Wetland lead to rigid interfaces between the city and the Xixi Wetland. Thus it hinders the extension of wetland benefit. From the perspective of urban design and for the consideration of Xixi Wetland, the landscape of Xixi wetland has been extended and implanted to the site by physical techniques. In the meantime, supported by behavioral experience, the interface between Xixi Wetland and the city is blurred which ensures maximum benefits of Xixi Wetland.

Design Feature

Following the natural texture and native landscape features of Xixi Wetland, the design has decided to re-shape Xixi through abstract processing and improving to the wetland landscape. It aims to create cozy, elegant and colorful life based on “Cool, Wild, Simple Elegant” artistic conception. Four innovative concepts — Raised Land, Tranquil Placement, Upright Standing and Mountain Piling — are applied in the design. Four steps including re-shape Xixi, erect buildings on wetlands, bring wetlands to buildings, and vertically extend wetland landscape, manage to create an integrated whole of buildings and wetlands as well as build distinctive Zen space.

C-31-2地块主要技术经济指标

建设用地面积		20644 平方米	建筑密度			33.0%
总建筑面积		40914.79平方米	绿地率			35.67%
a.地上总建筑面积		24772.7 平方米	机动车停车			192 辆
其中	酒店总面积	10511.3 平方米	其中	地面停车		20 辆
	商业面积	391.69 平方米		其中	普通泊位	13 辆
	物业管理及配套面积	168.43 平方米			出租车	4 辆
	办公面积	13687.2 平方米			装卸车	1 辆
	垃圾房	14.0 平方米			大巴车	2 辆
b.地下室总建筑面积		16142.09平方米				
	投影占地面积	6813.21平方米		地下停车		172 辆
	建筑容积率	1.20	其中	无障碍车位		4 辆
			非机动车停车面积			968.97 平方米
人防面积		12233 平方米				

C-31-1地块主要技术经济指标

西溪湿地公园

建设用地面积		26605 平方米
总建筑面积		52683.62平方米
a.地上总建筑面积		31922.6 平方米
其中	办公面积	25258.07平方米
	餐饮面积	6535.43 平方米
	物业管理及配套面积	129.13 平方米
b.地下室总建筑面积		20761.02平方米
	投影占地面积	9066.0 平方米
	建筑容积率	1.20

	34.08%	
	35.75%	
车	247	辆
停车	35	辆
普通泊位	30	辆
出租车	4	辆
装卸车	1	辆
停车	212	辆
障碍车位	4	辆
停车面积	2648.51平方米	

建筑高度统计表

楼号	室内外高差	±0.00黄海标高	最高点绝对标高
1#、27#	0.10	4.80	28.00
2#, 12#, 13#, 24#~26#	0.10	4.80	28.00
3#~8# 14#~20#	0.30	4.80	20.00
9#~11#, 21#~23#	0.30	4.80	13.50

代征用地面积指标

征地名称		征地面积
总代征用地		6320平方米
其中	代征道路	2681平方米
	代征绿地	2431平方米
	代征河道	1208平方米

图例:

- 人行出入口
- 消防车出入口
- 机动车出入口
- 3F 建筑地上自然层层数
- H=10.00 建筑室外地坪至女儿墙顶高度
- 4.60(WL) 景观水面标高
- 4.70 室外场地设计标高
- 4.65 道路设计标高（变坡点）

注:

1: 本总平面图标高按本地黄海高程定。
2: 图中建筑与建筑的间距为外墙之间的距离，以米为单位。
3: 本项目景观设计，建设方另行委托。

总平面图 1:500

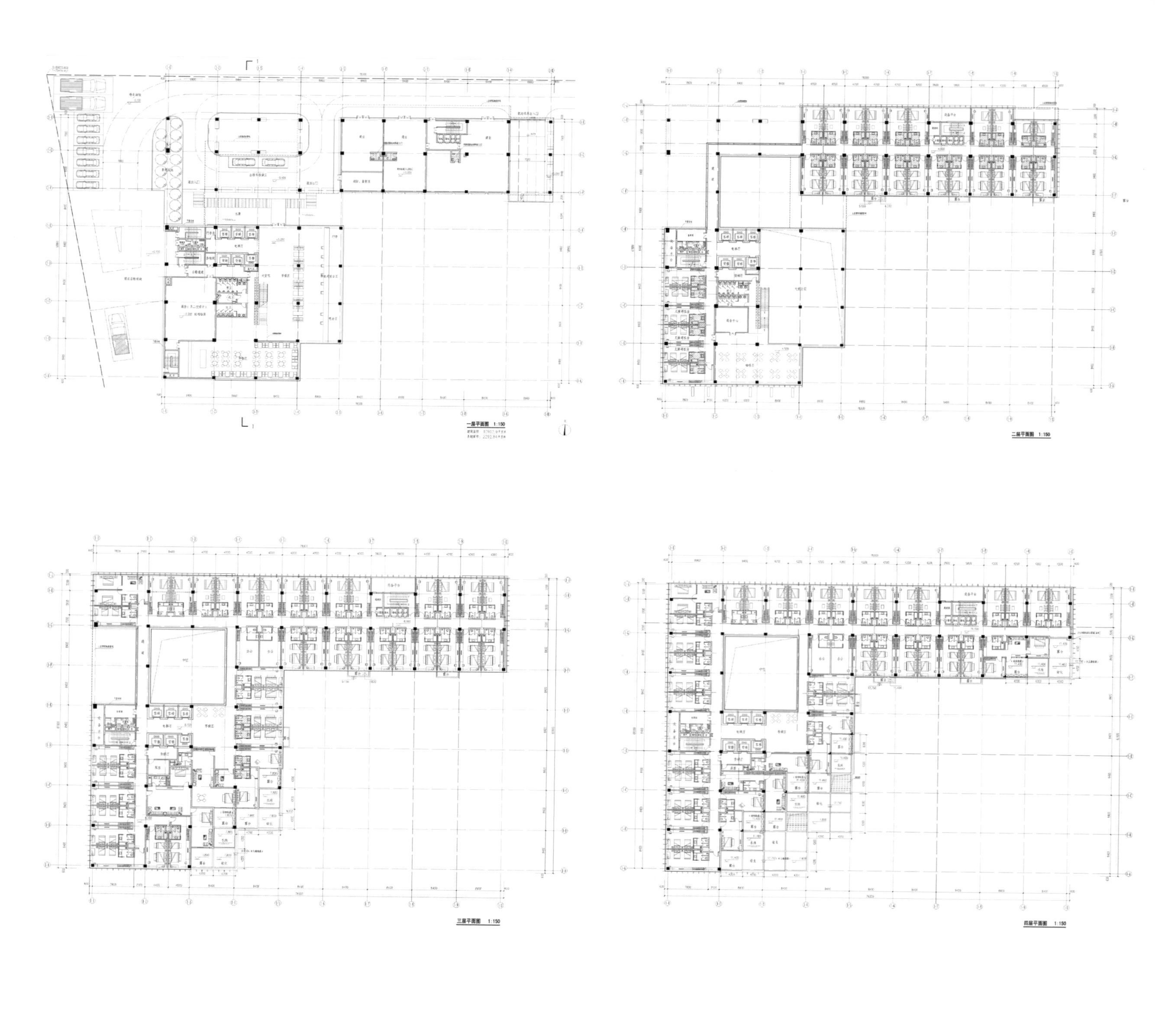

一层平面图 1:150

二层平面图 1:150

三层平面图 1:150

四层平面图 1:150

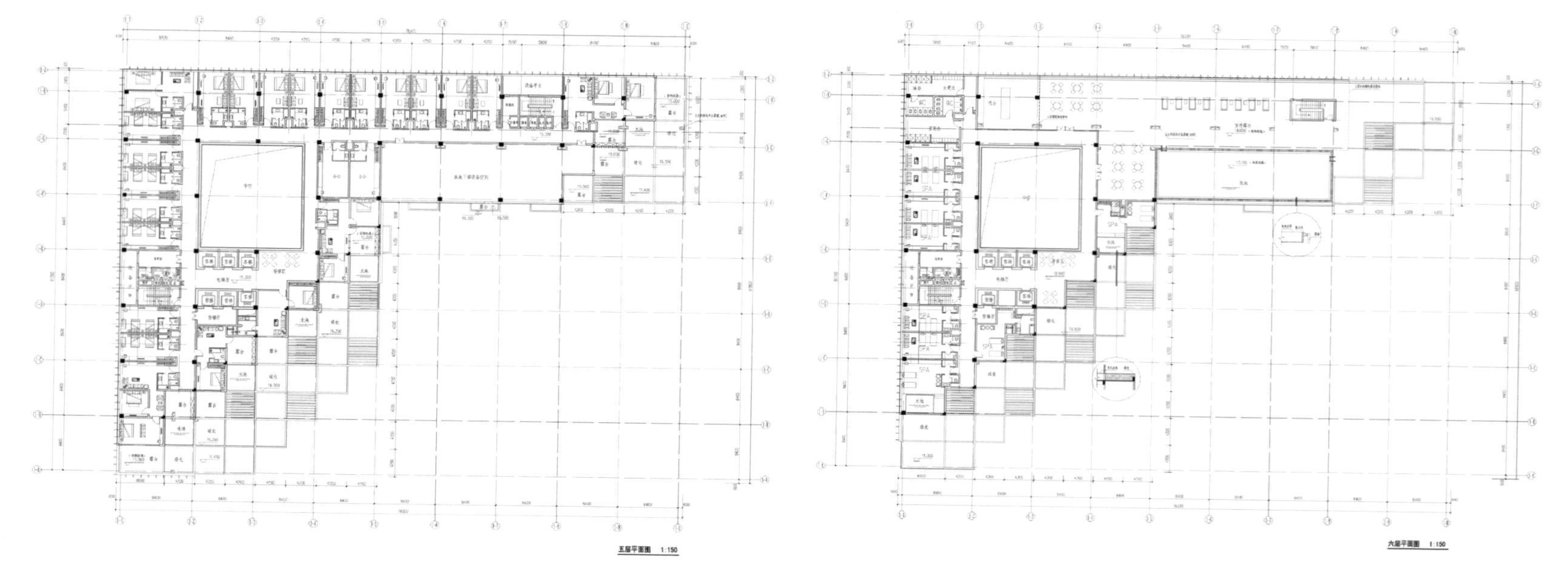

五层平面图 1:150

六层平面图 1:150

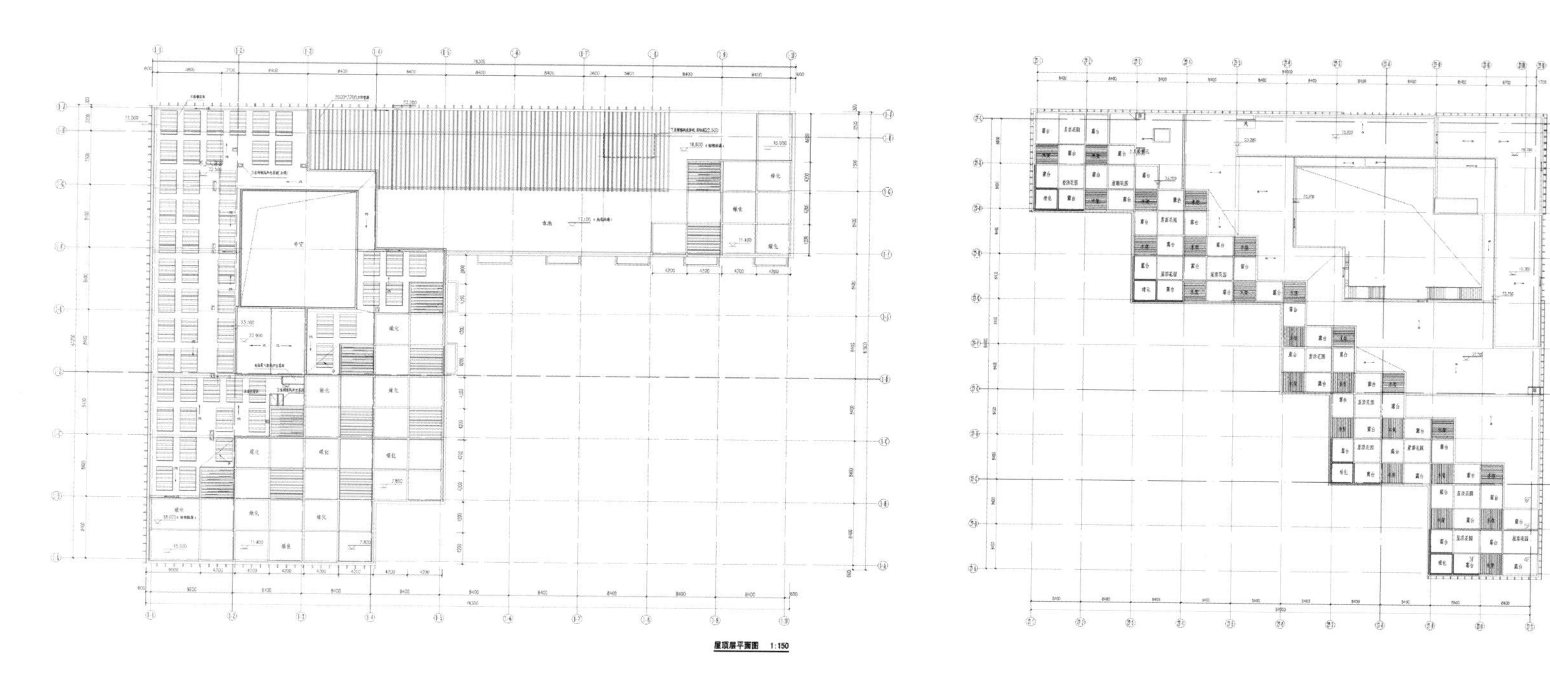

屋顶层平面图　1:150

屋顶平面图　1:150

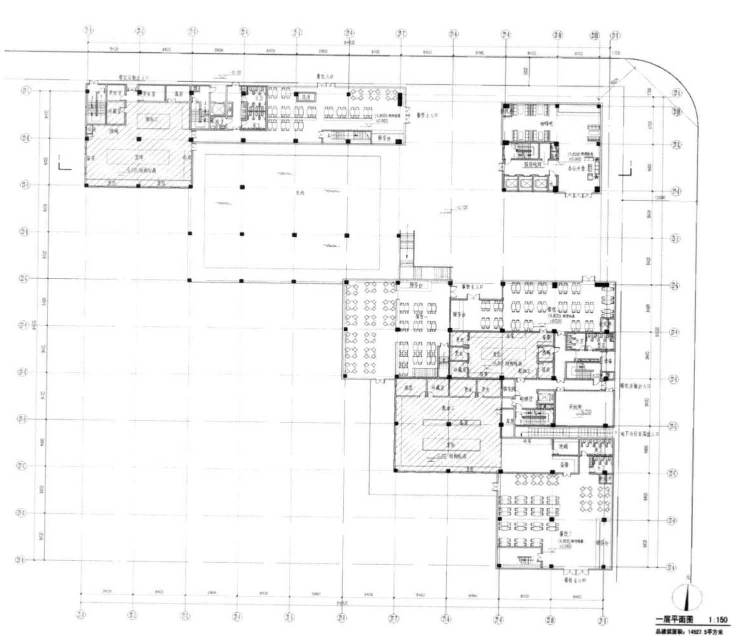

一层平面图　1:150

总建筑面积：14527.5平方米
占地面积：3552.32平方米

二层平面图　1:150

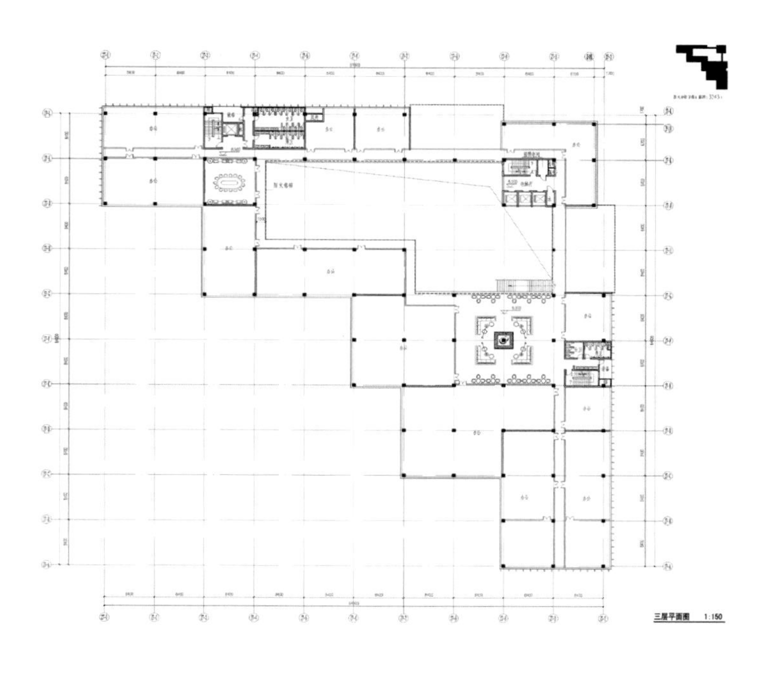

三层平面图 1:150

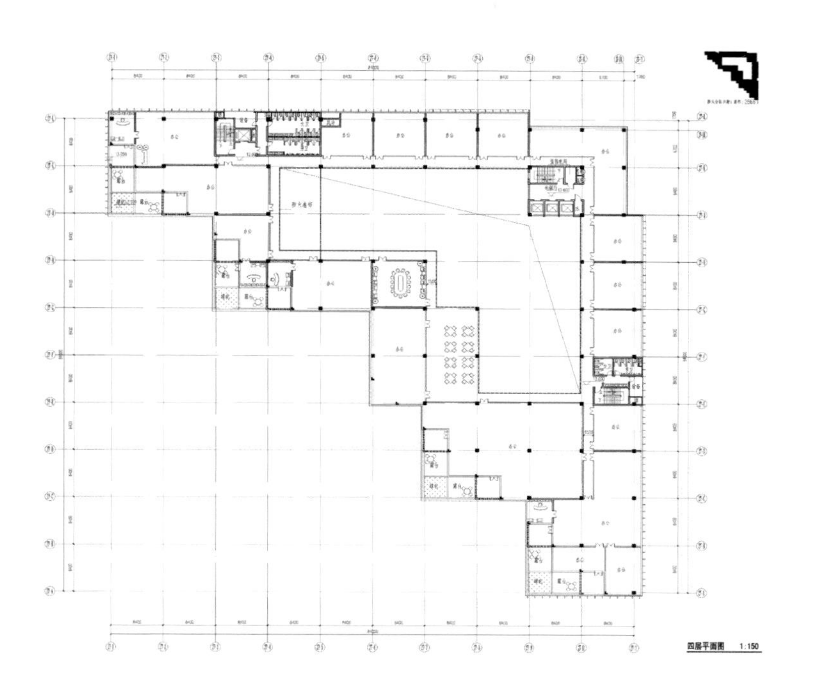

四层平面图 1:150

五层平面图 1:150

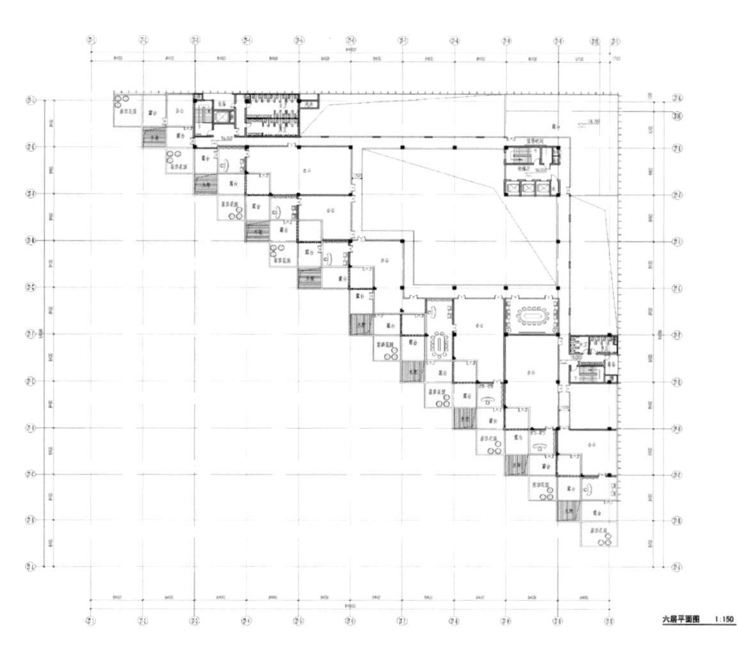

六层平面图 1:150

竖向装饰杆件
木百叶
上悬开启扇
隐框分缝
立面图例：
单片钢化玻璃 颜色：无色透明
玻璃幕墙 颜色：无色透明
干挂高级石材 颜色：深灰色
干挂石材 颜色：深灰色
双片中空 LOW-E 玻璃 颜色：无色透明
单片钢化玻璃 颜色：草绿色不透明
干挂陶板 颜色：咖啡色
干挂石材 1 颜色：深灰色
1-1 — 1-10 轴立面图 1:150
1-10 — 1-1 轴立面图 1:150

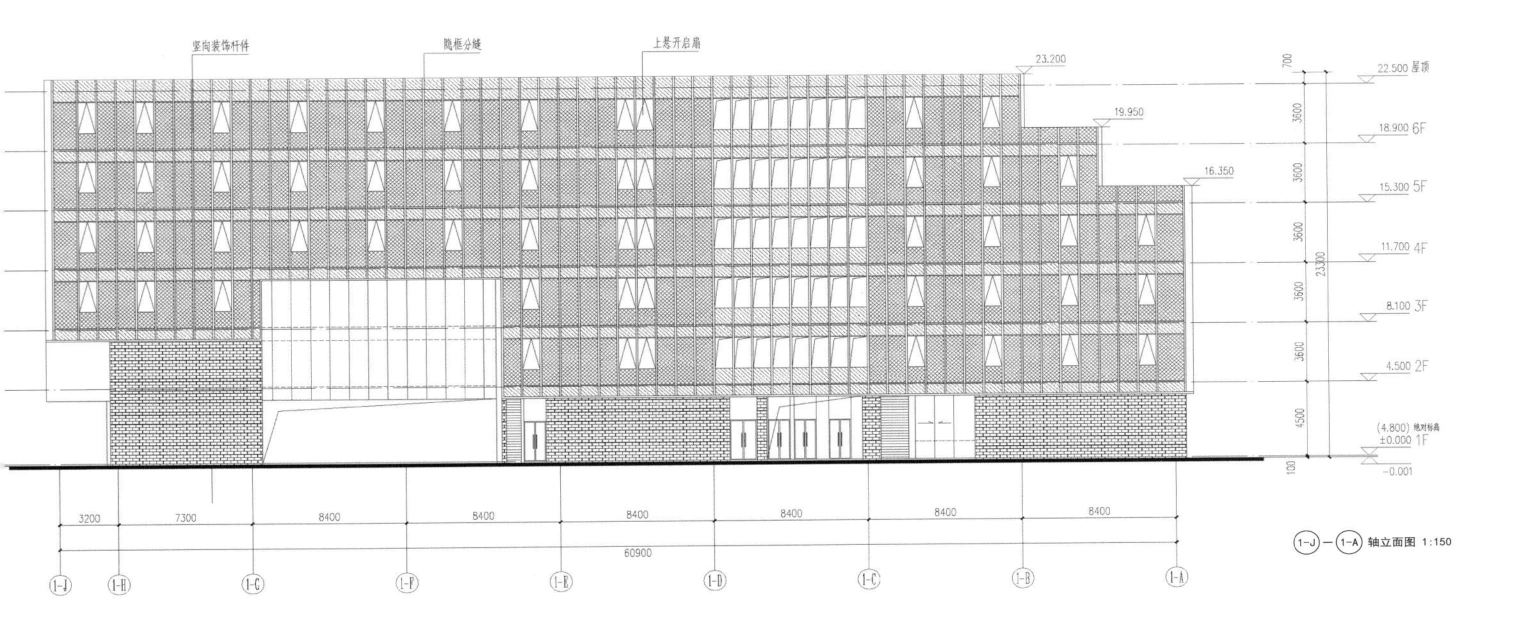
竖向装饰杆件
隐框分缝
上悬开启扇
23.200
19.950
16.350
22.500 屋顶
18.900 6F
15.300 5F
11.700 4F
8.100 3F
4.500 2F
(4.800) 绝对标高
±0.000 1F
-0.001
23300
3200
7300
8400
60900
(1-J)—(1-A) 轴立面图 1:150

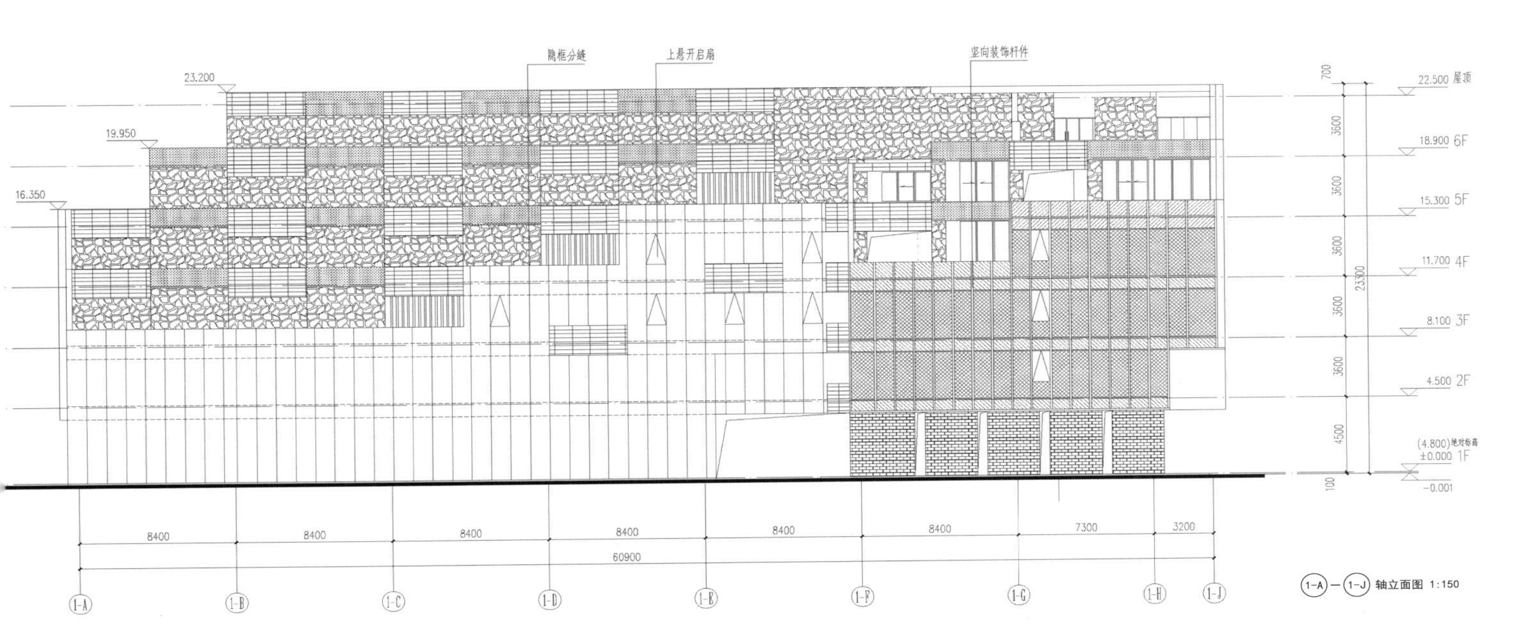
隐框分缝
上悬开启扇
竖向装饰杆件
23.200
19.950
16.350
22.500 屋顶
18.900 6F
15.300 5F
11.700 4F
8.100 3F
4.500 2F
(4.800) 绝对标高
±0.000 1F
-0.001
23300
8400
7300
3200
60900
(1-A)—(1-J) 轴立面图 1:150

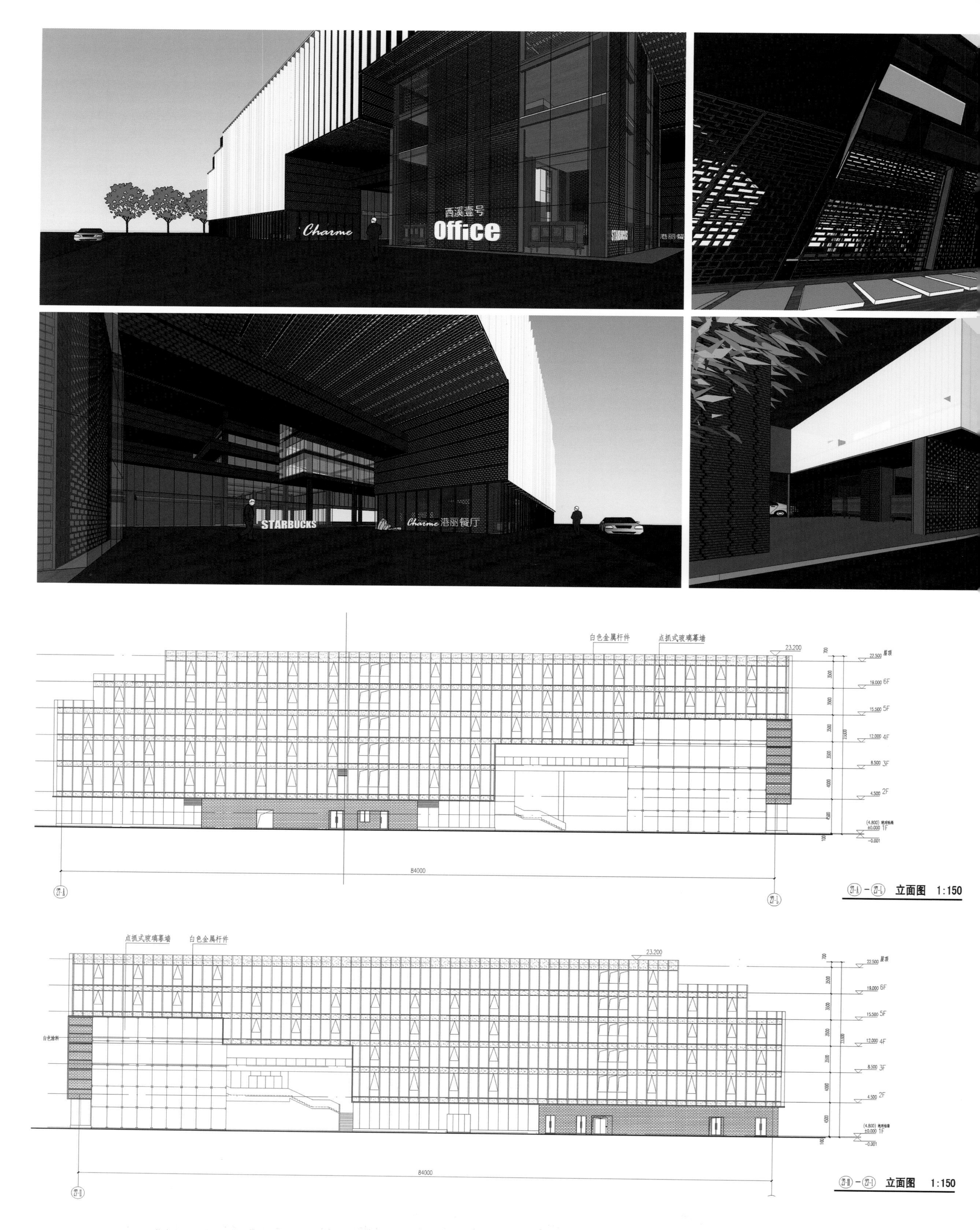
西溪壹号
Office
Charme
STARBUCKS
港丽餐厅
白色金属杆件
点抓式玻璃幕墙
23.200
84000
(27-A) — (27-L) 立面图 1:150
点抓式玻璃幕墙
白色金属杆件
23.200
84000
(27-H) — (27-1) 立面图 1:150

白色金属杆件
屋面
6F
5F
4F
3F
2F
1F
84000
(27-1)-(27-11) 立面图 1:150
白色金属杆件
84000
(27-L)-(27-A) 立面图 1:150

上海文新传媒谷

Wenxin Media Campus

设计单位：加拿大 CPC 建筑设计顾问公司
开发商：文汇新民联合报业集团
项目地址：中国上海市
总用地面积：85 000 ㎡
总建筑面积：262 000 ㎡
绿地率：25%
容积率：2.03
设计团队：邱 江　韩 强

Designed by: Coast Palisade Consulting Group
Developer: Wenhui Xinmin United Press Group
Location: Shanghai, China
Total Land Area: 85,000 m^2
Gross Floor Area: 262,000 m^2
Greening Ratio: 25%
Plot Ratio: 2.03
Design Team: Qiu Jiang, Han Qiang

方案一 Proposal I

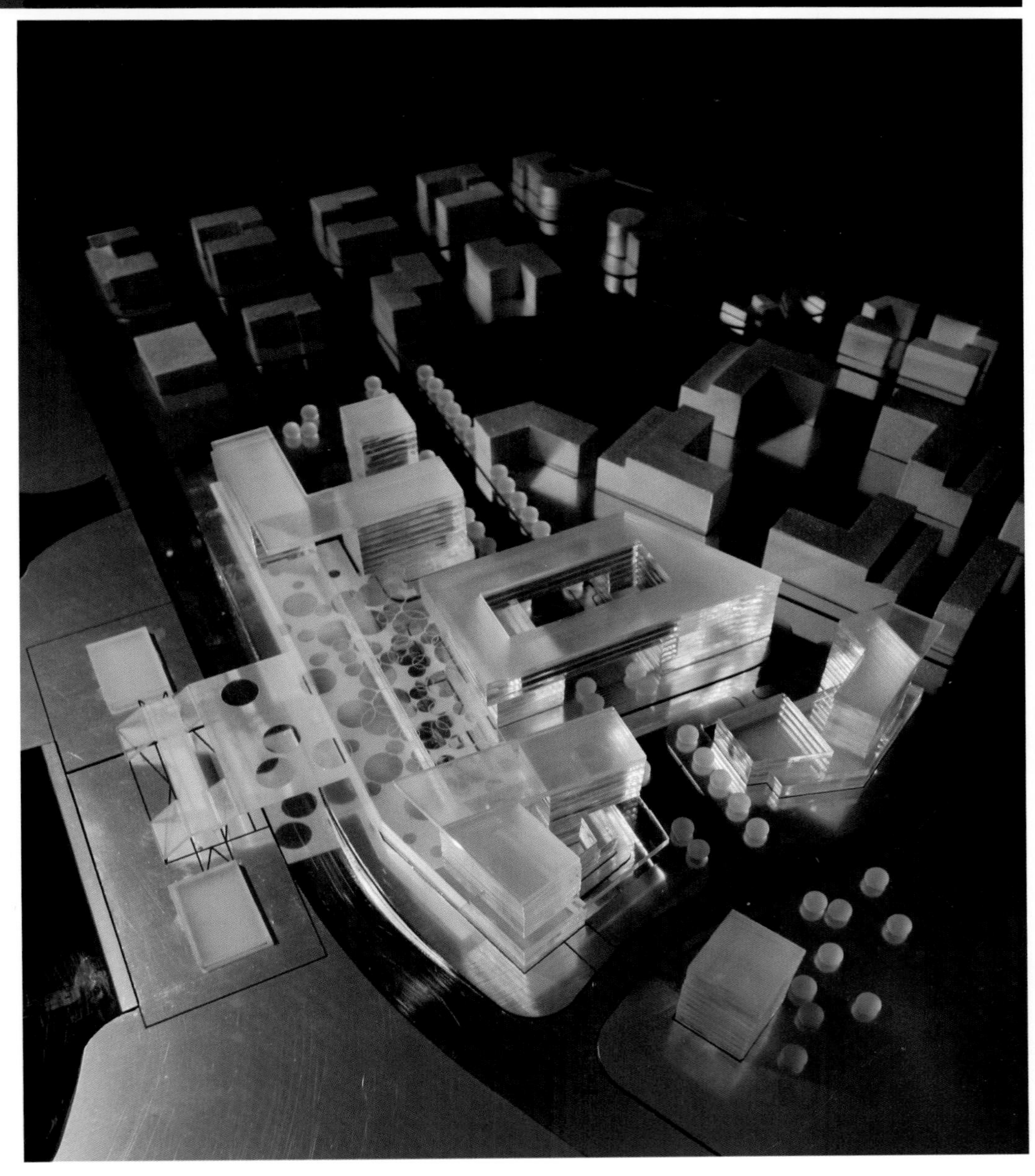

设计理念

项目的规模为构建一个媒体中心创造了良好的条件：它不仅仅是多个复合建筑体的集合，又或者是在建筑风格或表皮特征具有相关性的建筑集合，而是一个像纽约洛克菲勒中心、东京六本木山一样，是一个提倡创造性、多样性的媒体文化中心。在这里，人们可以自由地交流信息、观点和想法。

设计特色

考虑到场地的基本特征、面向邻近的高速公路和公园的广阔视野、由规划中的景观桥构成的行人轴线、通向场地的垂直车流流线等因素，设计在地块的东部边界上设置了一个朝向公园的立面和大型的广场。

这一布局达到了以下效果：创建了一个横向的地标，其立面的规模与邻近的高速公路和公园协调，使项目对高速公路和公园产生了强烈的视觉冲击力；将建筑的边界延伸至邻近的公园，使更多的办公空间可享有面向公园的视野；将规划中的景观桥与项目连接起来作为广场向公园的延展部分，这样设计比单单把建筑直接联系公园更有张力，同时也将公园的视野扩展到项目内；创建了一个重要的“核心”，综合地容纳了不同的设施和公共空间，从而使不同空间之间形成互动；由于广场被延伸至一层，交通更为便利，更重要的是，它打破了传统办公的约束。

设计通过将广场的水平高度抬升至交通流线之上，使连接了这个综合体大多数建筑的人行流线显得便捷流畅，还预留了更多的底层空间与车行流线、出入口、下客区、停车坪相接。这并不意味着广场以下的空间就成为了地下室，设计通过在地下空间设计大型的开口、引入自然光线、增加绿化和水景元素，使之成为一个明亮、开敞的空间，宛如一个朝向天空开放的空间。

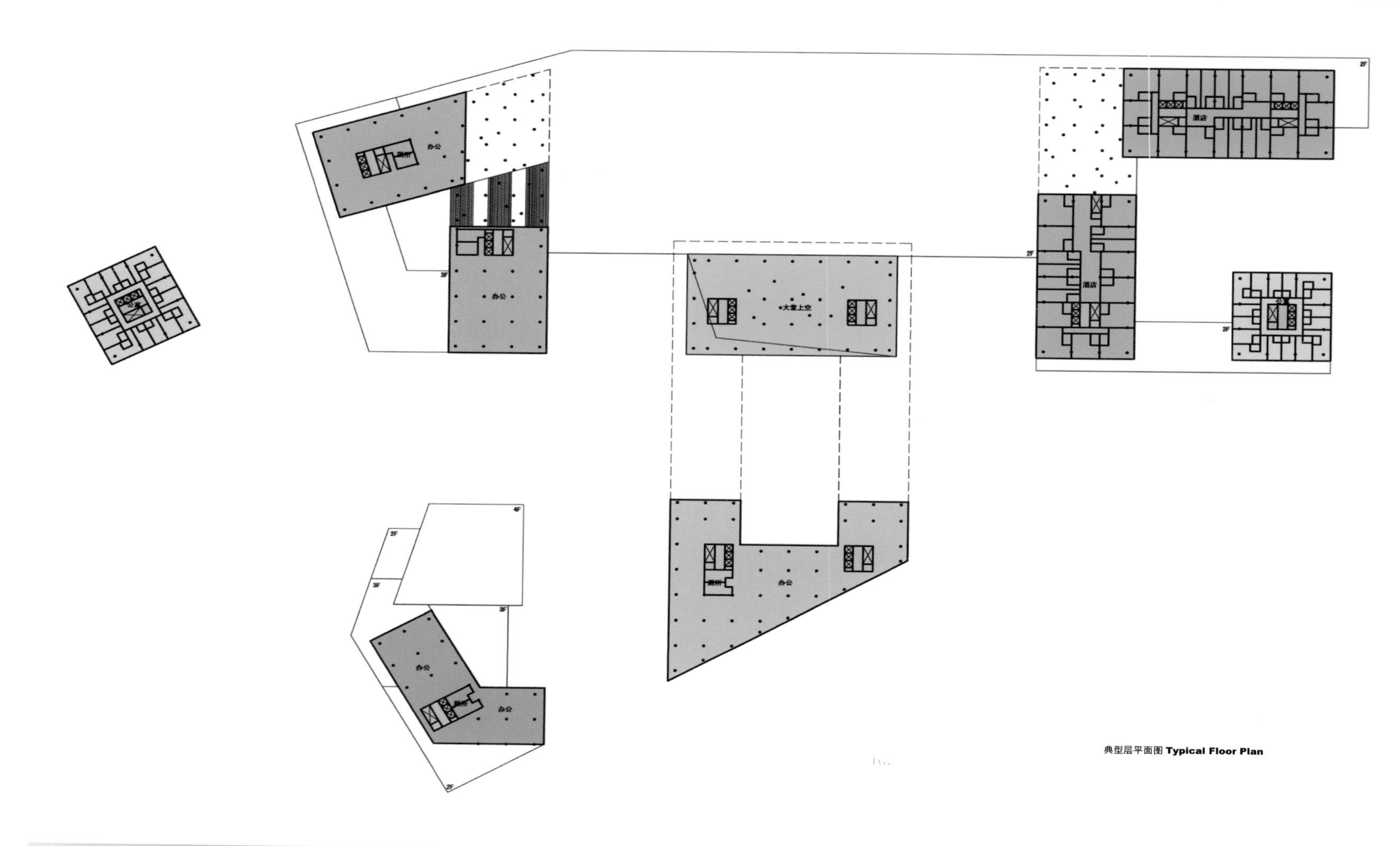

典型层平面图 **Typical Floor Plan**

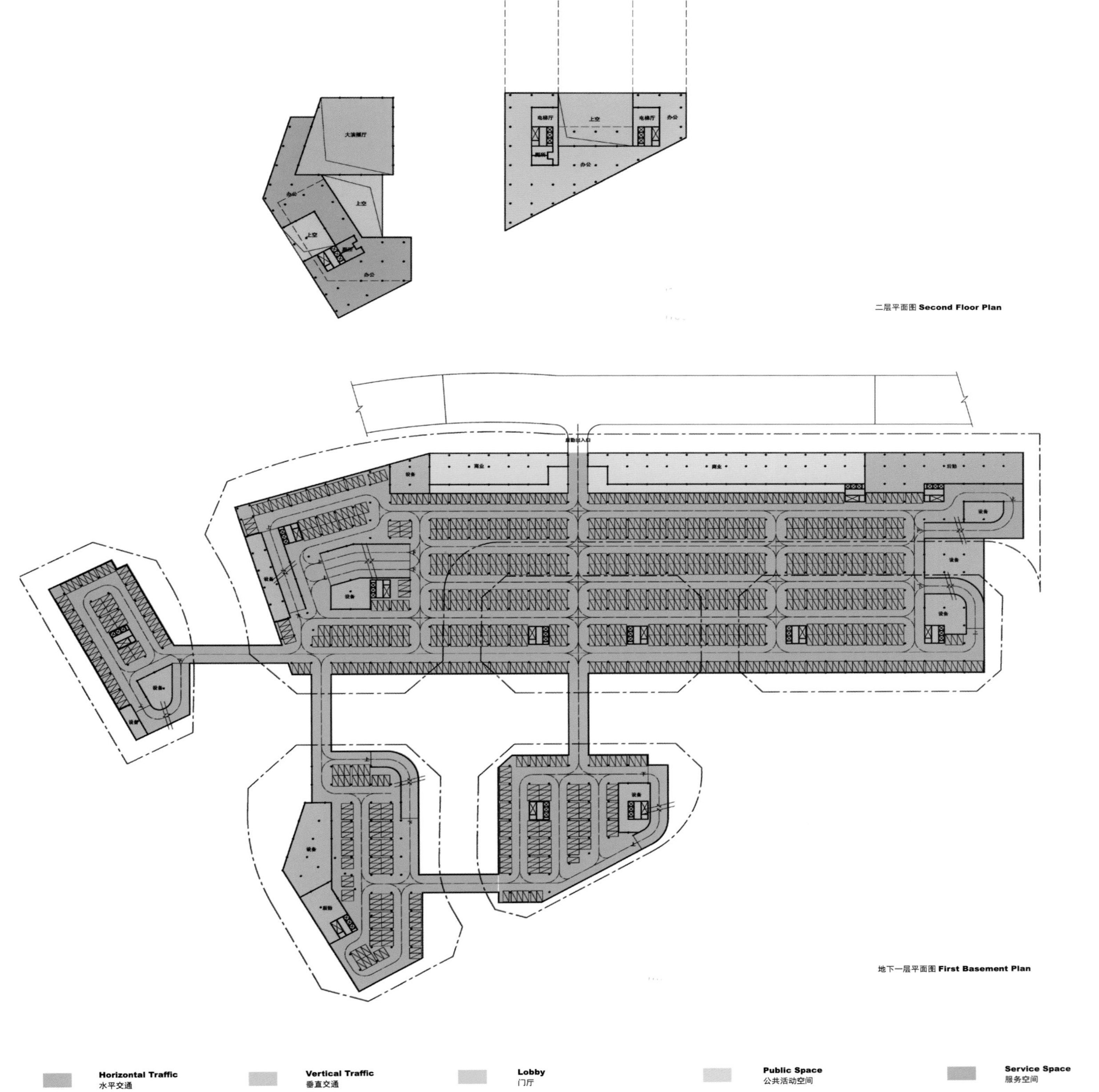

二层平面图 **Second Floor Plan**

地下一层平面图 **First Basement Plan**

Horizontal Traffic
水平交通

Vertical Traffic
垂直交通

Lobby
门厅

Public Space
公共活动空间

Service Space
服务空间

Corridors
露台

Headquter Office
文新总部办公

Conference
会议室

Exhibition
展览

New Media Center
新媒体中心

New Media Center
新媒体中心

HuaWen Media Center
华文媒体交流中心

Restaurant
餐厅

Minhang Media Center Office
闵行媒体中心办公

Boadcasting Room
闵行媒体中心演播厅

Commercial
商业

Car Parking
停车

Function Analysis
功能分析

- **Horizontal Traffic** 水平交通
- **Vertical Traffic** 垂直交通
- **Lobby** 门厅
- **Public Space** 公共活动空间
- **Service Space** 服务空间
- **Corridors** 露台
- **Headquter Office** 文新总部办公
- **Conference** 会议室
- **Exhibition** 展览
- **New Media Center** 新媒体中心
- **New Media Center** 新媒体中心
- **HuaWen Media Center** 华文媒体交流中心
- **Restaurant** 餐厅
- **Minhang Media Center Office** 闵行媒体中心办公
- **Boadcasting Room** 闵行媒体中心演播厅
- **Commercial** 商业
- **Car Parking** 停车

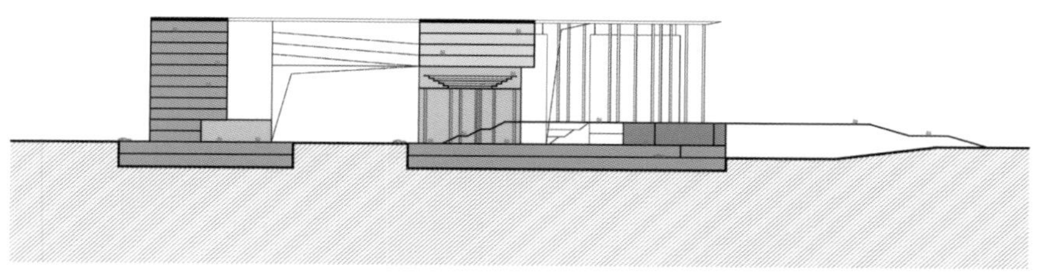

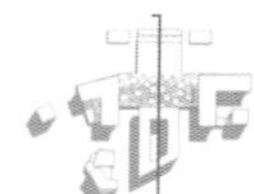

剖面图 **1-1 section 1-1**

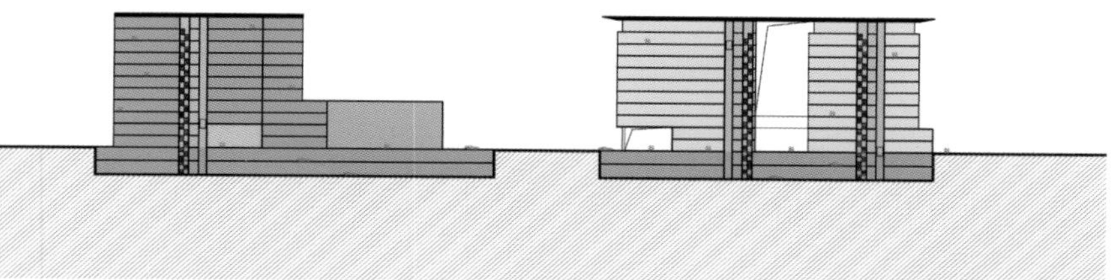

剖面图 **2-2 section 2-2**

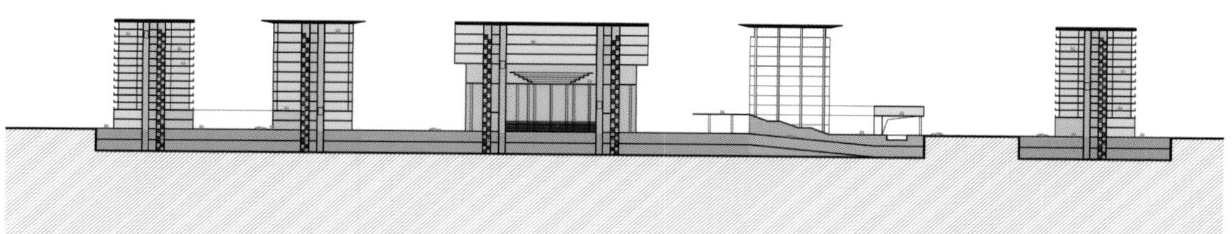

剖面图 **3-3 section 3-3**

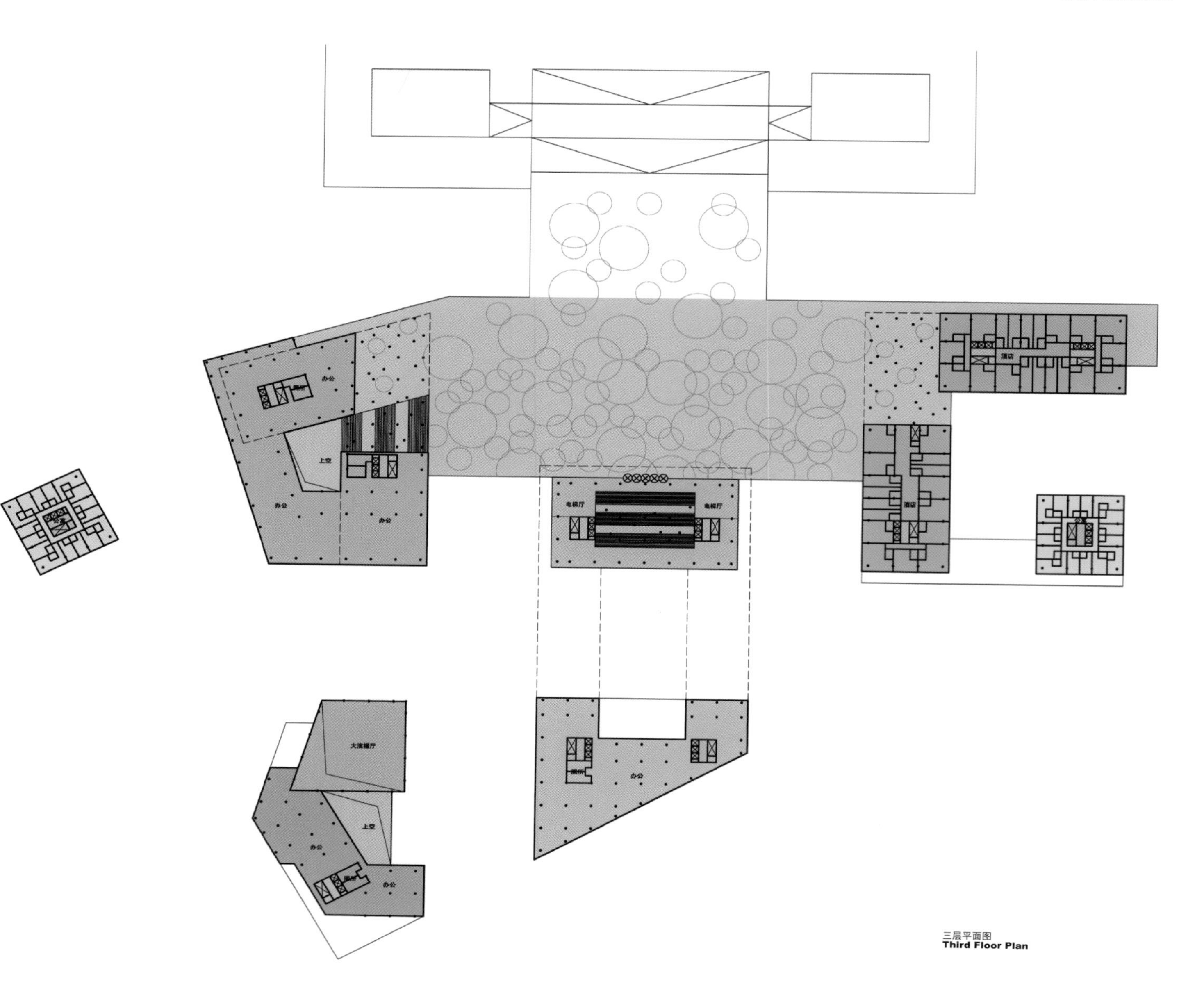

三层平面图
Third Floor Plan

东立面（面向文化公园）

西立面（面向商务区）

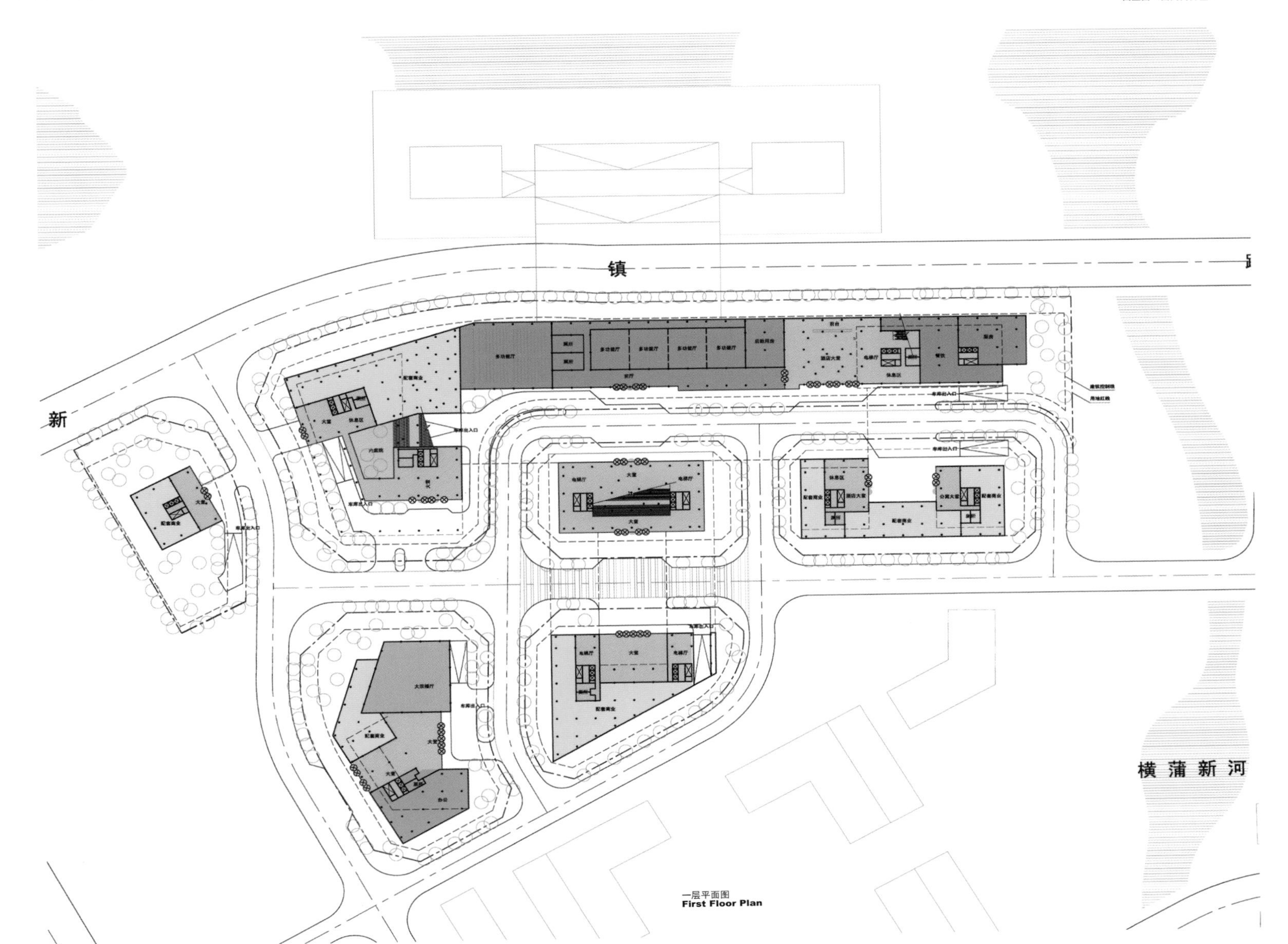

一层平面图
First Floor Plan

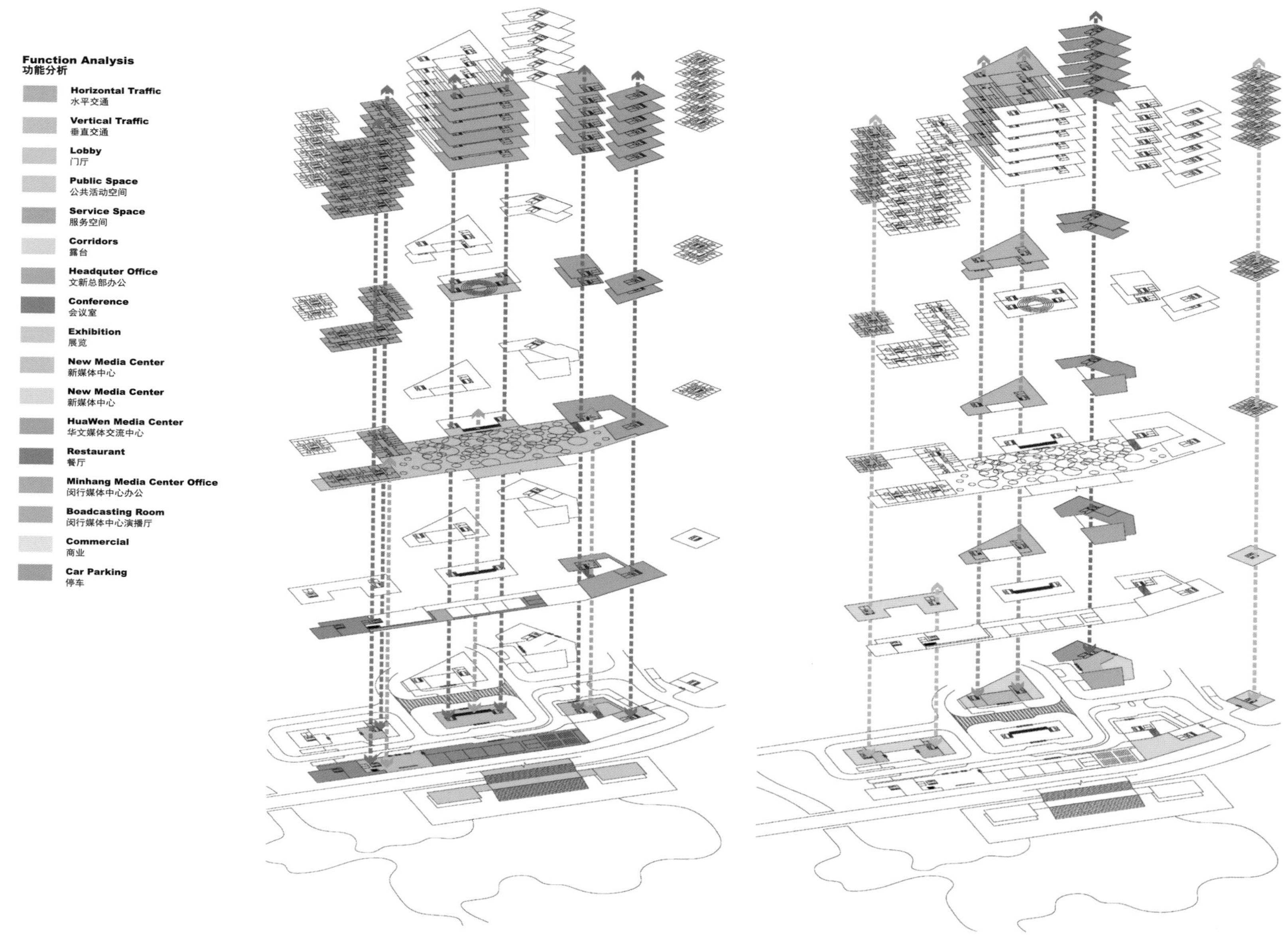

Function Analysis
功能分析
Horizontal Traffic
水平交通
Vertical Traffic
垂直交通
Lobby
门厅
Public Space
公共活动空间
Service Space
服务空间
Corridors
露台
Headquter Office
文新总部办公
Conference
会议室
Exhibition
展览
New Media Center
新媒体中心
New Media Center
新媒体中心
HuaWen Media Center
华文媒体交流中心
Restaurant
餐厅
Minhang Media Center Office
闵行媒体中心办公
Boadcasting Room
闵行媒体中心演播厅
Commercial
商业
Car Parking
停车

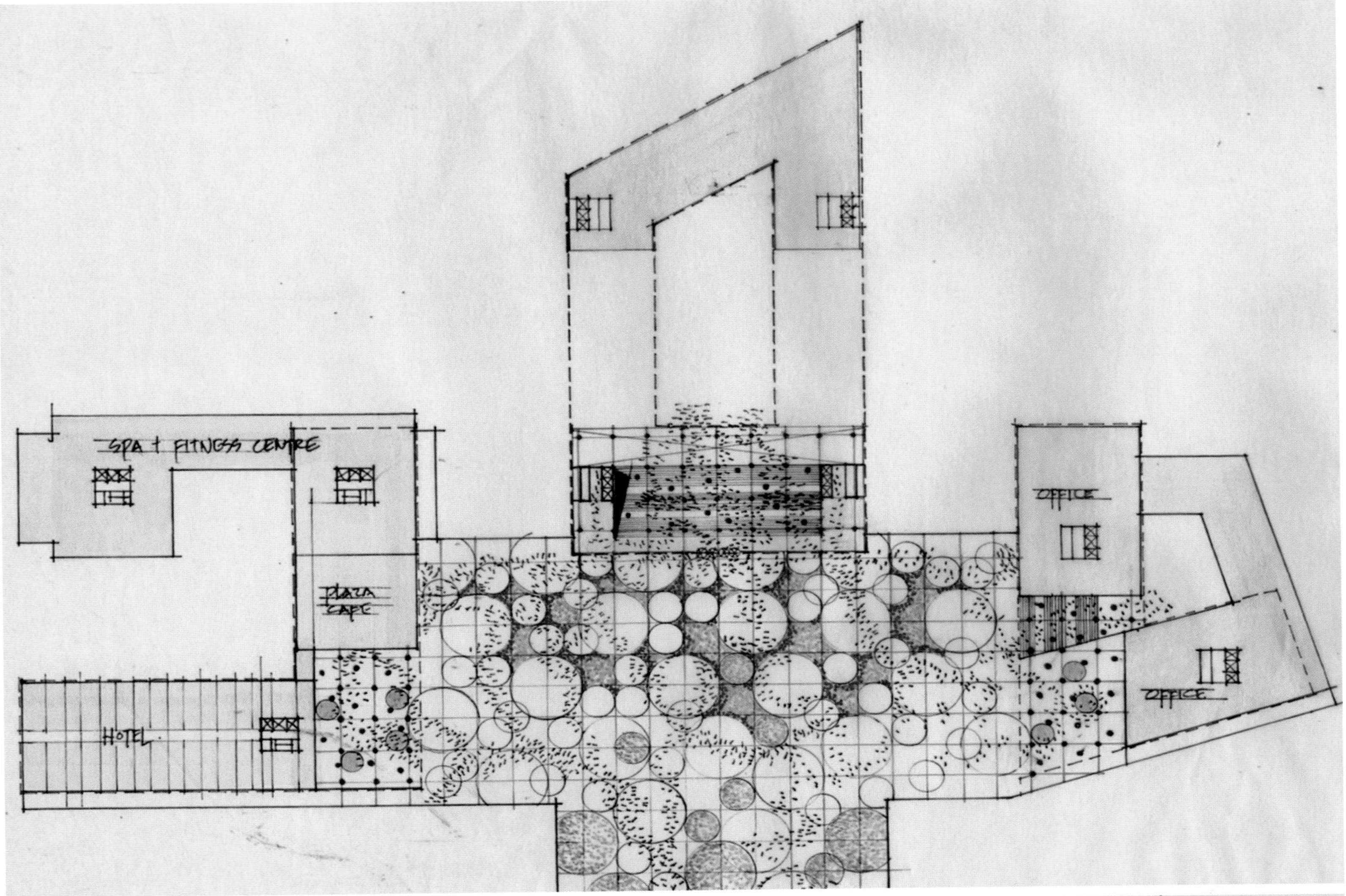

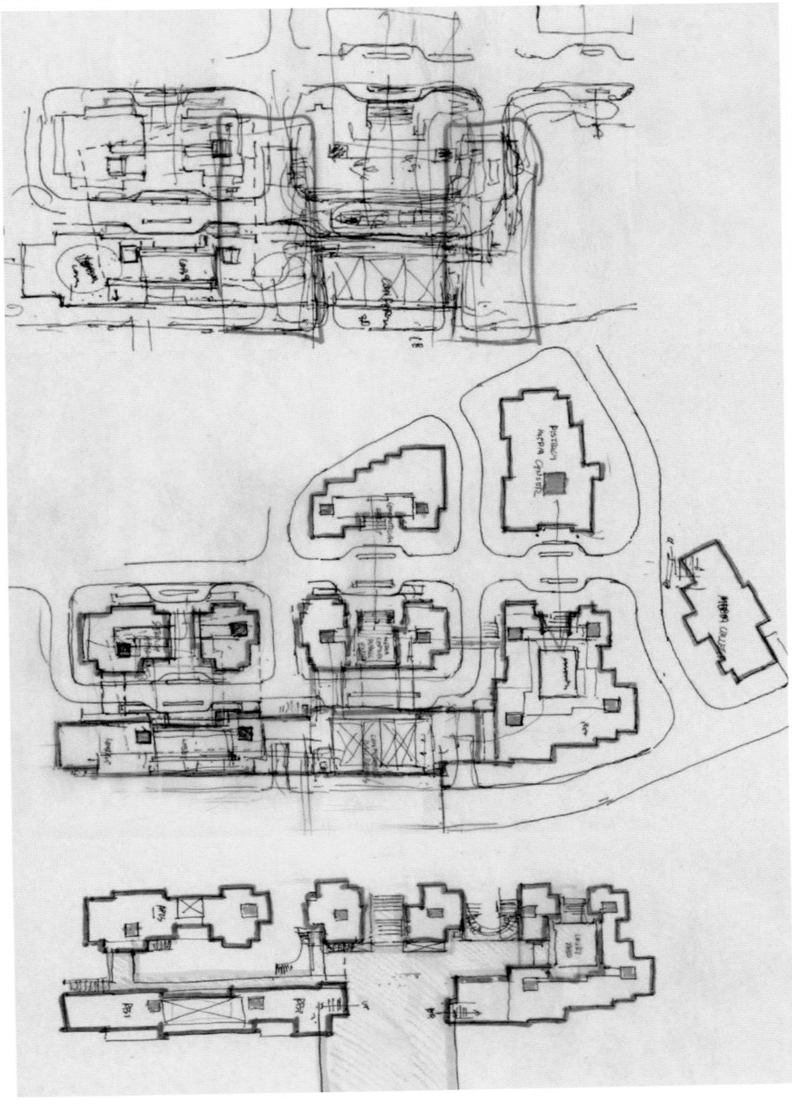

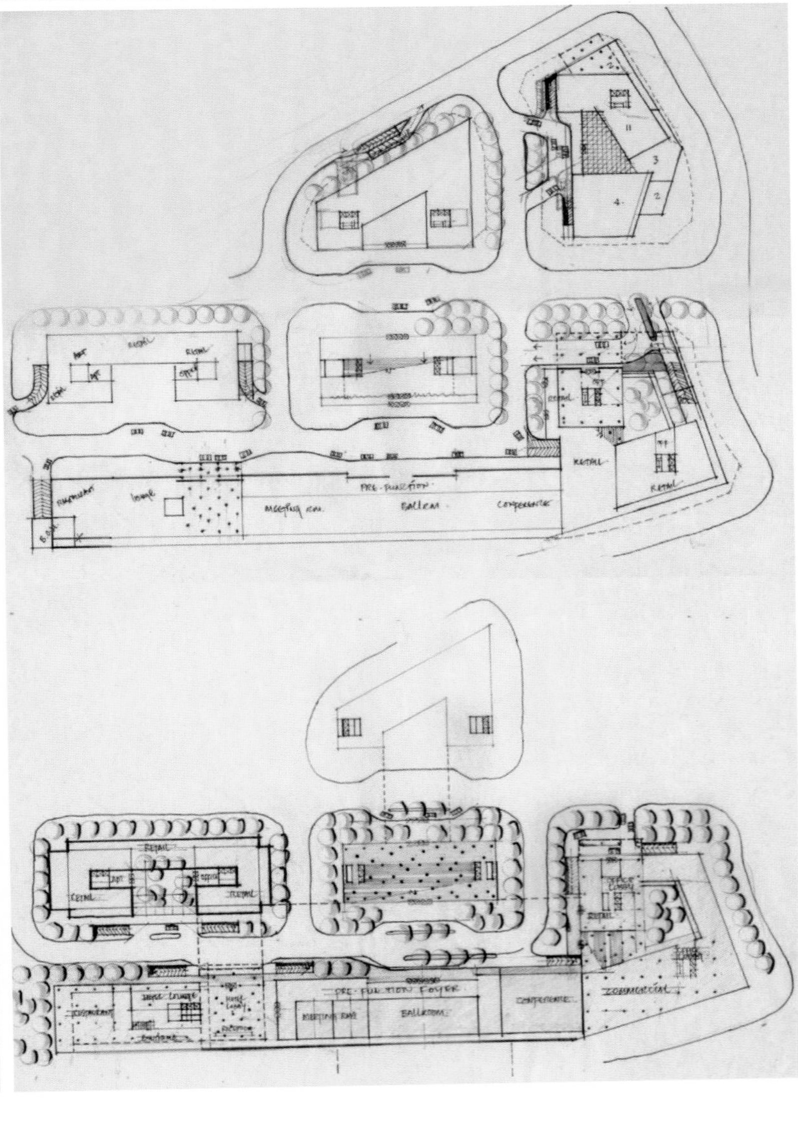

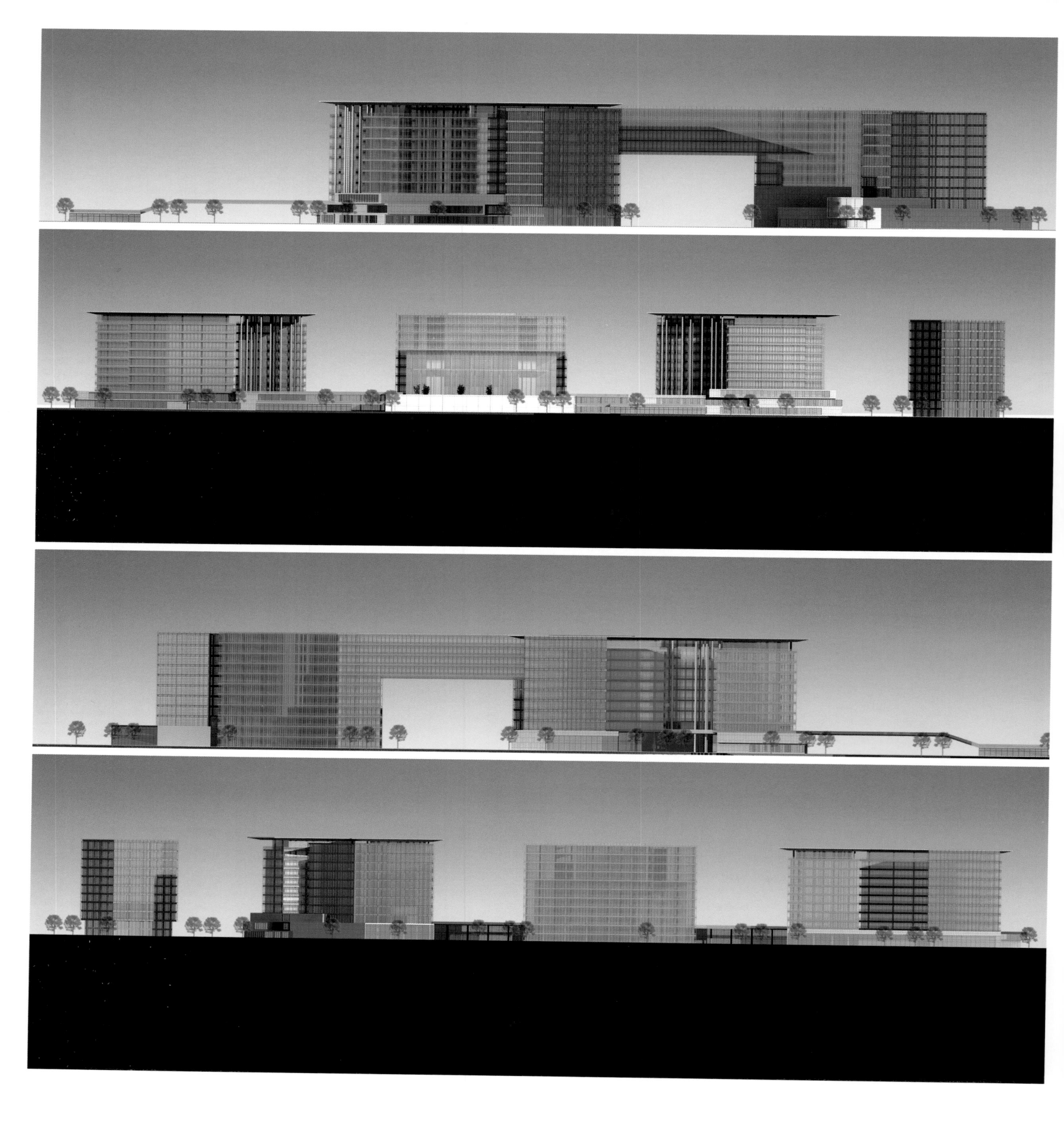

Design Concept

The scale of this project presents a rare opportunity to create a center – like Rockefeller Center in New York, or Roppongi Hills in Tokyo – that goes beyond being a mere collection of largely unrelated buildings, or related perhaps only by style or some sort of surface features. Instead the project would create a true media "Campus", a comprehensive center that encourages creativity, diversity, and the free interchange of information, ideas and opinions.

Design Feature

Recognizing the essential features of the context, the macro views from the adjacent highway and park, the pedestrian axis created by the planned landscape bridge leading into the site, and the vehicular aixs perpendicular to it, the entire eastern edge of the project, facing the park, has been set up as a formal elevation with a large plaza area at its centre.

This arrangement achieves a number of important results: creates a – necessarily horizontal – "landmark" elevation of an appropriate scale relative to the adjacent park and highway, directions from which the project may be viewed with maximum impact; extends the building perimeter adjoining the park, allowing more offices to have a view to the park; fully incorporates the planned landscape bridge with the design as a vital extension of the plaza into the adjacent park, which goes further than merely linking the project with the park, in effect it also extends the park into the project; establishes an essential "heart" to the complex where occupants of the various components of the facility and the general public may interact freely, free from vehicular traffic since the plaza has been extended beyond the original ground level, but more importantly, free from the psychological constraints set up by the standard office typology.

文汇新民集团新基地
新媒体中心
华文媒体中心
新媒体中心
闵行区梅地亚中心

经济技术指标—方案A

总用地面积(平方米)		85000
总建筑面积(平方米)		261678
地上总建筑面积(平方米)		172340
其中	文汇新民集团新基地	71090
	新媒体中心	39060
	闵行区梅地亚中心	20980
	华文媒体交流中心	41210
地下总建筑面积(平方米)		89338
首层建筑面积(平方米)		23159
覆盖率		27.2%
容积率		2.03
绿地率		25%

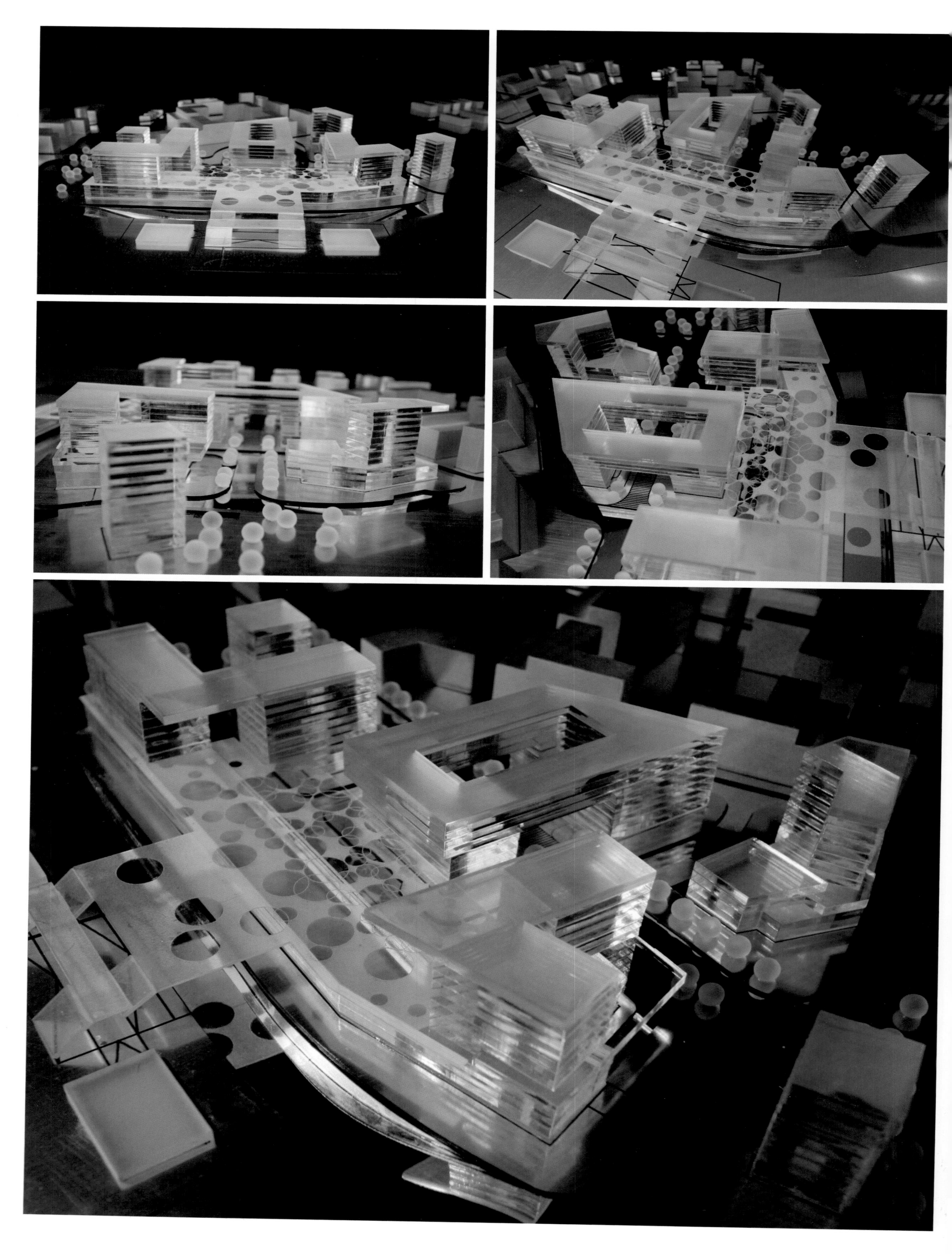

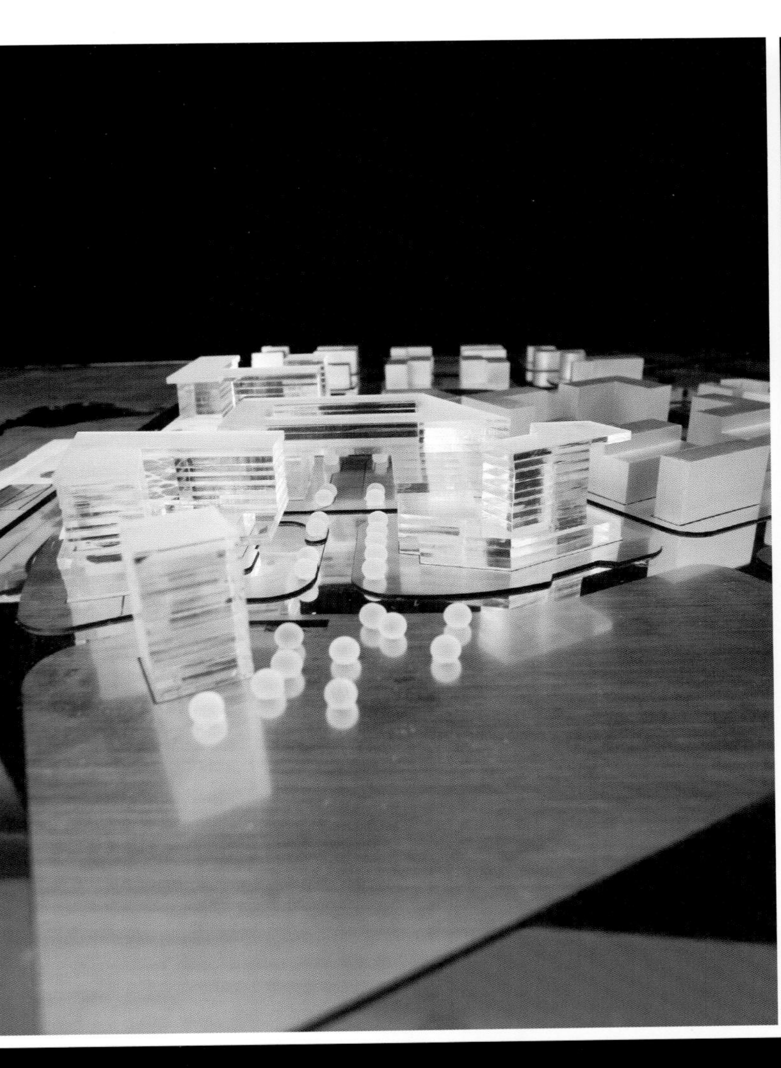

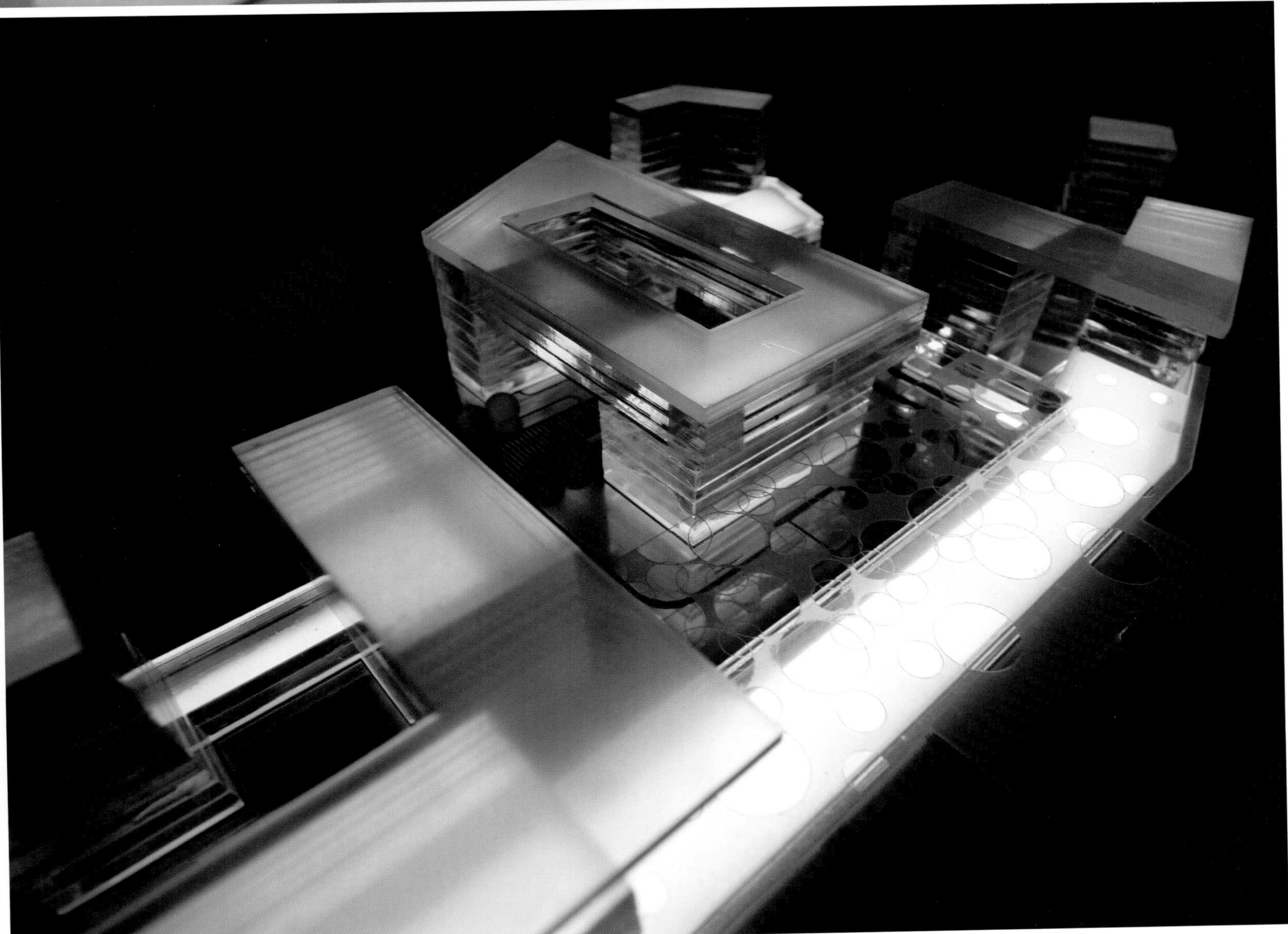

In a more practical sense, raising the plaza level above the traffic permits fully unencumbered pedestrian circulation between most of the buildings of the complex, while on the other hand, frees up the ground level to deal more directly with the vehicular traffic, the entrances and exits, the drop-offs and the parking ramps. But that does not mean that the area beneath the plaza would be by any means "basement-like" since through the creative use of large openings, skylights, landscaping and water features, it could be made into an area just as light, airy and vibrant as if it were completely open to the sky.

方案二 Proposal II

设计理念

设计师认为“空间”是人类的第一个住所，而“建筑”才是第二个住处。设计旨在构建一个集中的建筑体量，将之作为交流空间来承载人们对过往的记忆，同时也将人的感官带入未来。在这一方案中，设计师将“空间”诠释为“门”——可向里看与往外看的“门”，通过三道“门”再现了媒体发展史的三个阶段。

设计特色

三道“门”对称地朝向连接着公园和湖泊的景观桥，建筑也就自然地融入周围环境中，成为湖中央舞台的自然背景。设计克服了建筑高度的限制，在水平面上构建了一个横向的地标，一个引人注目的标志性建筑。

设计的重点在于表现了与地面衔接的和谐感。设计延续了原有的桥梁，以引导人们从公园进入中间的“门”，使建筑自然地成为景观的一部分。建筑的基座象征着起伏的地景，这一空间融合了不同的功能分区，可提供多样化的服务。

基层建筑从地面向天空生长，而大体量建筑从天而降，两者互相衔接，既形成了一种和谐的美感，又产生了强烈的视觉冲击力。中间办公建筑的楼顶将会是一个体现现代建筑艺术的“空中会堂”，也是媒体界普里兹奖的颁奖会堂。它是一个悬空 30 米、倒置的全透明玻璃结构，这一会场可直接通向 6 400 平方米的屋顶花园。

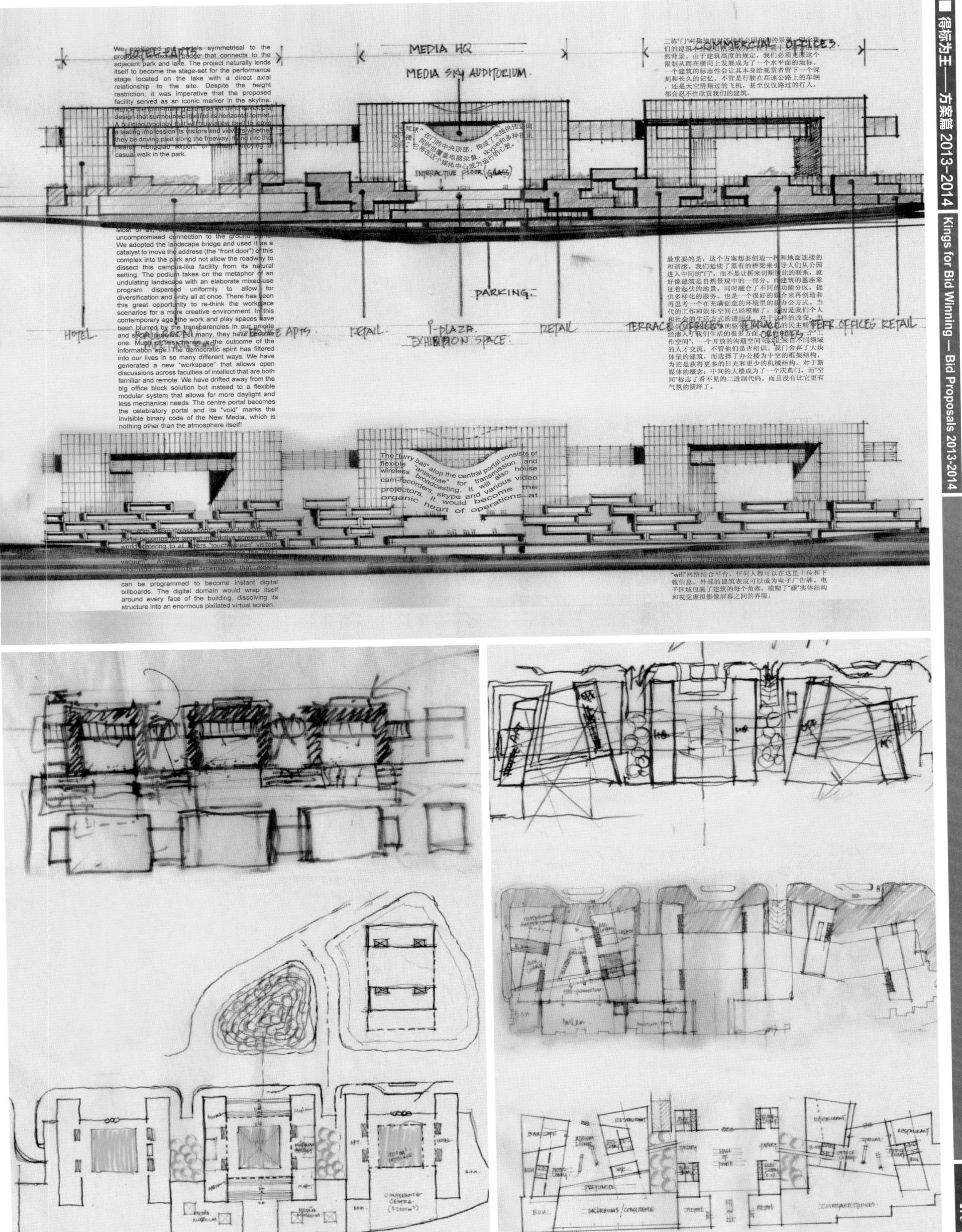
We positioned our portals symmetrical to the
proposed landscape bridge that connects to the
adjacent park and lake. The project naturally lends
itself to become the stage-set for the performance
stage located on the lake with a direct axial
relationship to the site. Despite the height
restriction, it was imperative that the proposed
facility served as an iconic marker in the skyline.
design that surmounted itself to its horizontal format.
a lasting impression its visitors and viewers whether
they be driving past along the freeway, flying into the
casual walk in the park.
HOTEL+APTS
MEDIA HQ
MEDIA SKY AUDITORIUM.
COMMERCIAL OFFICES.
三栋"门"对称地面对连接着公园和湖的景观，因此我
然背景。由于建筑高度的规定，我们必须克服这个
限制从而在横向上发展成为了一个水平面的地标。
一个建筑的标志性会让其本身给观赏者留下一个深
刻和长久的记忆。不管是行驶在高速公路上的车辆
，还是天空滑翔过的飞机，甚至仅仅路过的行人，
都会忍不住欣赏我们的建筑。
"毛茸球"在门的中央顶部，构成了天线供传送网
络广播。同时也覆盖电脑录像，skype和多种视讯
项目。它将在这个媒体中心成为运行的心脏。
INTERACTIVE FLOOR (GLASS)
uncompromised connection to the ground plane.
We adopted the landscape bridge and used it as a
catalyst to move the address (the "front door") of this
complex into the park and not allow the roadway to
dissect this campus-like facility from its natural
setting. The podium takes on the metaphor of an
undulating landscape with an elaborate mixed-use
program dispersed uniformly to allow for
diversification and unity all at once. There has been
this great opportunity to re-think the workplace
scenarios for a more creative environment. In this
contemporary age, the work and play spaces have
been blurred by the transparencies in our private
and social lifestyles. For many, they have become
one. Much of this change is the outcome of the
information age. The democratic spirit has filtered
into our lives in so many different ways. We have
generated a new "workspace" that allows open
discussions across faculties of intellect that are both
familiar and remote. We have drifted away from the
big office block solution but instead to a flexible
modular system that allows for more daylight and
less mechanical needs. The centre portal becomes
the celebratory portal and its "void" marks the
invisible binary code of the New Media, which is
nothing other than the atmosphere itself!
PARKING.
最重要的是，这个方案想要创造一种和地面连接的
和谐感。我们延续了原有的桥梁来引导人们从公园
进入中间的"门"，而不是让桥来切断彼此的联系，就
好像建筑是自然景观中的一部分。而建筑的基座象
征着起伏的地景，同时融合了不同的功能分区，提
供多样化的服务。也是一个很好的媒介来再创造和
再思考一个在充满创意的环境里的新办公方式。当
代的工作和娱乐空间已经模糊了，原因是我们个人
和社会的生活方式的透明化。对于这样的改变，也
作空间"，一个开放的沟通空间可以让来自不同领域
的人才交流，不管他们是否相识。我门舍弃了大块
体量的建筑，而选择了办公楼为中空的框架结构，
为的是获得更多的日光和更少的机械结构。对于新
媒体的概念：中间的大楼成为了一个庆典门，而"空
间"标志了看不见的二进制代码。而且没有比它更有
气氛的演绎了。
HOTEL.
RETAIL.
i-PLAZA.
EXHIBITION SPACE.
RETAIL
TERRACE OFFICES
TERRACE OFFICES
TERR. OFFICES RETAIL
The "furry ball" atop the central portal consists of
flexible "antennae" for transmission and
wireless broadcasting. It will also house
cam-recorders, skype and various video
projectors. It would become the
organic heart of operations at
can be programmed to become instant digital
billboards. The digital domain would wrap itself
around every face of the building, dissolving its
structure into an enormous pixilated virtual screen.
"wifi"网络结合平台。任何人都可以在这里上传和下
载信息。外部的建筑表皮可以成为电子广告牌。电
子区域包裹了建筑的每个角落，模糊了"碳"实体结构
和视觉虚拟影像屏幕之间的界限。
BALLROOMS / CONFERENCE
1ST FLR.

N
30
60
100m
新 镇 路
新 镇 路
2F
2F
2F
11F
11F
11F
1F
7F
路

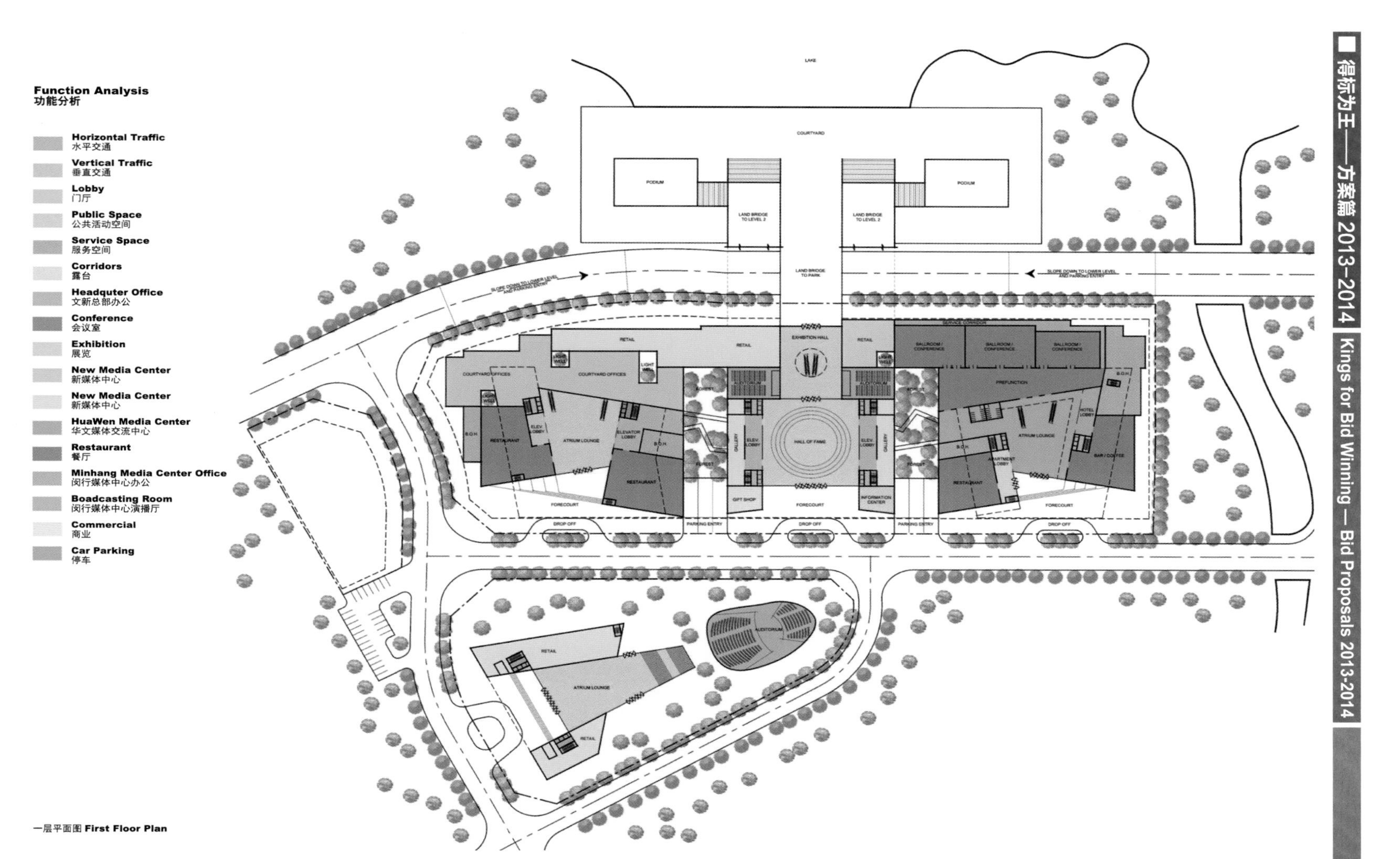

一层平面图 First Floor Plan

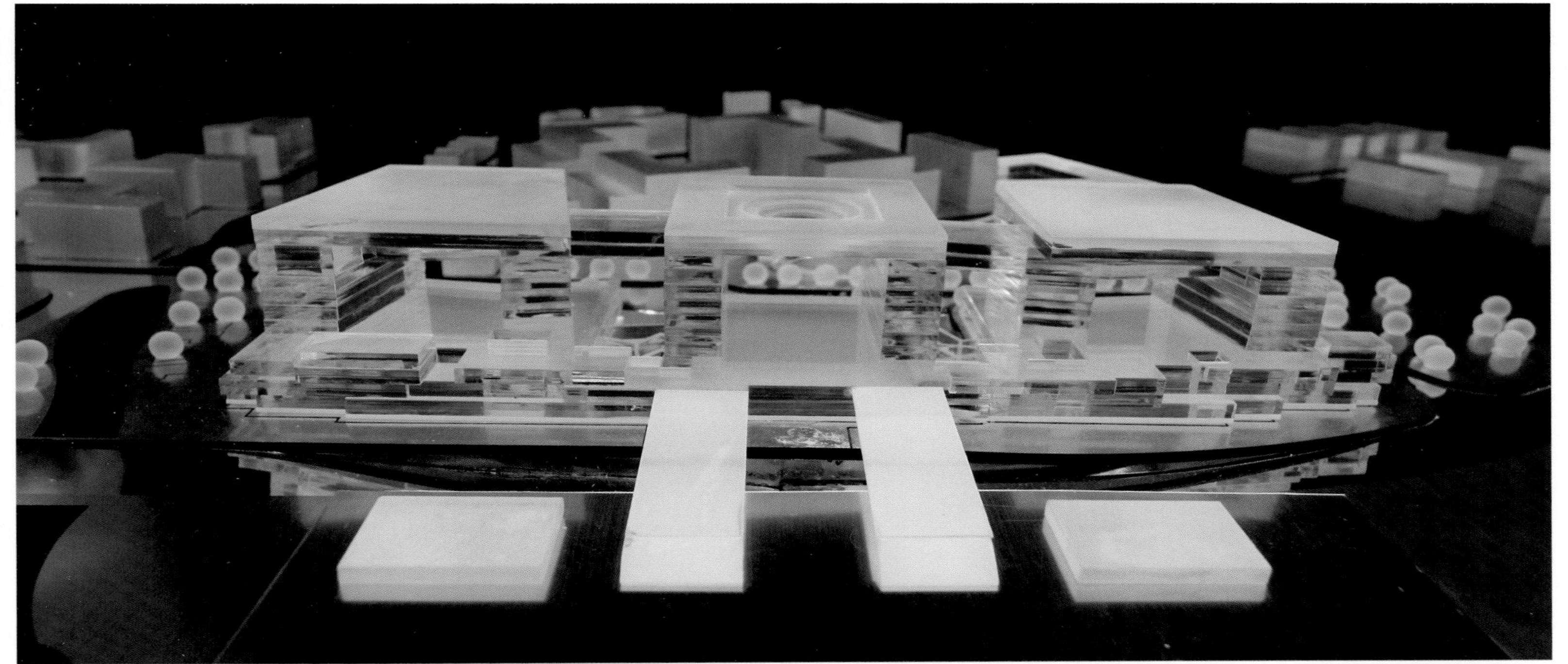

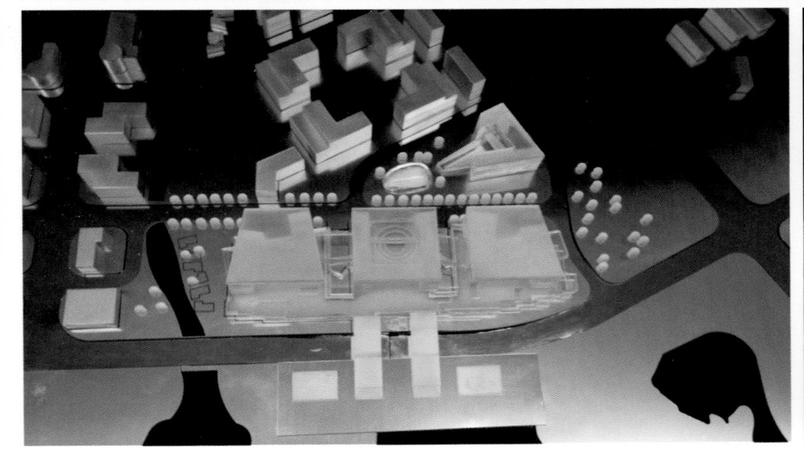

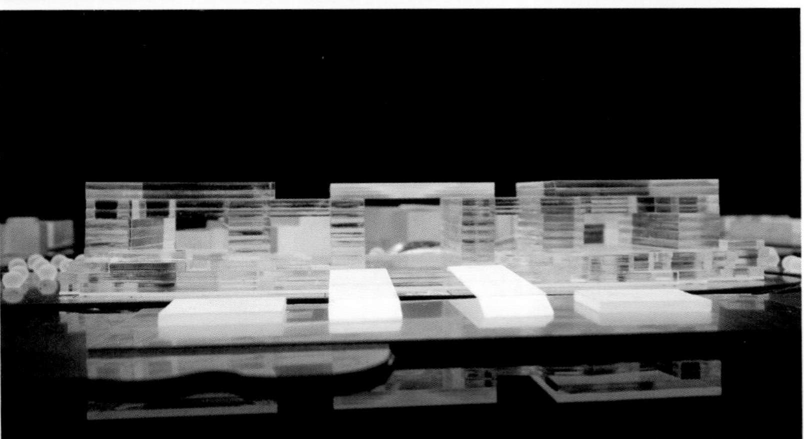

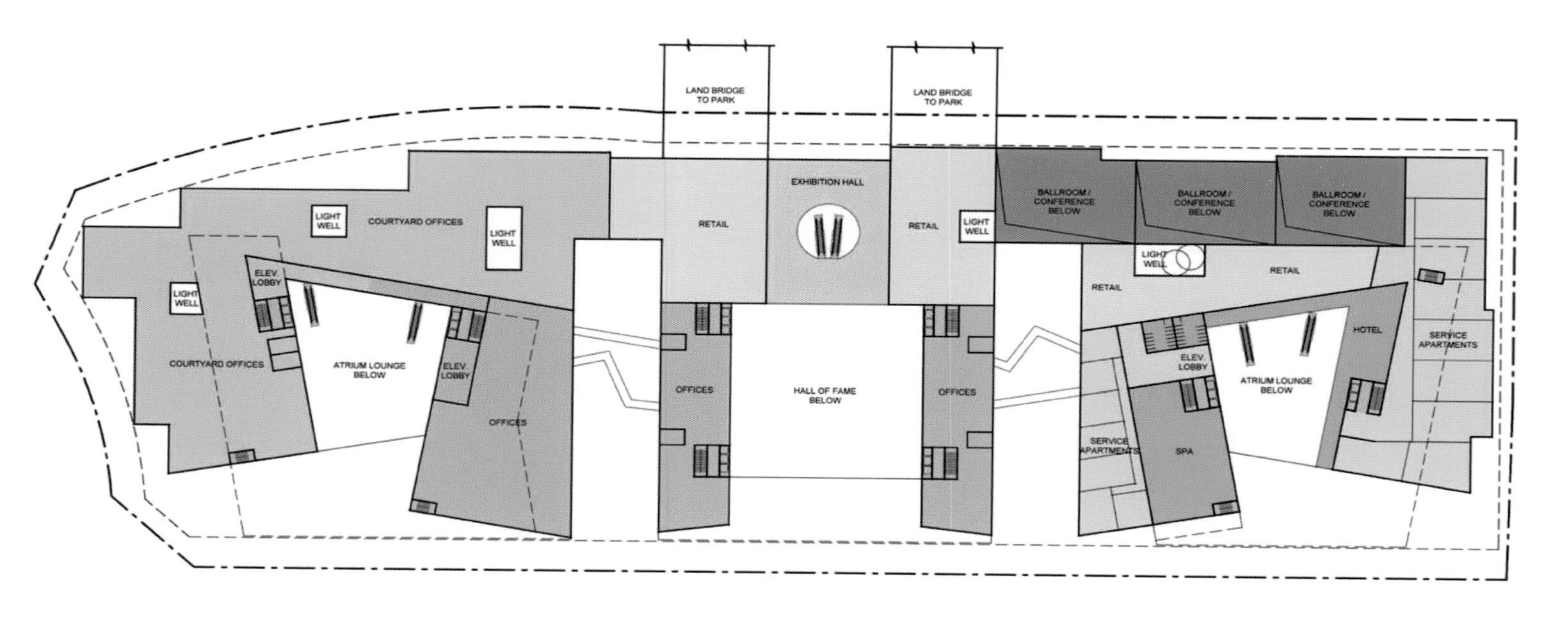

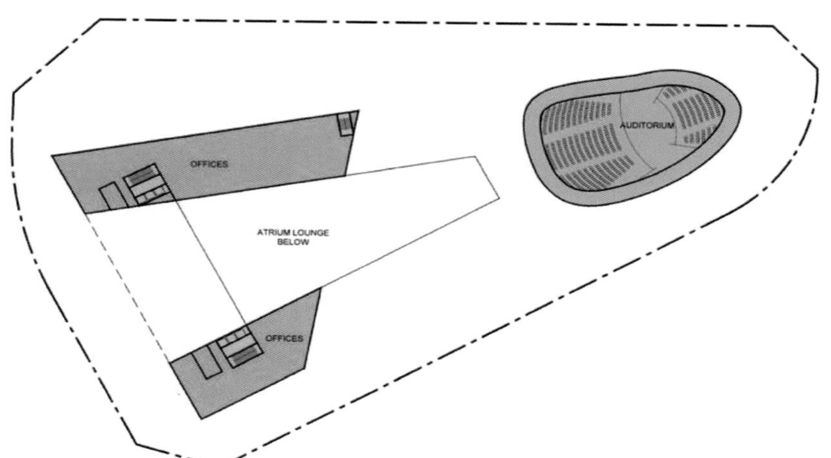

二层平面图 **Second Floor Plan**

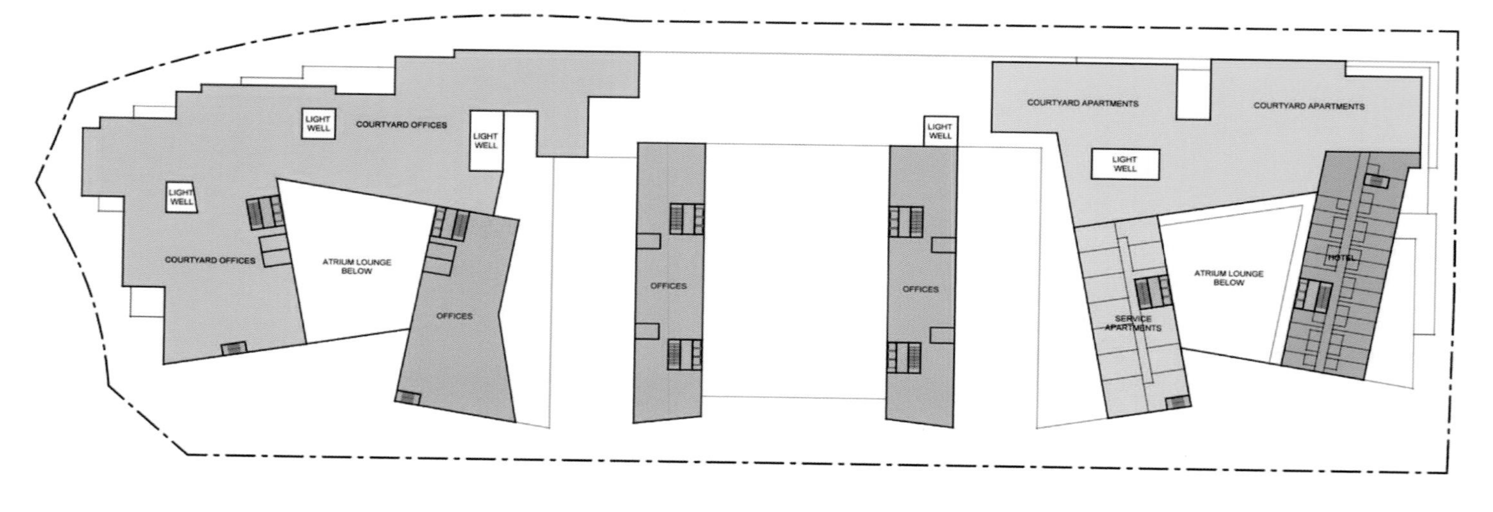

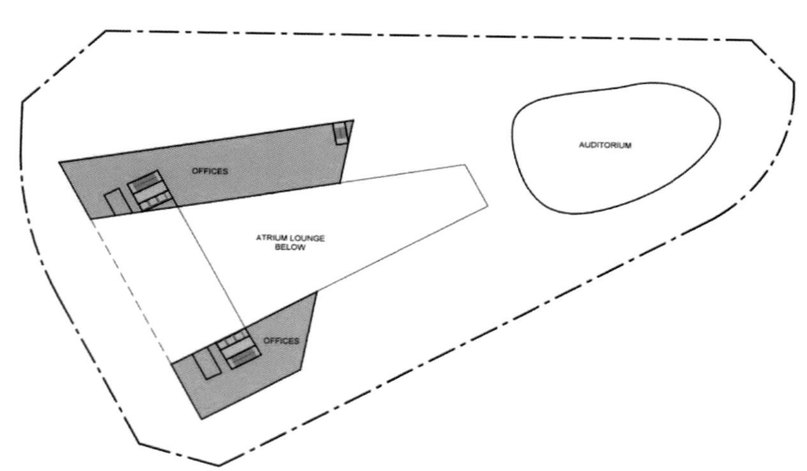

三层平面图 **Third Floor Plan**

Function Analysis
功能分析

Horizontal Traffic
水平交通

Vertical Traffic
垂直交通

Lobby
门厅

Public Space
公共活动空间

Service Space
服务空间

Corridors
露台

Headquter Office
文新总部办公

Conference
会议室

Exhibition
展览

New Media Center
新媒体中心

New Media Center
新媒体中心

HuaWen Media Center
华文媒体交流中心

Restaurant
餐厅

Minhang Media Center Office
闵行媒体中心办公

Boadcasting Room
闵行媒体中心演播厅

Commercial
商业

Car Parking
停车

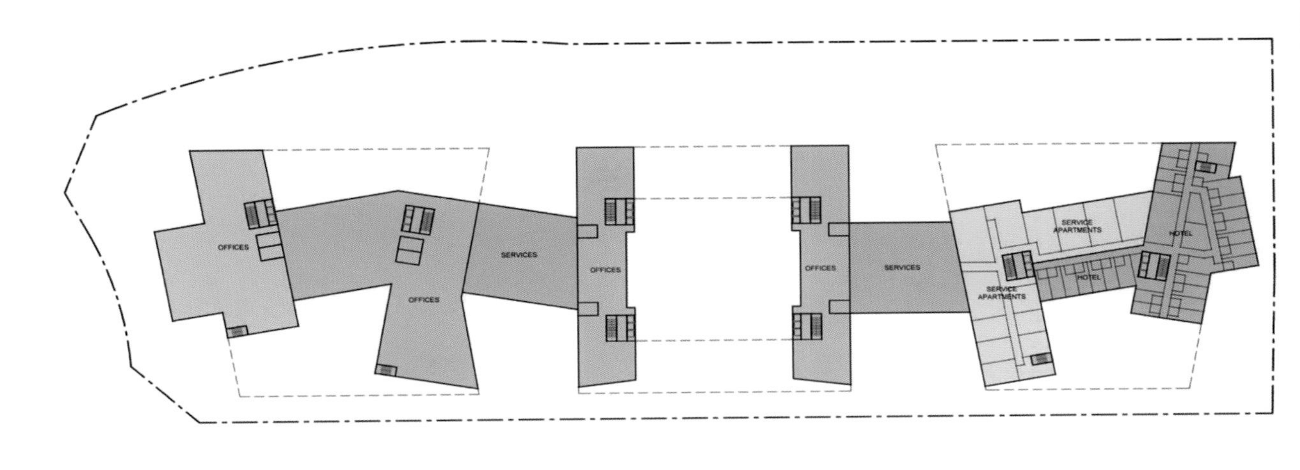

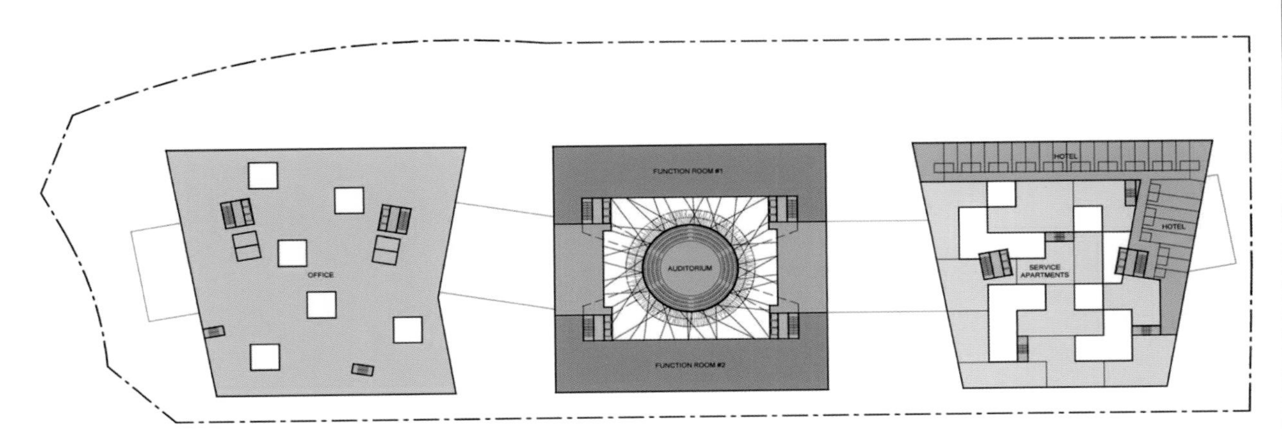

8-11 层平面图 8-11 Floor Plan

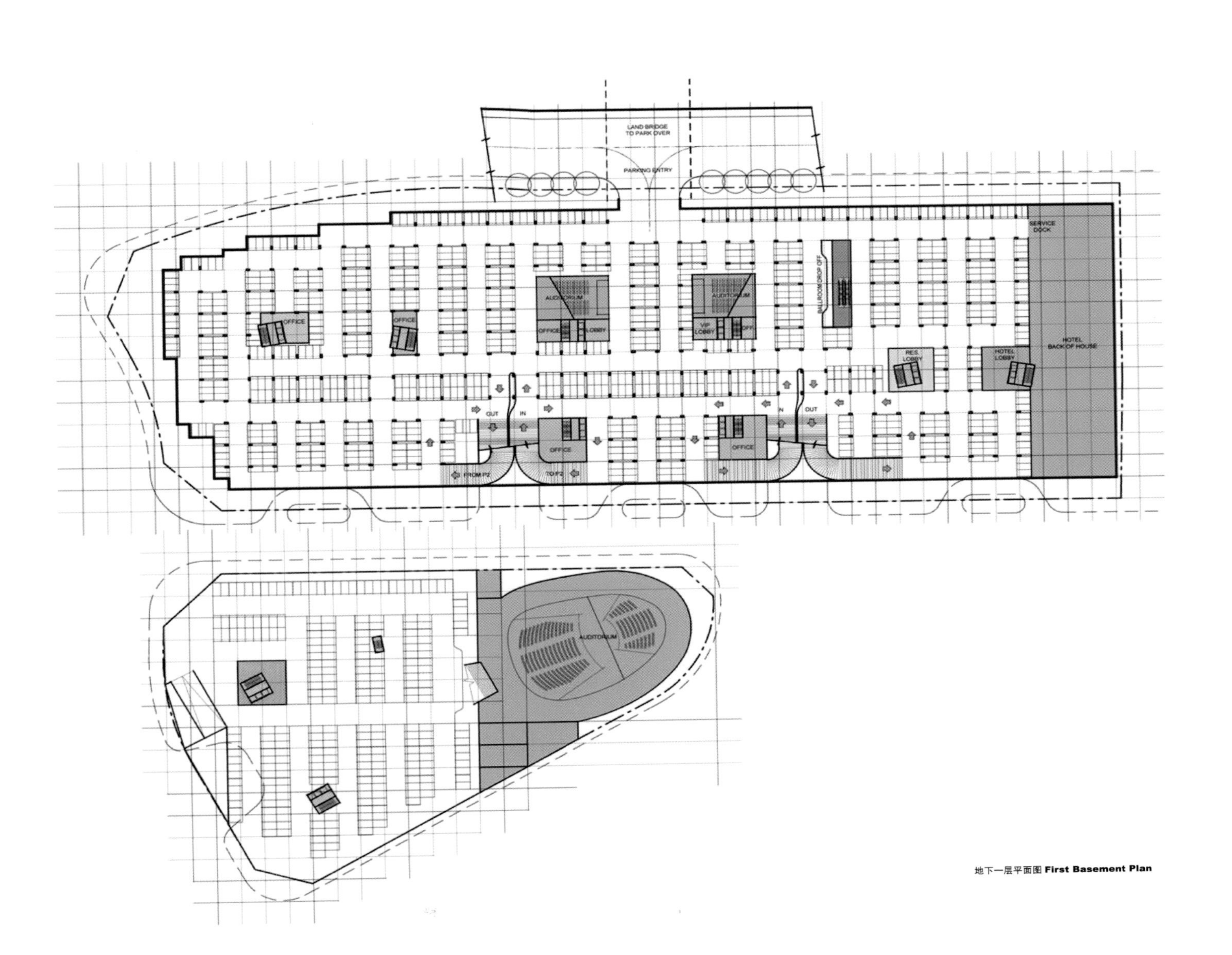

地下一层平面图 First Basement Plan

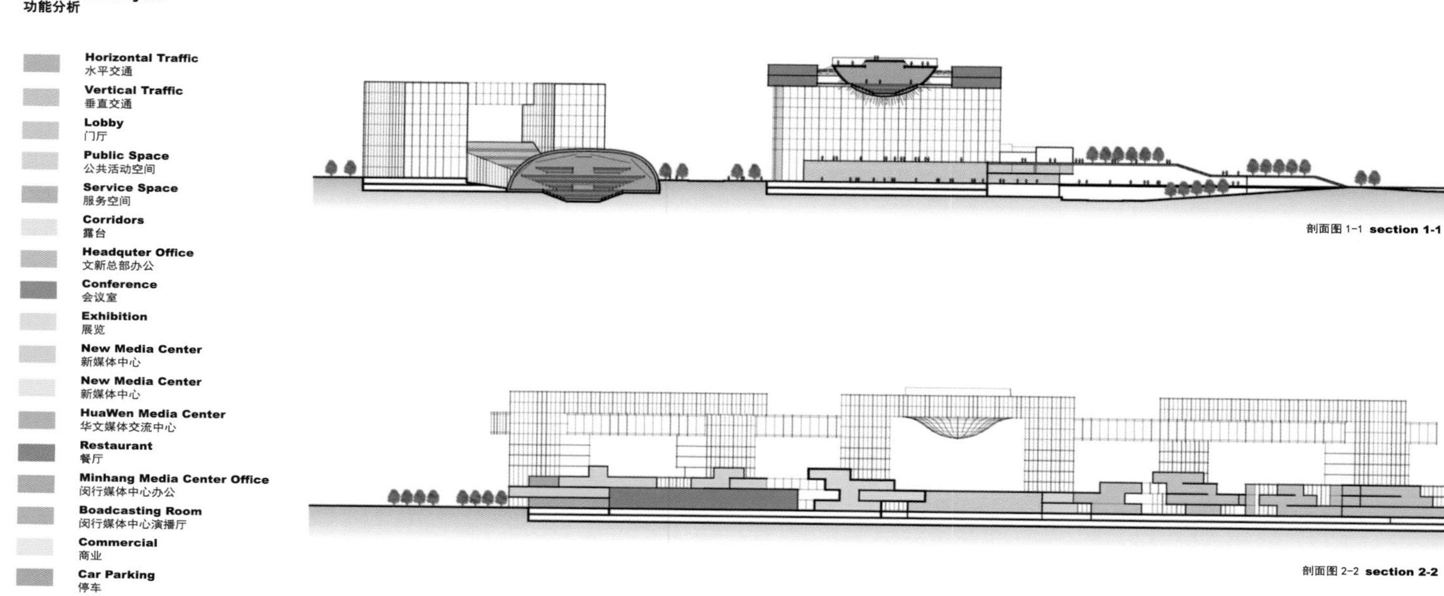

剖面图 1-1 **section 1-1**

剖面图 2-2 **section 2-2**

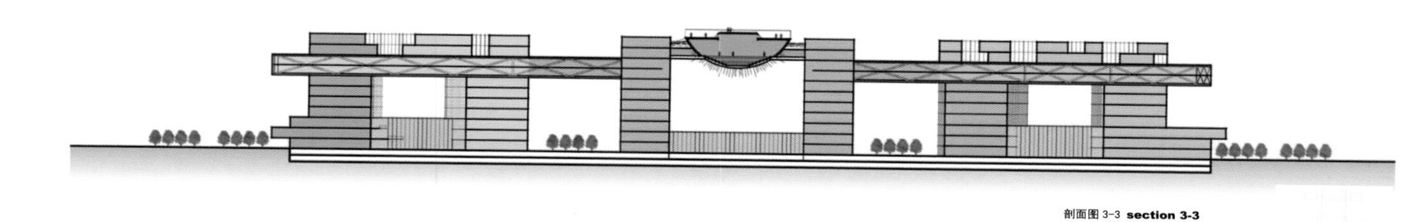

剖面图 3-3 **section 3-3**

东立面（面向文化公园）

西立面（面向商务区）

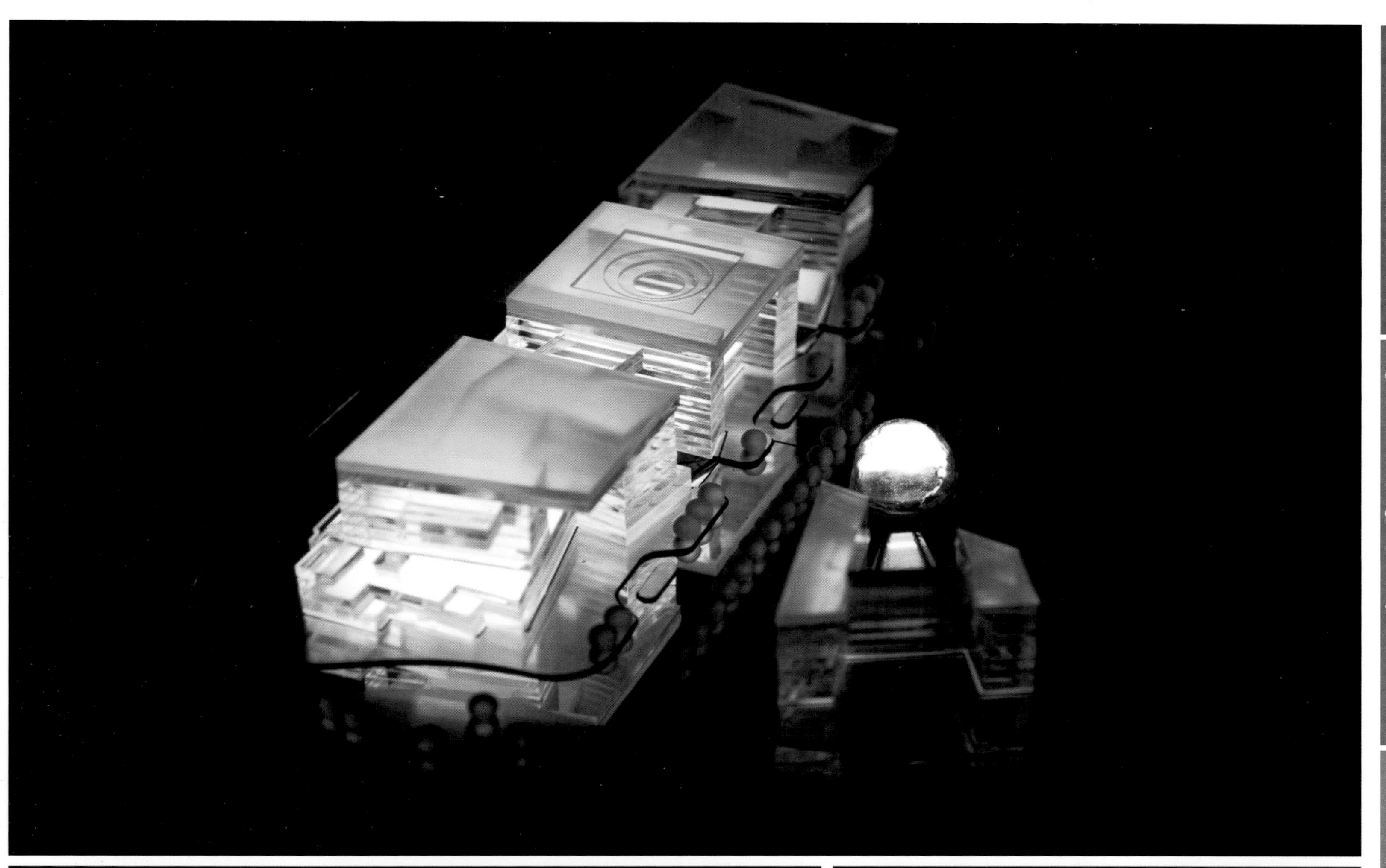

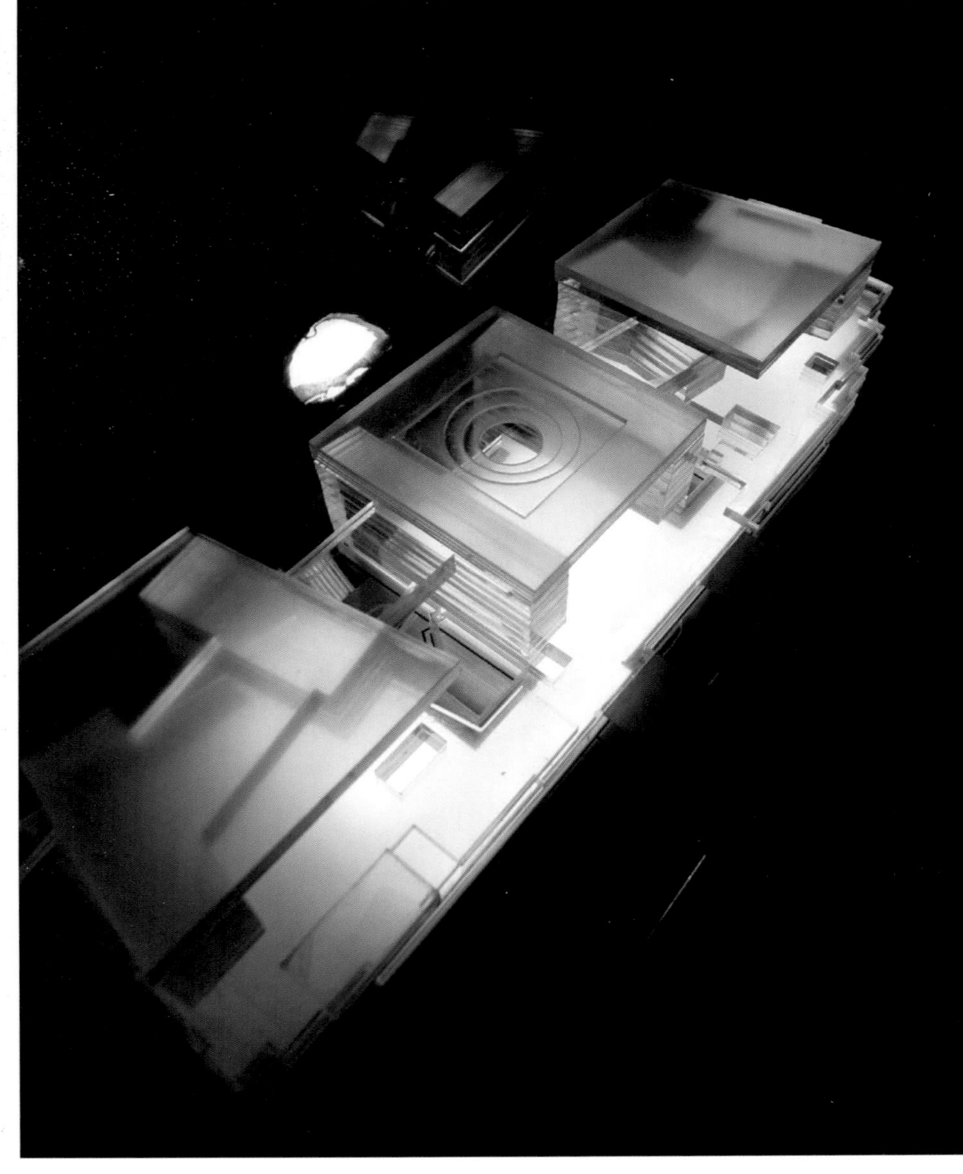

经济技术指标-方案B

项目		数值
总用地面积(平方米)		85000
总建筑面积(平方米)		262088
地上建筑面积(平方米)		172750
其中	文汇新民集团新基地	71304
	新媒体中心	41216
	闵行区梅地亚中心	20334
	华文媒体交流中心	39896
地下建筑面积(平方米)		89338
首层建筑面积(平方米)		30354
覆盖率		35.7%
容积率		2.03
绿地率		25%

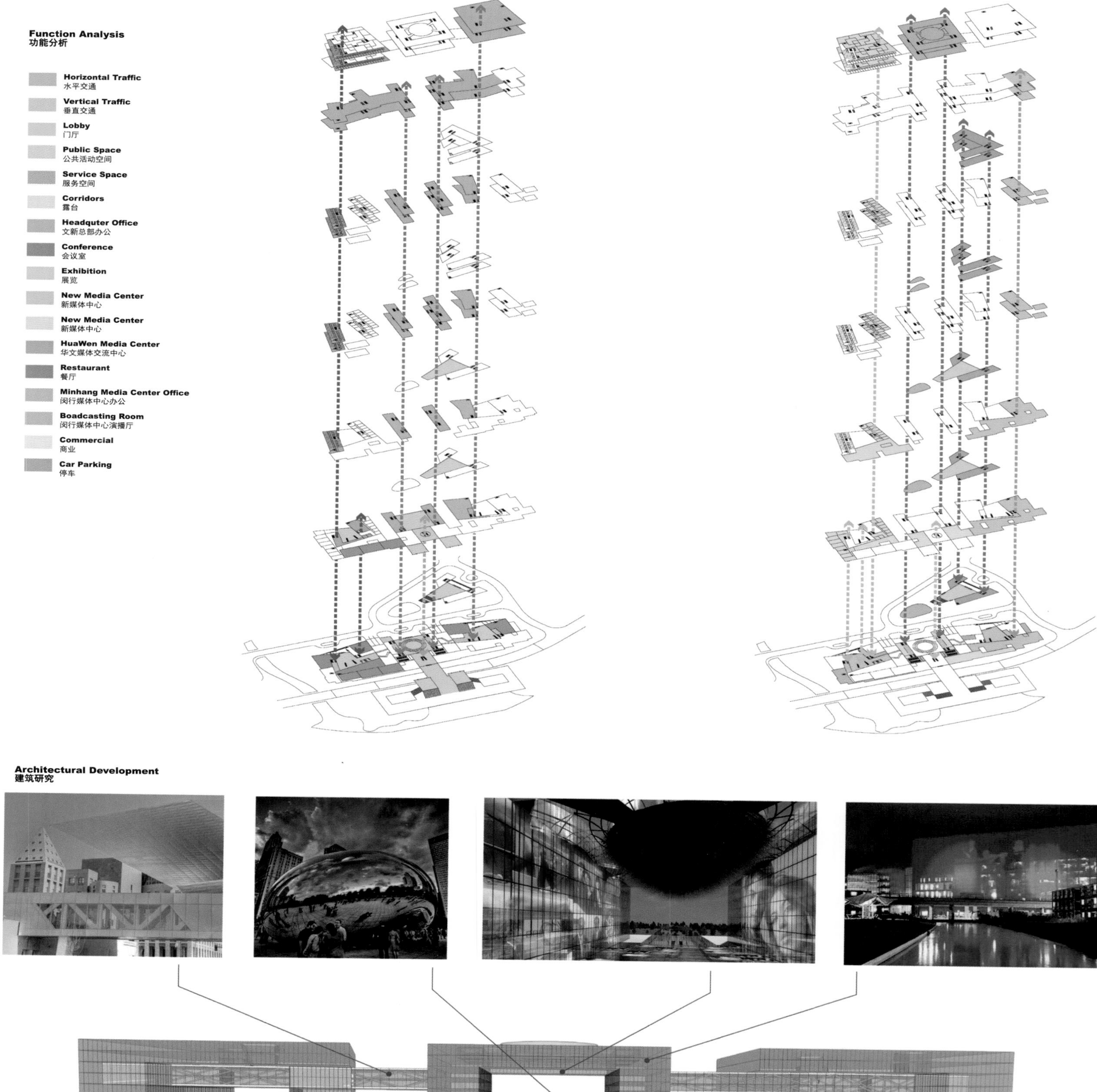

Architectural Development
建筑研究

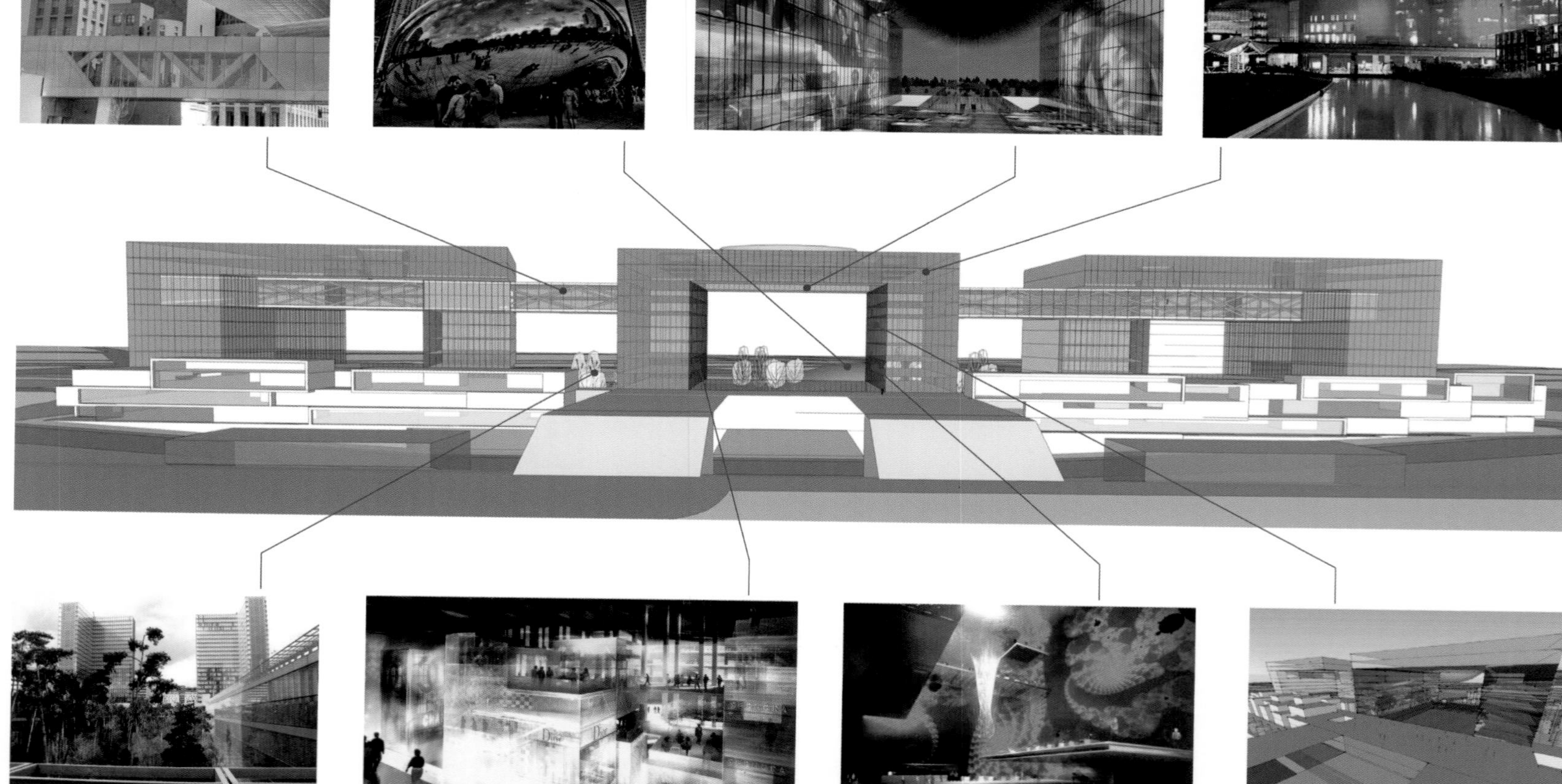

Design Concept

The "void" is our first habitation. The "building" is our second. The project has aimed to make a concentrated volume as a communicative space that carries the memory of people's past and simultaneously projects their senses into the future. The voids can be seen as "portals" for both "seeing out" and "looking back". The three portals represent the three stages of the historical development of media.

Design Feature

The portals are positioned symmetrical to the proposed landscape bridge that connects to the adjacent park and lake. The project naturally lends itself to become the stage-set for the performance stage located on the lake with a direct axial relationship to the site. Despite the height restriction, it was imperative that the proposed facility served as an iconic marker in the skyline.

Most of all, the proposal attempts to make an uncompromised connection to the ground plane. The project has adopted the landscape bridge and used it as a catalyst to move the address (the "front door") of this complex into the park. The podium takes on the metaphor of an undulating landscape with an elaborate mixed-use program dispersed uniformly to allow for diversification and unity all at once.

The low-rise buildings grow from the ground to the sky while large volume buildings seem like dropping from the sky. The connection of the two creates harmonious aesthetics as well as strong visual impact. The top part of the office building in the middle is an "air auditorium" reflecting modern architectural art. It is also the auditorium for awarding "Pritzker" of media field. The auditorium, a cantilevered 30-meter inverted transparent glass structure, can directly connect with the 6,400 square meters roof garden.

匈牙利布达佩斯 Ujpest 城市中心

Ujpest City Center

设计单位：3LHD
项目地址：匈牙利布达佩斯
基地面积：23 133 ㎡
总建筑面积：120 860 ㎡
设计团队：Sasa Begovic　Marko Dabrovic
Tatjana Grozdanic Begovic　Silvije Novak
Paula Kukuljica　Nives Krsnik Rister
Leon Lazaneo　Eugen Popovic
Vibor Granic　Zeljko Mohorovic

Designed by: 3LHD
Location: Budapest, Hungary
Site Area: 23,133 m^2
Gross Floor Area: 120,860 m^2
Project Team: Sasa Begovic, Marko Dabrovic, Tatjana Grozdanic Begovic, Silvije Novak, Paula Kukuljica, Nives Krsnik Rister, Leon Lazaneo, Eugen Popovic, Vibor Granic, Zeljko Mohorovic

项目概况

项目位于布达佩斯北部最重要的主要道路沿线，也位于多瑙河人行步道上的 Újpest 海湾。

WEST ELEVATION

SITE ENTRANCE

SITE ENTRANCE

SITE ENTRANCE

SITE

SITE ENTRANCE

SITE ENTRANCE

SITE ENTRANCE

SITE ENTRANCE

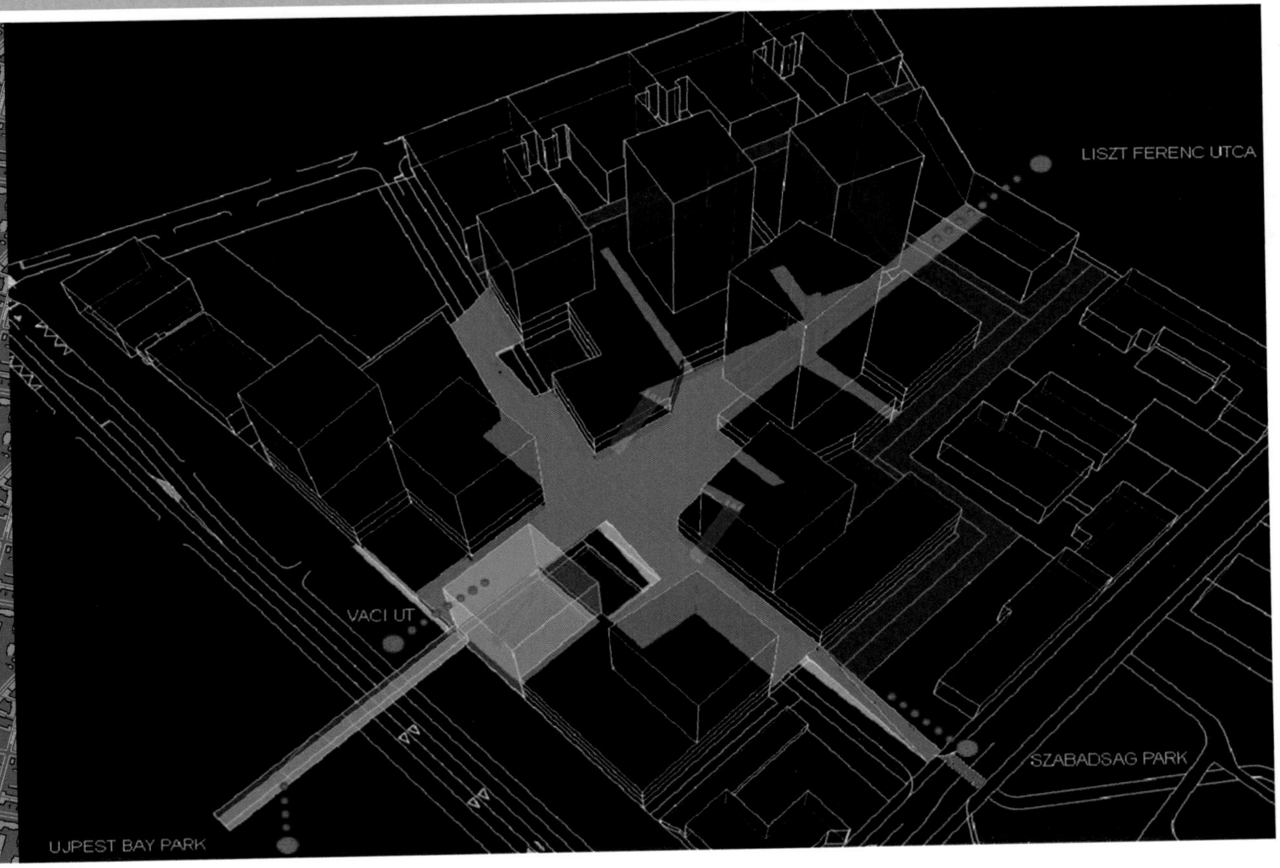

LISZT FERENC UTCA

VACI UT

SZABADSAG PARK

UJPEST BAY PARK

特色酒店布局
CHARACTERISTIC HOTEL LAYOUT

x5 55x25 20x28 5x32

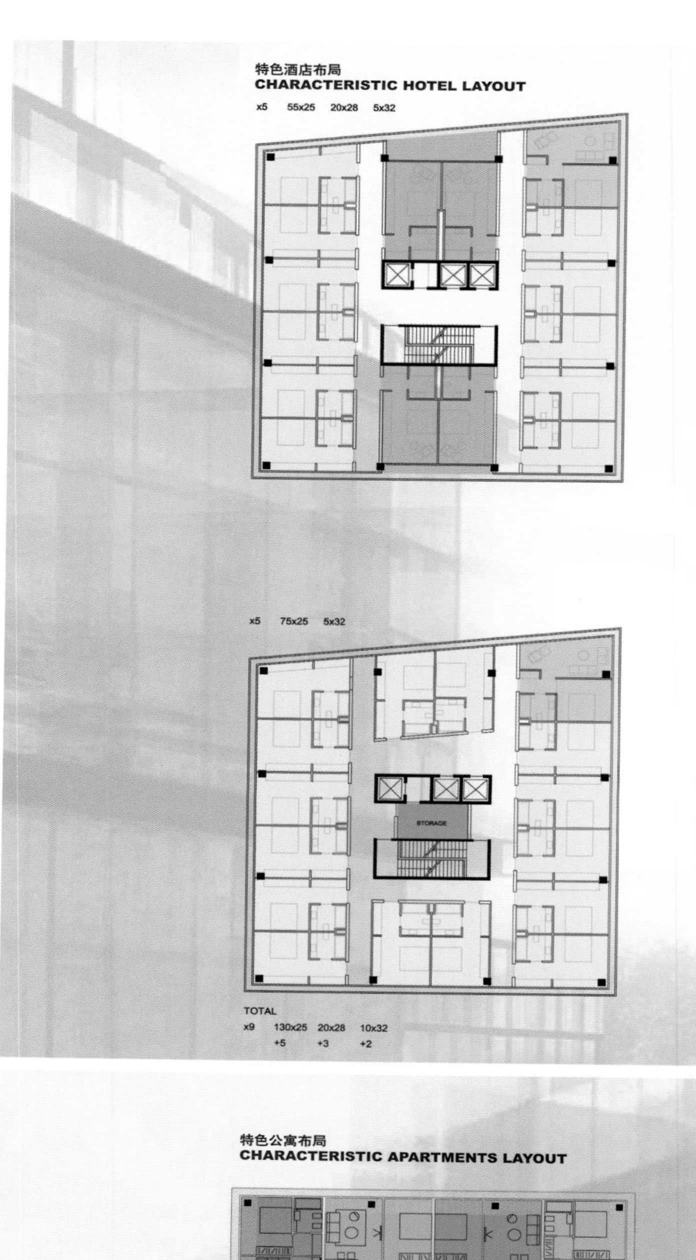

特色办公室布局
CHARACTERISTIC OFFICE LAYOUT

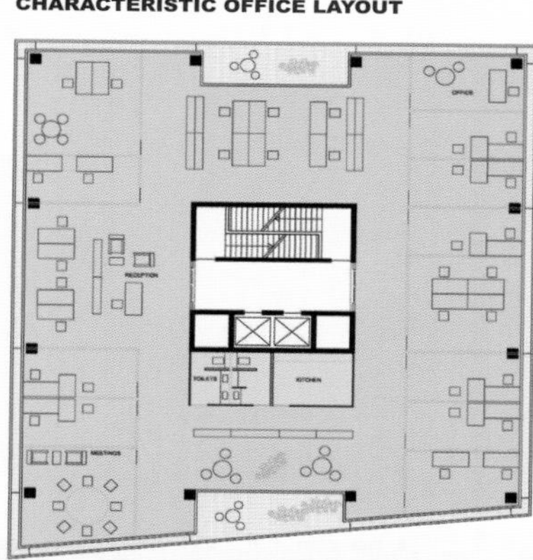

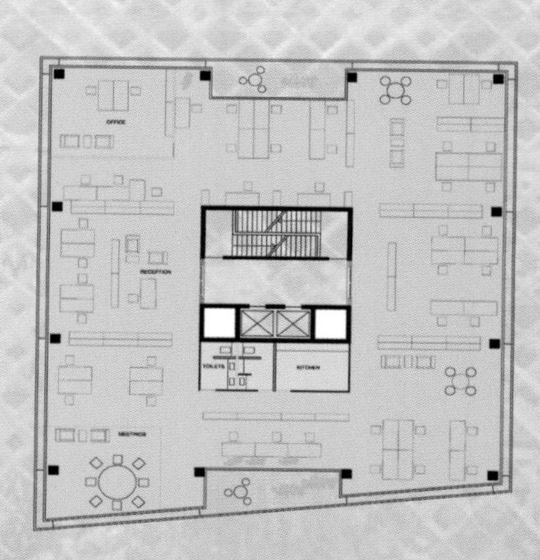

特色公寓布局
CHARACTERISTIC APARTMENTS LAYOUT

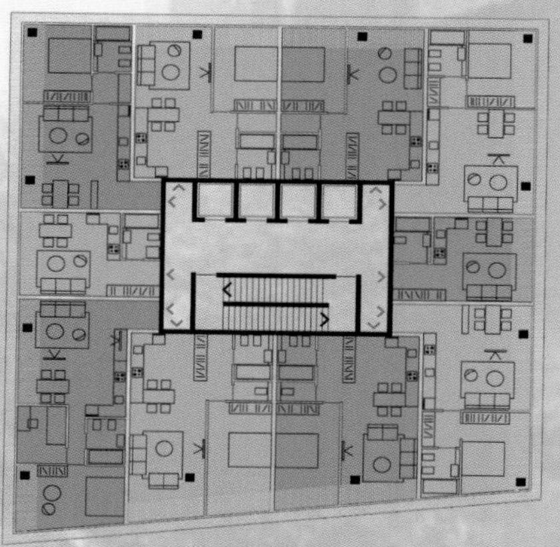

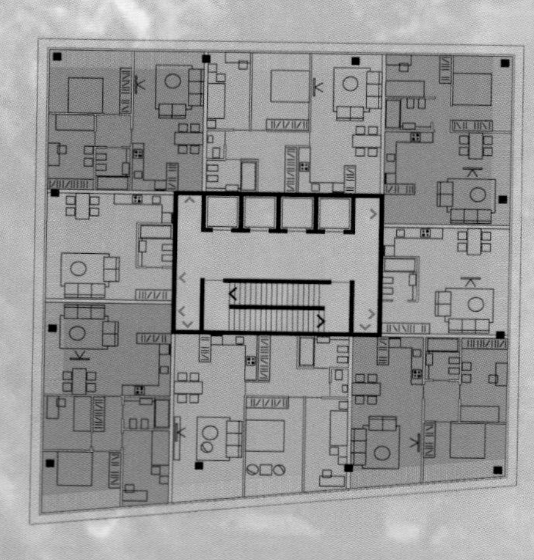

特色公寓布局
CHARACTERISTIC APARTMENTS LAYOUT

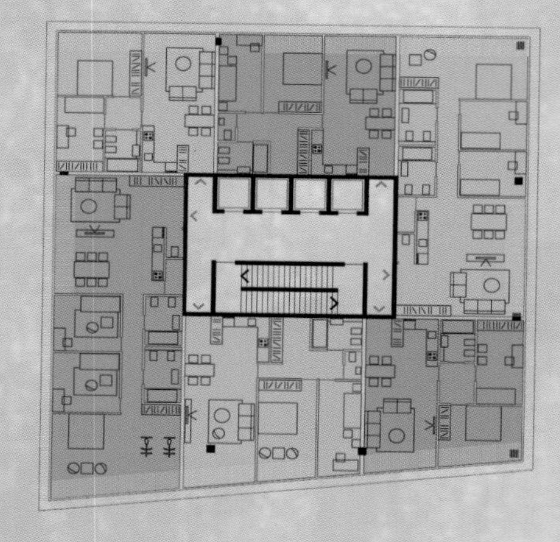

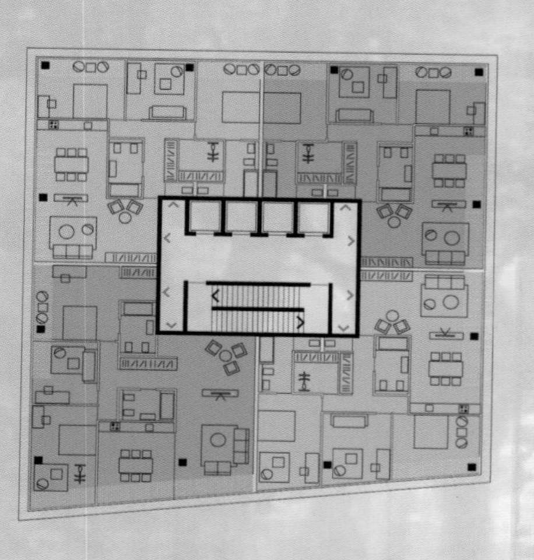

分区
DIV

11 层
11×8=
11×2=

6 层
6×6=36
6×2=12

26 层
26×4=1
26×2=52

3 层
3×4=12

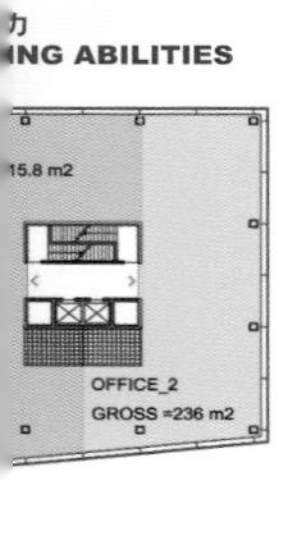

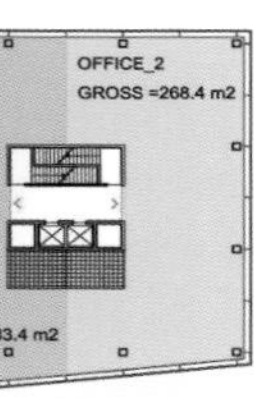

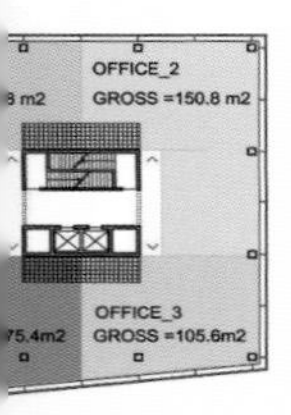

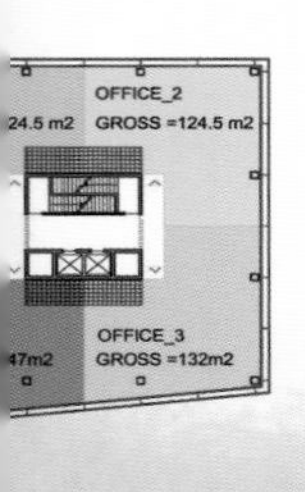

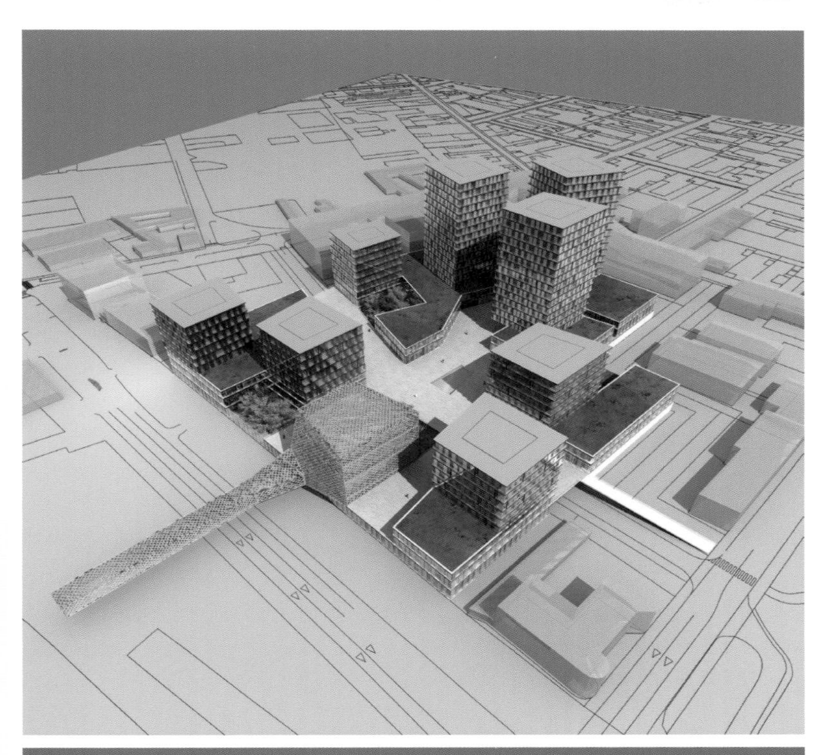

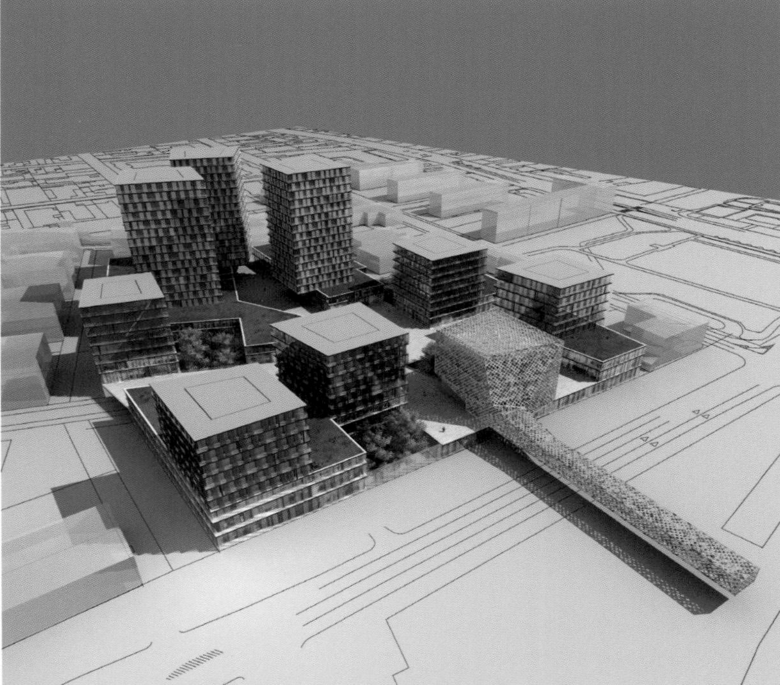

设计理念

基于对城市模型和当地已有建筑的了解，设计方案试图建立起与邻近场地的联系。设计师认为交通流线、步行流线，这些已有的元素都是与建筑相关的，它们不仅是与底层和地下层空间衔接的主要通道，也是构成场地垂直景观的元素。

建筑设计

整个项目由 9 栋不同功能和高度的建筑组成，配备有下沉广场和小型广场，设置在一条抬高的街道上，将周边的场地和建筑联系了起来。建筑根据交通线路和街景来布局排列，并考虑了建筑遮阳的问题。位于场地中心的是三栋住宅楼，边缘部分是商业大楼和酒店，其他的则是以展览、餐厅等公共职能为主的建筑。

在海拔 104 米的水平面上，即 Vaci 道路水平面上，设有该综合体的主要入口、健身中心和游泳池等体育设施，也可经由此处到达酒店和会议中心。酒店、体育设施以及零售区的下面都设有两层的地下停车场，以方便人员的出入。

Profile

The project is positioned along the most important main road of the north Pest, and the Újpest Bay on Danube promenade.

Design Concept

Based on careful analysis of all urban matrices and existing buildings, the scheme tries to find all possible connections to the neighboring sites. Designers believe circulation lines, pedestrian routes and other existing elements are related with buildings. They are not only main channels connecting the ground floor and basement, but also vertical landscape elements of the site.

Architectural Design

Nine buildings, with different functions and heights, compose this project which is also complemented with sunken plazas and small squares. It locates along elevated street connecting surrounding sites with the project. The nine elements are carefully positioned according to the pedestrian corridors and vistas, with taking care to the shadowing of the existing buildings. In the middle of the site, three verticals are housing facilities, and on the two edges verticals are business tower and hotel, other elements are dedicated to the public functions, exhibitions, restaurants and etc.

On level 104 meters over the sea level (Vaci road level) is main entrance in the complex with sport facilities including fitness center and swimming pool. The hotel and conference center can be reached via this entrance. Two-floor garage below hotel, sport facilities and retail district is set to facilitate access.

天津大岛写字楼、五星级酒店及商业街区

Commercial District and Five-Star Hotel & Office Complex on Grand Island, Tianjin

设计单位：ZPLUS 普瑞思国际
主设计师：刘顺校 Julliette.M.S
周湘津 Jan Menges
开发商：天津滨海投资发展有限公司
项目地址：中国天津市
总用地面积：180 000 ㎡
总建筑面积：98 000 ㎡
建筑密度：31%
绿化率：22%
容积率：2.48

Designed by: ZPLUS
Chief Designer: Liu Shunxiao, Julliette.M.S,
Zhou Xiangjin, Jan Menges
Developer: Tianjin Binhai Investment Development Co., Ltd.
Location: Tianjin, China
Land Area: 180,000 m^2
Gross Floor Area: 98,000 m^2
Building Density: 31%
Greening Ratio: 22%
Plot Ratio: 2.48

项目概况

项目位于天津市河西区梅江南生态居住区的公共中心，东侧为梅江南国际俱乐部，西侧是友谊南路，北部为梅江南大岛酒楼，优越的地理位置使得项目的战略地位日益明显。

分区设计

大岛写字楼和五星级酒店

建筑由裙房和 A、B 两栋塔楼构成，A 座为办公楼，B 座为酒店。高层综合体建筑的 1-4 层是连接两栋塔楼的裙房，以弧线形为造型主题，与临近的大岛酒楼的形态相呼应，两者之间形成对话和联系。建筑外檐选用石材、玻璃等现代材质，通过分格和百叶构成协调多样的肌理变化。立面以竖向线条为主，白色的天然石材幕墙赋予两座塔楼端庄、沉稳的气质和雕塑般的立体感。

商业街区

商业街区包括集中商业、商业街和酒店式公寓，设计通过整合现有地块的功能和城市空间，使建筑形态服从城市设计的整体要求，并结合公共绿化带，创造舒适宜人的外部环境。

商业娱乐建筑采取集中与分散相结合的布局方式，相对集中的部分采用方形，沿基地西北角布置，形体虽然整齐，但在立面上做出些许挖空和挑出的处理，丰富主干道沿街立面。内部商业街区采用三层小体量式分散布局，形成尺度宜人的步行环境，并设置独立、单独的楼梯和出入口。

在造型上，商业街区以竖向彩条玻璃为韵律，局部点缀凸出和凹入的体量，形成时尚、酷感的商业街景。后方的公寓则较为规整，设计采用竖向线条形成韵律，加上彩釉玻璃，使之与商业部分浑然一体，构成关联度较高的外立面表皮，达到商业建筑应有的视觉丰富度和包容度。

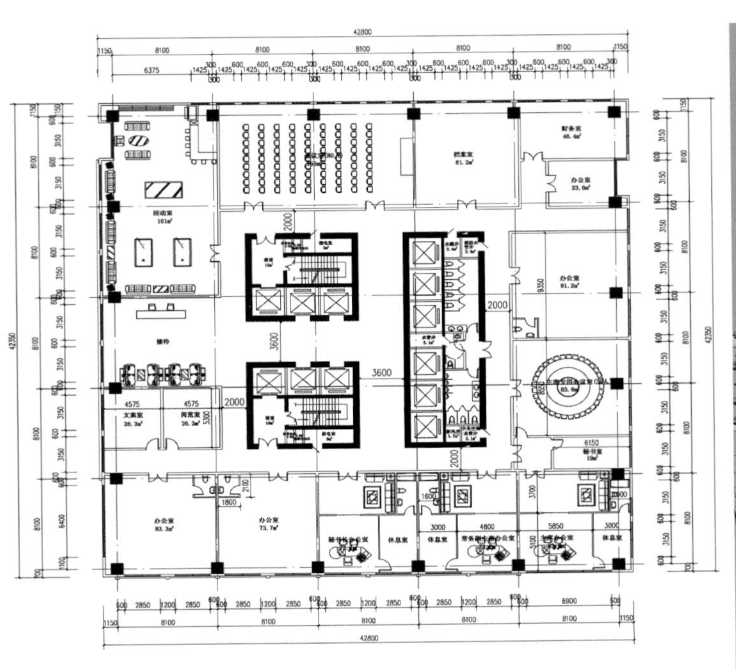

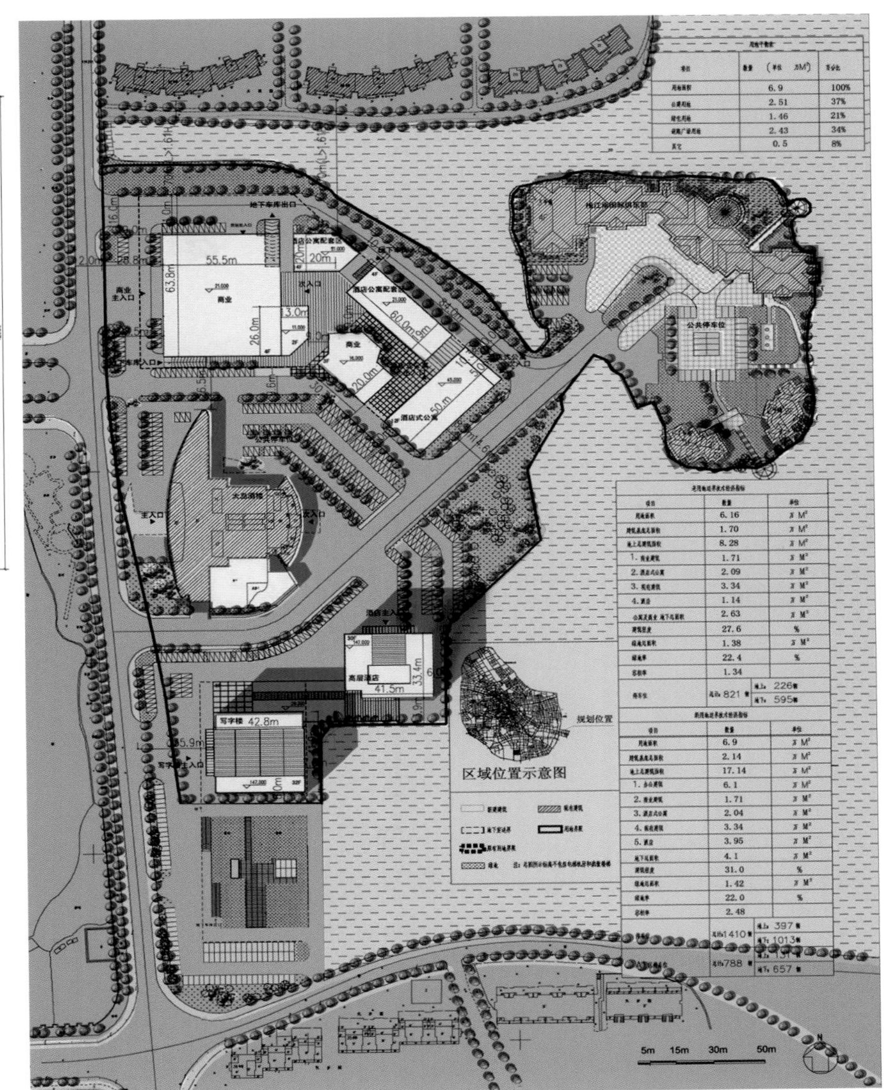

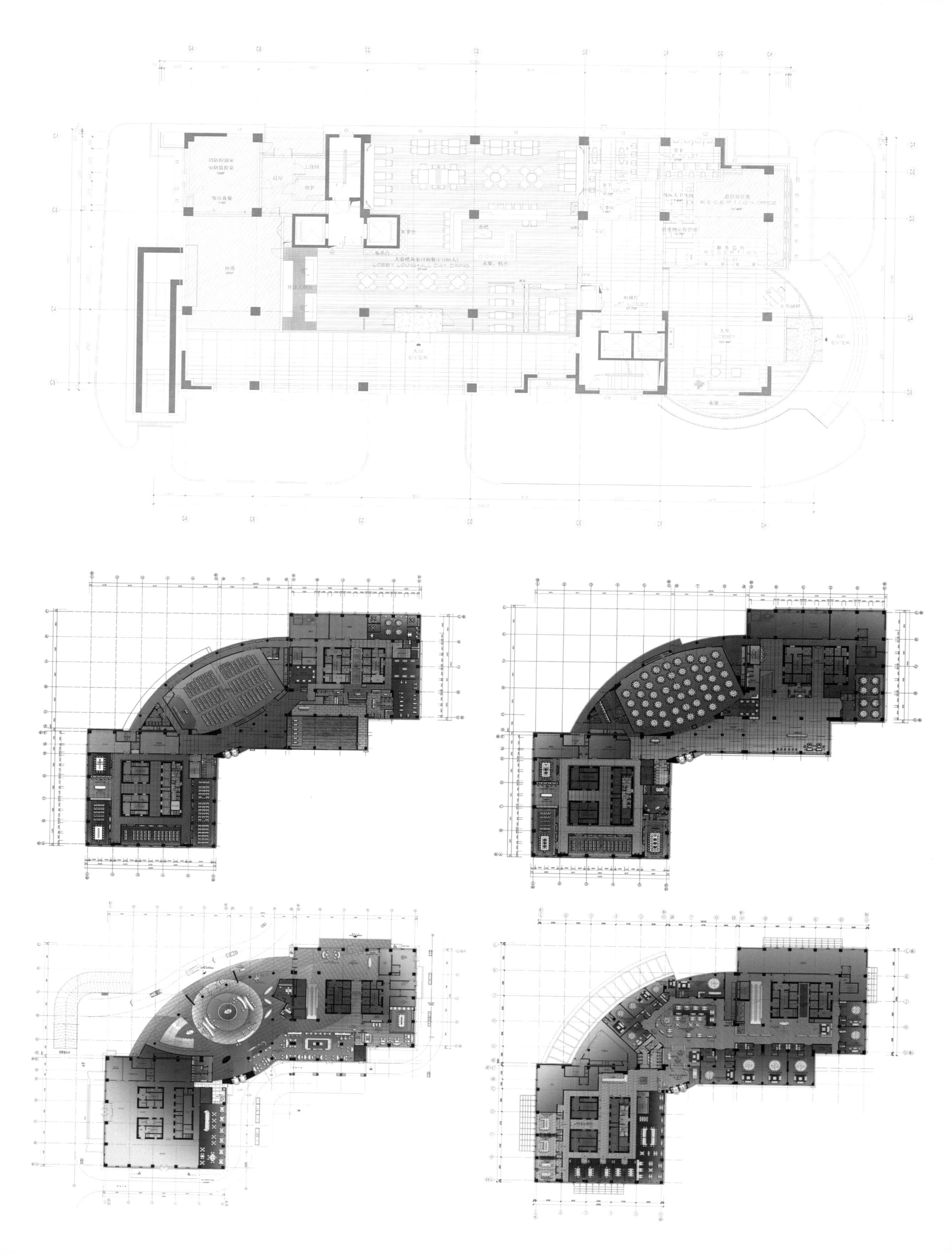

消防控制室
安防监控室
客房备餐
过厅
电井
厨房
开放式厨房
备餐台
大堂吧及全日制餐厅(101人)
LOBBY LOUNGE/ALL DAY DINING
水吧
水果、糕点
VIP
入口
ENTER
男卫
MALE TOILET
女卫
FEMALE TOILET
行李间
总台办公室
RECEPTION OFFICE
贵重物品保管室
服务总台
RECEPTION
电梯厅
LIFT LOBBY
大堂
LOBBY
水景

Profile

The project is located in public center of Meijiangnan Ecological Residential Area of Hexi District, Tianjin City. With Meijiangnan International Club to the east, Youyi South Road to the west and Jiangnan Grand Island Restaurant to the north, the project has an increasingly prominent strategic position due to its superior geographical location.

Subarea Design

Office Complex & Five Star Hotel on Grand Island

Building complex comprises of Tower A, Tower B, Office Building A and Hotel B. The 1-4 floors of high-rise building complex form the podium connecting these two towers. Taking arcs as theme of architectural image, these two towers echo to nearby Grand Island Restaurant and create dialogue and connection between the two. Stone, glass and other modern materials applied in external eaves constitute coordinated diversified texture through grids and louvers. Building elevation is dominated by vertical lines. White stone facades give these two towers grand dignified temperament as well as sculpture-like stereoscopic impression.

Commercial District

Commercial district involves commercial buildings, commercial streets and serviced apartments. By integrating existing functions and urban spaces of the site, the project manages to comply with overall requirements of urban design. Thanks to public green land, an enjoyable external environment is created.

Commercial & entertainment buildings have a layout of combining the centralized with the decentralized. Relatively centralized parts are arranged along the northwest of the site in a square shape. Even though they have regular forms, certain carved and cantilevered parts in the facades have enriched street fronts of the main street. Internal commercial block has a scattered layout of small three-storey volumes which creates fine pedestrian space. There are also independent stairs and entrances there.

Commercial block, taking vertical colored ribbon-shaped glass as rhythm, is locally decorated by concaved and protruded volumes, which forms fashionable cool commercial street landscape. Apartments at the back of commercial block have regular forms. Vertical rhythmic lines and colored glazing glass of these apartments coordinate with the commercial block, making mutually related facades and achieving expected visual richness and tolerance of commercial buildings.

澳大利亚悉尼 The Entrance

The Entrance

设计单位：Tony Owen Partners
项目地址：澳大利亚悉尼

Designed by: Tony Owen Partners
Location: Sydney, Australia

项目概况

The Entrance 是悉尼北部的一个滨海旅游度假小镇，这个多功能的项目构建在小镇主要道路与海滨拐角处。

建筑设计

项目由 12 000 平方米的零售裙楼和其上的豪华住宅塔楼组成。三层的裙楼包括一家超市、滨海餐厅、百货商店、多功能中心以及体育馆。住宅塔楼设想的阳台边缘游泳池和大型居住空间将为当地的住宅设计提供新的参照。

设计采用的几何体源自河口的沙滩因受沿海潮汐的影响而形成的曲折形态。交叉的鸡蛋形建筑结构，既赋予了建筑独特的外观，同时也确保了所有的阳台都有双倍的高度，可最大限度地吸收太阳光。

在设计过程中，设计师运用了立体建模软件，以了解这个复杂几何体的结构与金属外壳之间的相互作用。设计师通过对结构和立面的元素建模，精准地把握空间中的任一元素与其他元素之间的关系，从而将整个建筑变为一个统一、简单明了的整体。

Profile

The Entrance is a coastal resort town north of Sydney. This mixed use development occupies a site known as the “Key Site” as it is the focal site in the town located on the corner of the main street and the scenic waterfront.

Architectural Design

The development consists of 12,000 square meters of tourist related retail podium and a luxury apartment tower above. The 3 storey podium contains a supermarket, specialty stores, a waterfront restaurant level, as well as department store, function centre and gym. The residential tower will set new standards for the area with proposed balcony edge swimming pools and grand living spaces.

The geometry of the deign is derived from the curved patterns created in the estuarine marsh sands due to the effects of the coastal tidal flows. The unique cross over egg shaped design ensures that all balconies have a double height space to maximize sun penetration.

The project has utilised 3-D modelling software in the design to know how the structure interacted with the metal cladding in this complex geometry. By modelling every element of structure and façade designers know exactly how each piece is related to another in space. In this way, the building becomes a unified and quite straightforward whole.

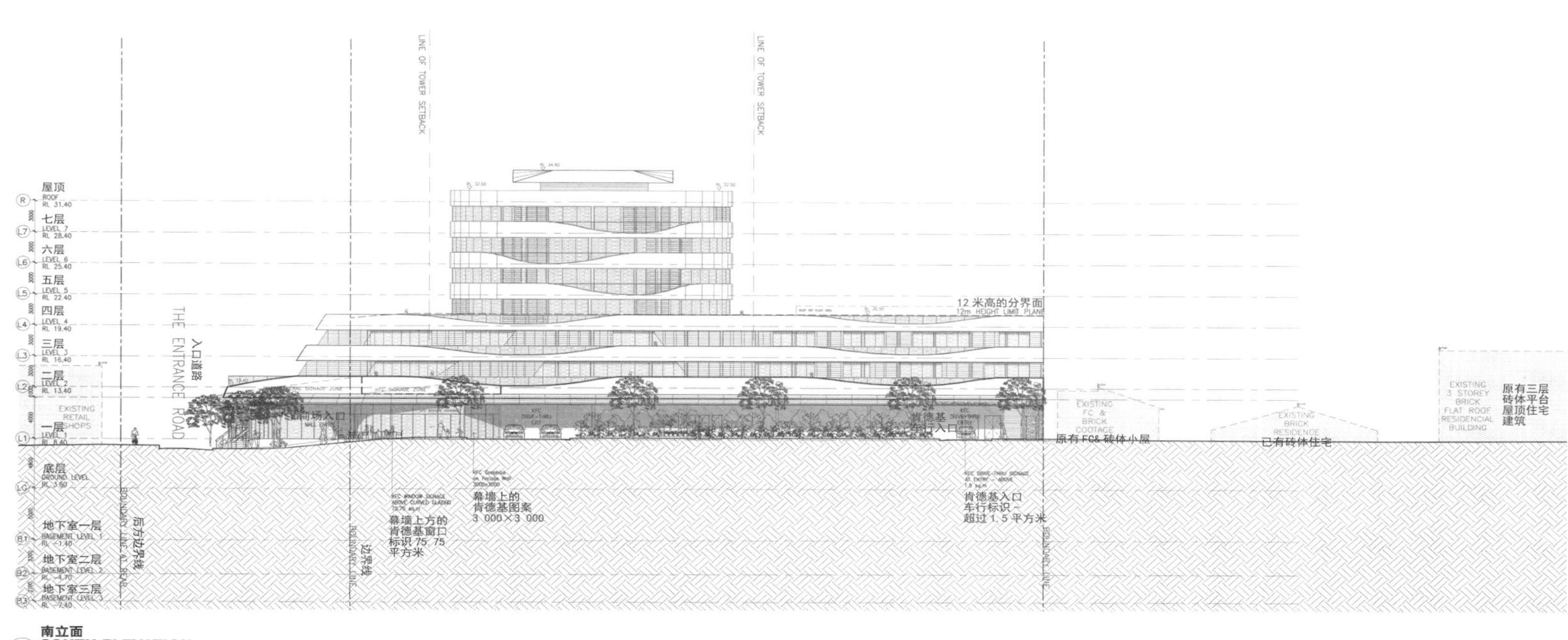

南立面
SOUTH ELEVATION
从滨海大道观测
VIEW FROM OCEAN PARADE

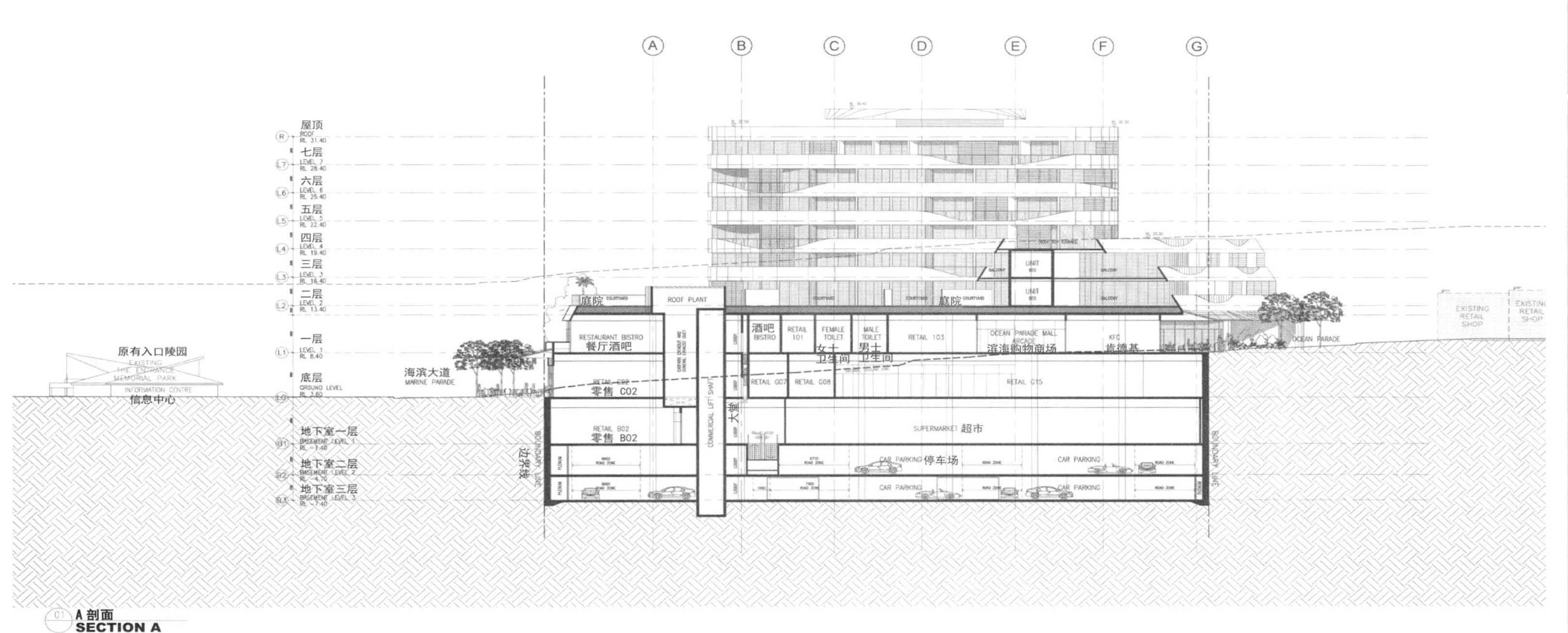

A 剖面
SECTION A

屋顶 ROOF RL 30.40
七层 LEVEL 7 RL 27.40
六层 LEVEL 6 RL 24.40
五层 LEVEL 5 RL 21.40
四层 LEVEL 4 RL 18.40
三层 LEVEL 3 RL 15.40
二层 LEVEL 2 RL 11.90
一层 LEVEL 1 RL 8.40
底层 GROUND LEVEL RL 3.60
地下室一层 BASEMENT LEVEL 1 RL -1.40
地下室二层 BASEMENT LEVEL 2 RL -4.70
地下室三层 BASEMENT LEVEL 3 RL -7.40
地下室四层 BASEMENT LEVEL 4 RL -10.10
地下室五层 BASEMENT LEVEL 5 RL -12.80
地下室六层 BASEMENT LEVEL 6 RL -15.50
地下室七层 BASEMENT LEVEL 7 RL -18.20

24 米高分界面 24m HEIGHT LIMIT PLANE
12 米高分界面 12m HEIGHT LIMIT PLANE
信息中心 INFORMATION CENTRE
海滨大道 MARINE PARADE
滨海大道 OCEAN PARADE
露台 TERRACE 景观区 LANDSCAPED AREA
单元生活/就餐 UNIT LIVING/DINING
机房 PLANT ROOM
厨房 KITCHEN 浴室 BATH
UNIT
消防楼梯井 FIRE STAIR
LIFT SHAFT 电梯井
COMMUNITY SPACE 社区空间
LOBBY
RESIDENTIAL LOBBY
COMMERCIAL OFFICE 04 商业办公室 04
RETAIL 108
DRIVE THRU
RESIDENTIAL LOBBY 住宅大厅
D.D.S.
DIRECT DISCOUNT STORE
花盆 PLANTER
RETAIL B06
超市 SUPERMARKET
房屋后面的超市 BACK OF HOUSE
停车场 CAR PARKING
ROAD ZONE
后边界线 REAR BOUNDARY LINE
边界线 BOUNDARY LINE

B 剖面 SECTION B
比例 1:200@A1 SCALE 1:200@A

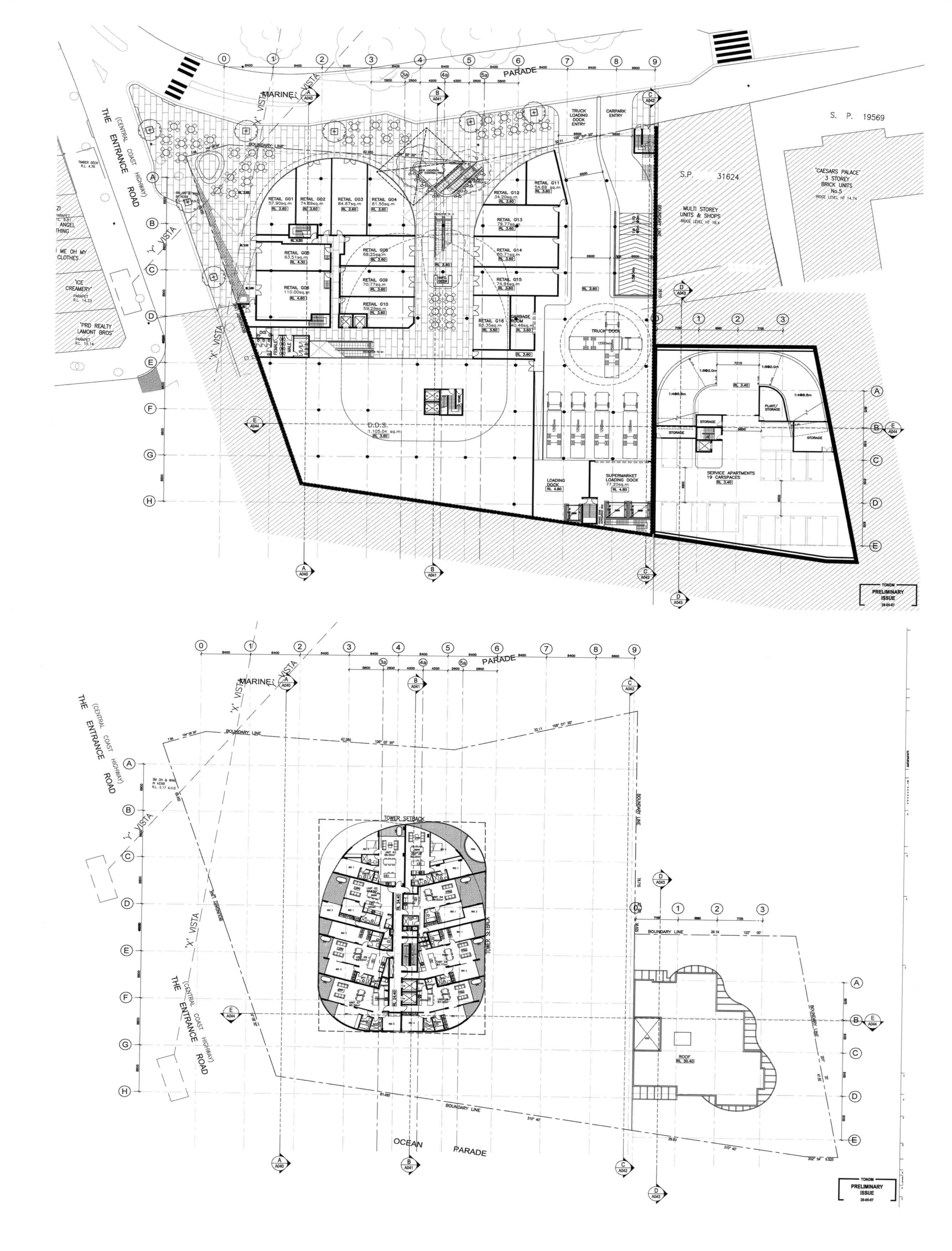

MARINE PARADE
THE ENTRANCE ROAD
(CENTRAL COAST HIGHWAY)
'X' VISTA
'Y' VISTA
BOUNDARY LINE
TRUCK LOADING DOCK ENTRY
CARPARK ENTRY
S. P. 19569
S.P. 31624
MULTI STOREY UNITS & SHOPS
'CAESARS PALACE' 3 STOREY BRICK UNITS No.5
RETAIL G01 57.90sq.m
RETAIL G02 74.69sq.m
RETAIL G03 84.67sq.m
RETAIL G04 81.55sq.m
RETAIL G05 63.51sq.m
RETAIL G06 110.00sq.m
RETAIL G08 68.25sq.m
RETAIL G09 70.77sq.m
RETAIL G10 59.29sq.m
RETAIL G11 54.69 sq.m
RETAIL G12 34.20sq.m
RETAIL G13 79.77sq.m
RETAIL G14 80.71sq.m
RETAIL G15 74.94sq.m
RETAIL G16 68.35sq.m
GARBAGE ROOM 40.46sq.m
TRUCK DOCK
D.D.S. 1,105.04 sq.m
LOADING DOCK
SUPERMARKET LOADING DOCK 77.20sq.m
FEMALE
MALE
STORAGE
PLANT/ STORAGE
SERVICE APARTMENTS 19 CARSPACES
'ICE CREAMERY'
'PRD REALTY LAMONT BROS'
ME OH MY CLOTHES
PRELIMINARY ISSUE
TOWER SETBACK
ROOF RL 30.40
OCEAN PARADE
PRELIMINARY ISSUE

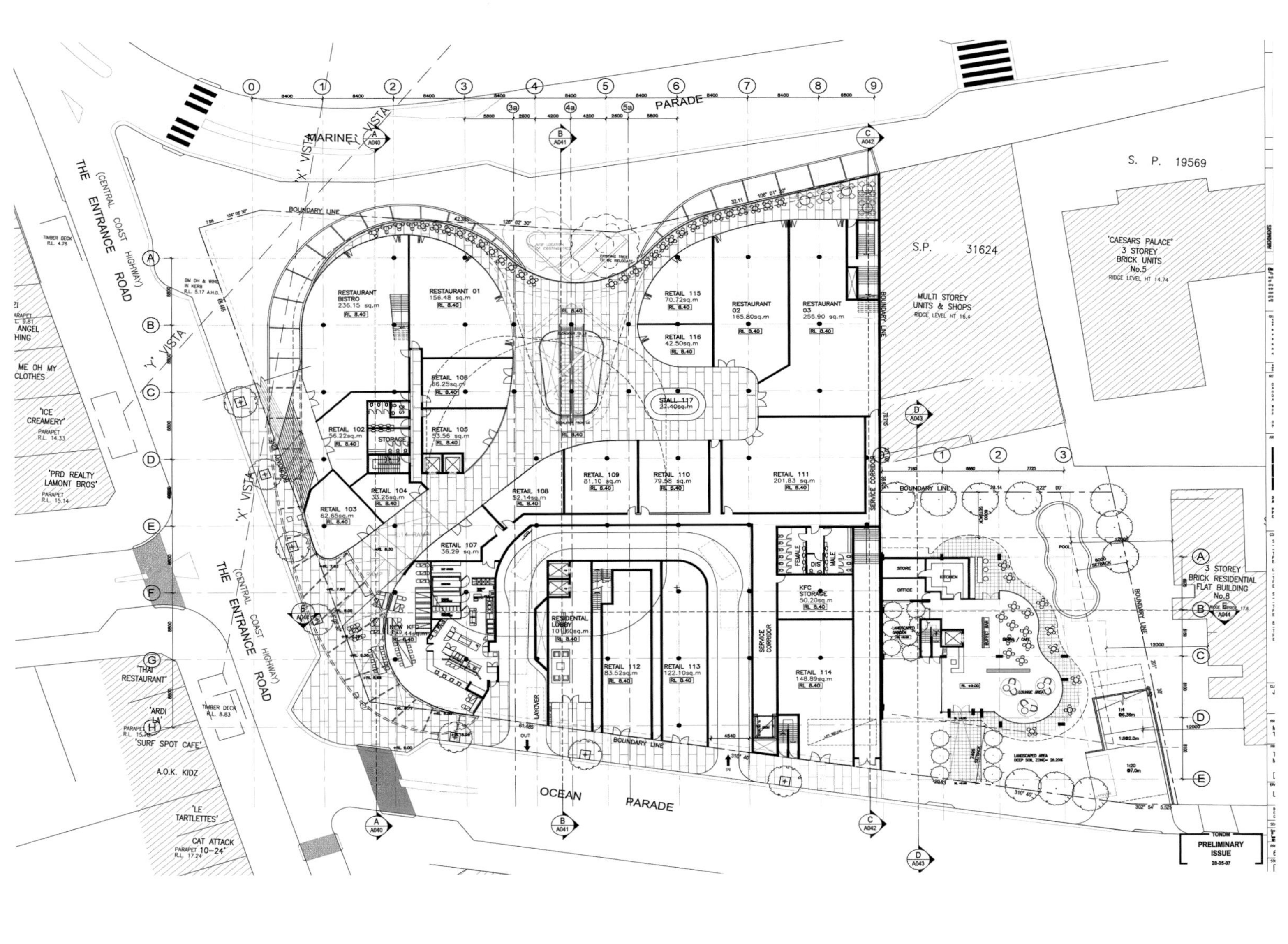
MARINE PARADE
OCEAN PARADE
THE ENTRANCE ROAD (CENTRAL COAST HIGHWAY)
'X' VISTA
'Y' VISTA
BOUNDARY LINE
S. P. 19569
S.P. 31624
'CAESARS PALACE' 3 STOREY BRICK UNITS No.5
MULTI STOREY UNITS & SHOPS
RESTAURANT BISTRO 236.15 sq.m
RESTAURANT 01 156.48 sq.m
RESTAURANT 02 165.80sq.m
RESTAURANT 03 255.90 sq.m
RETAIL 115 70.72sq.m
RETAIL 116 42.50sq.m
RETAIL 106 66.25sq.m
RETAIL 105 53.56 sq.m
RETAIL 102 56.22sq.m
RETAIL 103 62.65sq.m
RETAIL 104 33.26sq.m
RETAIL 108 52.14sq.m
RETAIL 107 36.29 sq.m
RETAIL 109 81.10 sq.m
RETAIL 110 79.58 sq.m
RETAIL 111 201.83 sq.m
STALL 117 32.40sq.m
RETAIL 112 83.52sq.m
RETAIL 113 122.10sq.m
RETAIL 114 148.89sq.m
RESIDENTAL LOBBY
NEW KFC
KFC STORAGE 50.20sq.m
SERVICE CORRIDOR
STORE
OFFICE
KITCHEN
POOL
LOUNGE AREA
3 STOREY BRICK RESIDENTIAL FLAT BUILDING No.8
'THAI RESTAURANT'
'SURF SPOT CAFE'
A.O.K. KIDZ
'LE TARTLETTES'
CAT ATTACK
'ICE CREAMERY'
'PRD REALTY LAMONT BROS'
ME OH MY CLOTHES
ANGEL HING
PRELIMINARY ISSUE

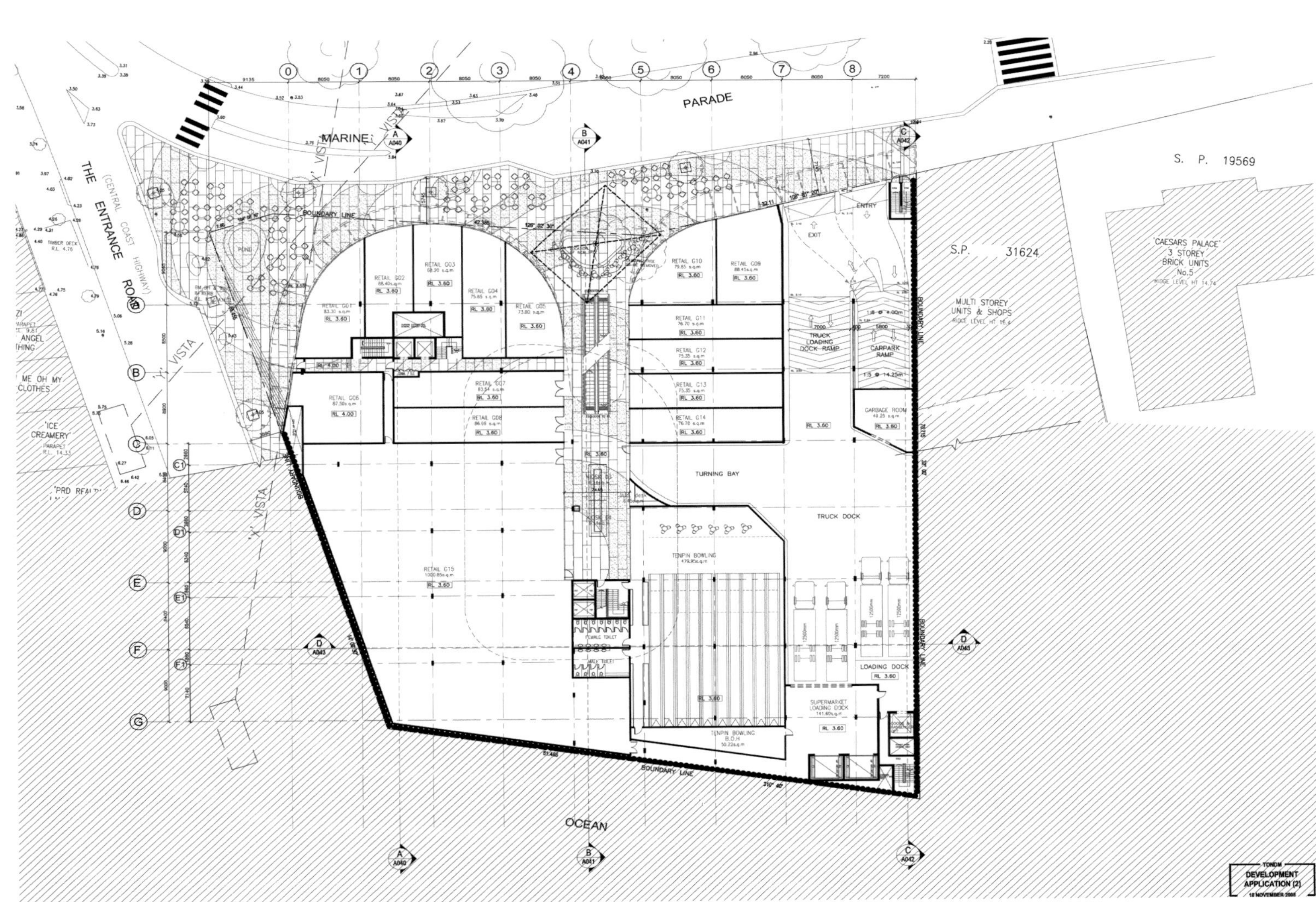
MARINE PARADE
OCEAN
THE ENTRANCE ROAD (CENTRAL COAST HIGHWAY)
'X' VISTA
'Y' VISTA
BOUNDARY LINE
S. P. 19569
S.P. 31624
'CAESARS PALACE' 3 STOREY BRICK UNITS No.5
MULTI STOREY UNITS & SHOPS
POND
EXIT
ENTRY
TRUCK LOADING DOCK RAMP
CARPARK RAMP
GARBAGE ROOM
TURNING BAY
TRUCK DOCK
TENPIN BOWLING
RETAIL C15
LOADING DOCK
SUPERMARKET LOADING DOCK
TENPIN BOWLING B.O.H
'ICE CREAMERY'
ME OH MY CLOTHES
ANGEL HING
DEVELOPMENT APPLICATION [2]